Control in System Dynamics

Series Editor
Jean-Paul Bourrières

Control in System Dynamics

Comparative Analysis of Feedback Strategies

Alain Oustaloup

iSTE

WILEY

First published 2024 in Great Britain and the United States by ISTE Ltd and John Wiley & Sons, Inc.

ISTE Ltd
27-37 St George's Road
London SW19 4EU
UK

www.iste.co.uk

John Wiley & Sons, Inc.
111 River Street
Hoboken, NJ 07030
USA

www.wiley.com

Library of Congress Control Number: 2024943874

British Library Cataloguing-in-Publication Data
A CIP record for this book is available from the British Library
ISBN 978-1-78630-022-5

Contents

Preface

Complemented with Some Background to the Closed Loop

The subject matter discussed in this book concerns automatic control, or more specifically, control in system dynamics. It is an educational work, both theoretical and applicative in nature, featuring a number of solved problems.

The personalization of the presented content results from a well-known past (which nourishes the present), a desire for innovation and simplifcation, and a long-standing deep thinking based on experience. It is true to say that thinking on this subject has always been conducted based on acquired experience in electronics and automation, through teaching, research and technological-transfer activities within the framework of partnerships with the industrial sector. Driven by a concern for balance, these activities can also be credited for favorably supporting my career, first as an assistant in electronics and automation at the University of Bordeaux from 1978 to 1990, and then as an automation professor at the engineering schools of ENSEIRB (École Nationale Supérieure d'Électronique et de Radioélectricité de Bordeaux) and the Polytechnic Institute of Bordeaux, from 1991 to 2018.

Having had the privilege of presenting the foundations of the control loop at the prestigious Les Houches summer school in physics (France), I had the opportunity very early on to become aware of the favorable perception of our discipline by physicists, who indeed concluded my presentation by interpreting automation as a veritable science, even if my talk remained limited to the subject of the control loop itself. But in the context of the control loop, is automation not considered consistent with automation building based on the famous *closed loop*, better known by electronics engineers for its feedback (or negative reaction), and the *open loop*, which simultaneously determines stability and performance in terms of frequency?

The history of the closed loop answers this question, providing testimony through founding works, their context and their development, associated with the main dates. In fact, several elements are drawn in large part from one synthesis document produced by Jean-Claude Trigeassou, based on a contribution by Bennett, a science historian [BEN 79, BEN 93].

In its most widely accepted version, the closed loop can take different forms, notably that of the control loop (with controller), as studied in tracking and regulation in the automation control domain. The concept of the closed loop appears to date back to the 18th century, with the introduction of the mechanical device known as the Watt regulator, by Boulton and Watt in 1788. This device was regularly improved upon, notably by Siemens, and analyzed by physicists interested in its operation: Young in 1807, Airy in 1840 and 1851, and Maxwell in 1868 in his renowned article, "On Governors", the first ever article on the subject of automation.

As the transcontinental telephone between the East and West Coasts of the United States required reamplification of the signal by *repeaters* (relay amplifiers for telephone lines), "Bell Laboratories" worked to replace the first (magnetic) amplifiers with low-frequency electronic amplifiers, made possible by the invention of the triode. Black was recruited to Bell in 1925 to improve on electronic tube repeaters, whose main shortcoming was their narrow bandwidth, associated with significant distortion. His idea of introducing a negative reaction to widen the bandwidth and reduce distortion was patented in 1927 and published in 1934 in the *Bell System Technical Journal*. Even if the concept of a closed loop appears to date back more than a century prior, it is Black specifically who can be credited for introducing the closed loop into the electronics field, through the shaping of feedback, as it is still used today.

But the feedback that could be at the origin of an oscillation, or indeed an instability, led Nyquist, also at Bell, to develop his famous criterion established based on Cauchy's theory and published in 1932 under the title, "Regeneration Theory". In the case of a stable control loop in open loop, where the Nyquist criterion required the critical point (–1) to not be surrounded for closed-loop stability, a significant decrease in gain accompanied by an (illusory) low phase rotation was then sought.

In 1934, the problem was put to Bode, who demonstrated, for minimal phase-shifting systems (or systems with minimum phase), that a gain curve corresponded to a single phase curve, with this link between gain and phase already presenting a naturally admitted constraint in terms of synthesis and therefore performance: complementary to the Bode diagrams dating back to 1938, this contribution, associated with that of Nyquist, was published in 1940, and went on to be included in a book, published in 1945. Furthermore, the first proposition for the systematic adjustment of the P, I and D actions of the PID regulator should be attributed to Ziegler and Nichols, who published their widely known method in an article back in 1942.

Finally, the control loop was also used for temperature and pressure regulation, as well as for servomechanisms (notably for speed and position), according to the principles set out by Hazen in 1934, very similar to those of Black, although not inspired by them. Research into servomechanisms, which was developed significantly to support the armament effort in the Second World War, was given particular impetus in areas that were a priority for armament, including the domain of control and guidance of anti-aircraft defense guns. In 1940, Brown specified the performances of servo-controls, most specifically in transient state, through the damping coefficient and natural frequency. One of Brown's researchers, Hall (thesis published in 1943), constructed a chart, known as the Hall chart, plotted in the complex plane (also known as the Nyquist plane), to change from the frequency response in open loop to the frequency response in closed loop. To change directly to closed loop based on the Bode diagrams in open loop, Nichols had the idea of transposing the Hall chart into a new plane, known as the Nichols plane, whose gain in decibels is applied on the ordinate, and whose phase in degrees is applied on the abscissa [JAM 47]. An even older origin of the Nichols plane and chart can be found in the Black plane and chart, dating back to 1934, in the sense that the Nichols plane results from the permutation of the Black gain and phase axes, accompanied by the expression of the gain in decibels (in compliance with the Bode representation). Yet these works, which contributed to the precision (in position and speed) of servomechanisms, were not enough to address the more general problem of *pursuit of target*, which also required the anticipation of the movements of the target and a prediction of its future behavior. So Wiener developed his famous filter, known as the Wiener filter (1942), used to guide DCA artillery based on radar information, thus replacing manual guiding. Section 1.10.4 provides additional details of a more technical nature.

Having gone through these major points in its history, let us now present the control loop in the context of the present book, specifying that by studying all aspects of the control loop, we are able to present, in a clear, structured manner, all of the functions, properties and performances that can characterize all other controls.

Moreover, the unification exercise considered in Chapter 4, which consists of developing a control loop equivalent to any other control among a set of controls, makes it possible to demonstrate an absence of specific contribution in terms of performances, even with the synthesis method differing, depending on the control.

Certainly expressing a change with respect to the control loop, this methodological contribution, which, in principle, allows us to assume a bonus in terms of performance, and which, subsequently, retains this in the best case scenario, leads us at the very least to think of the principle, according to which it is advisable to change the form (here, the methodology) to preserve the substance (here, performance). It is true that evolution is *de rigueur*, so if a change is inevitable when it is not necessary, is it not preferable to opt for a change of form that is fundamentally the least detrimental?

With the control loop thus all the more reinforced, as indeed confirmed by its usage in all robustness approaches, it is widely developed in the first two chapters of this book, which accord it due credit. This didactic choice should help the reader to obtain a measure of the major interest of the command loop, in the sense of its own interest and the interest it represents for other controls through the often-implicit place that it occupies, and which is highlighted by establishing the equivalent control loop. This same loop, through the nature of its regulator, makes it possible to objectively appreciate the richness of control strategy, with the control loop thus appearing as an absolute reference. This means that the control loop, which is the proven substrate of automation, must be very widely known to all, including specialists of other controls, at the risk of rediscovering lukewarm water, or even making cold water from it, by achieving less success than the control loop.

The innovative contribution in terms of performance undoubtedly lies in the so-called robust controls, including the CRONE control (the French abbreviation for the Robust Control of Non-Integer Order), which combines three significant advantages in terms of robustness: the choice of dynamic performance to be robustified; taking genuine uncertainty domains into account (unlike the H_∞ approach); and minimal parameterization of the transmittance to be optimized (unlike the QFT approach). Based on the structure of the control loop, it owes its specific nature to a regulator enriched by a non-integer, real or complex order. This order ensures the robustness of the control's stability degree with respect to the (parametric) plant uncertainties.

Represented in the Nichols plane by a straight-line segment whose description transmittance is based on complex, non-integer integration, the generalized template of the third-generation CRONE control is used in the covering technique to build iso-overshoot and isodamping contours. It transpires that the iso-overshoot contours are the Nichols amplitude contours, which are thus validated as iso-overshoot contours. The symmetry, in relation to the abscissa axis of the Nichols plane, of the *isodamping contours* (of Oustaloup), is a remarkable property that reinforces the relevance of this working plan in terms of analysis and synthesis.

Such contours, called performance contours, significantly enrich the well-known arsenal of tools and methods for frequency synthesis of the regulator (whichever it is) of the control loop. These performance contours are indeed tools of great interest, in the sense that they make it possible to respond, in a deterministic manner, to the performance specifications related to the overshoot and damping of control loop dynamics: surely the presentation of this new bonus in favor of the control loop, offering it an indisputable argument in its favor, is the best way to conclude this preface or, given the context, if we can make room for a little humor: to close the loop?

February 2022

Book Structure and Content

Along with the three appendices covering active noise control, non-integer control and performance contours, respectively, this book consists of five chapters and 13 solved problems, the content of which is discussed here by way of an introduction.

Chapter 1 starts by demonstrating that the plant output disturbance is none other than the effect on its output of the set of so-called "original" disturbances. Given this disturbance, it then develops *tracking-regulation duality* in the operational and frequency domains, applying the most dual approach possible in order to best structure the presentation of the control loop. After a very thorough and wide-ranging definition of transmittance in open loop, which responds to all of the linear systems forming part of engineering science, the open loop is used to physically present the stable, unstable or barely oscillating nature of the control loop, as well as the left-hand criterion, which governs this character for a stable open loop with minimum phase shifting. Transmittances in closed loop, which are the tracking and regulation transmittances, as well as the resulting tracking and regulation functions, which introduce *reference tracking* and *disturbance rejection*, respectively, constitute the substrate of tracking-regulation duality. The structural nature imposed by such a duality is then used to study, in turn, the input sensitivity (of the plant), which results in replacing the academic PID with the non-academic PID, the behavior and performances of the control loop, and its dynamics, before ending with the charts in tracking and regulation, whose duality speaks for itself.

Chapter 2 begins by defining stability, and presenting the fundamental stability condition, which can be verified using algebraic or graphical criteria. Next, the frequency stability margins, that is, the gain and phase margins, are defined. The phase margin is examined in more detail via the determination of an optimal phase margin, in the sense of the minimization of a quadratic criterion based on the steepness time and the first overshoot of the step response in tracking or in regulation. The stability degree, an essential concept in system dynamics, perfectly exemplified by the control loop owing to the coupling between open loop and closed loop, is presented in the time domain, then in the frequency domain, where particular use is made of this coupling by means of the amplitude contours presented in Chapter 1. This approach demonstrates, *inter alia*, that while the circle centered at (–1) in the complex plane offers (in the mathematical sense) a true measurement of the distance to the critical point, it is nevertheless not significant for a time dynamic performance able to quantify the stability degree in the time domain. When compared with pole-placement control, the frequency approach stands out as favorable in that it satisfies both a reduced first overshoot specification and a damping ratio specification. After defining the precision based on the difference between the sought and actual values of the control loop output, precision is presented in accordance with tracking-regulation duality and with the steady states corresponding to step and ramp tests. Once the stability degree–precision dilemma has been defined, it is highlighted in the Nichols plane, then solved by an open-loop gain, which provides a compromise between stability degree and precision. Next, this compromise is improved and the dilemma is brought into question by a non-integer order. A pragmatic definition of dynamics is provided, after which a "determination of dynamics" section identifies the frequency dynamic performances significant to time dynamic performances, which thus determine the dynamics. The controller synthesis thus forms part of a straightforward, systematic methodological approach that offers the user effecient methods, in user-guide form.

Chapter 3 results from an educational exercise aiming to extract the idea of different linearizing approaches using strategies that stand out because of their simplicity. Forming part of engineering sciences, these strategies, which enrich basic automation and can be applied to the majority of plants, are conducive to favoring the application of linearizing approaches in the most diverse sectors, notably in industry. More precisely, the presentation of each approach is structured by a principle reduced to its simplest expression, and by the application of this principle to one or more nonlinear plants, one of which is used for all approaches; this nonlinear plant, common to all monovariable applications, is a motor shaft that rotates a small mass secured to the shaft by means of a straight arm. In linearization by immersion, extended to a multivariable plant, the principle consists of linearizing the plant with respect to new inputs. In linearization by high gain, the principle is based on a high-gain control used for its disturbance rejection. Although the basic principle of linearization by high gain expresses the idea of a total disturbance rejection, linearization by disturbance rejection expresses the idea of a partial rejection, which is more realistic in terms of the plant input sensitivity. Indeed, its principle is based on an elementary control loop, whose gain in open loop remains compatible with solid plant input immunity. In fact, through its ability to reject disturbance, which in this case represents the plant's nonlinearity, the control loop has a linearizing character that has come to be frequently used in industry to overcome nonlinearities. While the linearization of the plant around a nominal trajectory constitutes a well-known, relatively widely used linearizing approach, the use made of it here stands out with the introduction of uncertainty domains, with the nonlinearity indeed represented frequentially in the form of uncertainty domains. In fact, this strategy, which displaces uncertainty domains such that they present as little as possible overlap of the areas presenting a low degree of stability, is able to specify the way in which the nonlinearities are taken into account in third-generation CRONE control (see Chapter 5). Finally, the chapter concludes with a rigorous contribution from Brigitte d'Andréa-Novel, which compensates for the extreme simplicity of the strategies proposed.

Chapter 4 presents different control schemes, highlighting their *point in common*, notably by showing that they are implicitly based on an elementary control loop (the regulator of which is then analyzed). Given that the study domain, frequency or time, differs depending on the nature of the control, it is difficult to conduct a comparative study of the main control principles. More generally, this difficulty is exacerbated by the difference in the *study framework*, which incorporates both the study domain (frequency or time) and the structure of the control scheme. As such, this chapter aims for a *common study framework*, seeking to develop a unified approach to control. As developed in the first two chapters, the in-depth study of the control loop, notably in the frequency domain, has seen the development of a tried-and-tested control methodology over time, capable of meeting several temporal dynamic performance specifications, via frequency dynamic performances, significant for these performances. Moreover, such a methodology ought to be interpreted as a *reference* among the main control methods proposed in this chapter, and indeed this book. It is true that the idea aims to reduce each control scheme to an *equivalent control loop*, so that they come under the same study framework (that of the control loop), and thus conduct an objective comparative study of the different control strategies: indeed, the nature of the regulator of each equivalent control loop allows an objective appreciation of the richness of each control strategy. One initial conclusion is that *the sophistication of the control scheme is not always synonymous with the enriching of control strategy*.

While a proportional, high-gain regulator has the benefit of rapidly showing the interest of the high-gain control through its insensitivity (or robustness), ensured by a transfer imposed by that of the reaction chain, this case is nevertheless not realistic since high gain comes at the cost of a large input sensitivity, and therefore low input immunity. In order to make judicious use of a high-gain principle such as this, it is then appropriate to aim, not for an insensitization of the control, but simply for a desensitization, by choosing an acceptable gain. As for the change to the control loop, rewriting the error signal of the high-gain control in an appropriate form makes it possible to transform its functional diagram in accordance with a transfer in cascade with an elementary control loop.

By assuring a unit transfer in tracking, the feedforward control enables differentiation between the dynamics in tracking and regulation, as indeed a reference prefilter does. It is true that appropriately rewriting the plant input enables the functional diagram of the control to be transformed in accordance with a reference prefilter in cascade with an elementary control loop.

In the absence of a disturbance, and in the case where the plant is perfectly modeled, the internal-model control is conducted without feedback and the controller synthesis is then remarkably simple. Appropriately rewriting the plant input enables the internal model scheme to be restructured in the form of an elementary control loop, the regulator of which is expressed. In the case of a "robustification", which consists of ensuring a reference-output transfer independent of the plant for the control, the (equivalent) elementary control loop thus obtained has the advantage of recognizing the *infinite character* underlying the control, not only through the regulator gain, but also through the input sensitivity in tracking and regulation.

As a preamble to quadratic and predictive controls, which are based on an optimization with respect to the input, $u(t)$, of the plant, a property that is essential for this optimization is established, a property according to which the output $y(t)$, and its derivatives depend on the past of the input $u(t)$, and not its instantaneous value (such a property resulting from the convolution).

The quadratic-criterion control is presented based on a second-order study plant, with input $u(t)$ or $u(k)$, depending on whether the model is continuous or discrete. The synthesis of the regulator is performed through the minimization of a quadratic criterion, continuous then discrete, which introduces a control weighting factor, δ. A continuous, direct approach based on continuous control and a discrete then continuous approach based on discrete control make it possible to result in the same elementary control loop. One result that is remarkable to say the least, is that the action of the regulator of this loop is reduced to a *proportional action*, $1/\delta$, which is none other than the inverse of the control weighting factor.

Predictive control, the principle of which is presented here in a continuous-time approach, makes a significant contribution through the highlighting of an elementary control loop equipped with a phase-lead regulator in the operational domain. The study of the regulator frequency placement with respect to unit gain frequency in open loop shows that the regulator cannot respond to the optimal frequency placement ensured by the direct frequency synthesis of a phase-lead regulator (as discussed in Chapter 2). The problem of regulator frequency placement is at the origin of two problems, for which the resolution principles are presented.

Chapter 5 examines the CRONE control (*Commande Robuste d'Ordre Non Entier*, robust non-integer order control). The robustness in question is that of the stability degree vis-à-vis the plant uncertainties. The stability degree is measured by the resonance ratio in tracking, or the damping ratio in tracking and regulation: the first ratio is indeed significant to the reduced first overshoot of the step response in tracking or in regulation; the second is significant to the damping of its oscillatory mode. Uncertainties are taken into account via the *genuine uncertainty domains* that they define, unlike the case of the H_∞ approach, where the *overestimation* of the uncertainty domains does not enable the best results to be obtained. With regard to the time dynamic performance, which is indissociable from the damping ratio, namely the natural frequency, the non-integer approach in no case seeks to ensure the robustness of this frequency. Indeed, allowing the control natural frequency to vary freely enables better "compliance" with the plant, notably by avoiding forcing its input when its dynamics become much slower than that imposed on the control. In accordance with the strategies used by the three generations of CRONE control, for the first generation, a *non-integer differentiation order*, which is real, makes it possible to parameterize the *transmittance of the regulator* with few parameters. For the second and third generations, a *non-integer integration order*, which is real or complex according to the generation (real for the second and complex for the third), makes it possible to parameterize the *open-loop transmittance* with few parameters, and thus reduce its optimization: *the synthesis of the regulator in the form of a rational fraction is then conducted uniquely for optimal transmittance in open loop*, thus avoiding the iterative synthesis of a regulator with numerous parameters (required, notably, by the QFT approach). In the case of the relaxation of water on a porous dyke, where the fractal nature of the porosity enables the concept of *fractal robustness* to be introduced, the robustness in question is that of relaxation damping with respect to the mass of water in motion, thus bringing into question the *mass-damping dilemma* in mechanics. The transposition of this type of robustness to the control domain can be a means of defining the initial approach of the CRONE control. Its first generations indeed result from two legitimate interpretations of the model governing the relaxation of water on a porous dyke, these two interpretations defining, respectively, the first-generation CRONE control through a *phase locking of the regulator* (at a non-integer multiple of $\pi/2$), around the unit gain frequency in open loop, ω_u, and the second-generation CRONE control through a *phase locking in open loop* (at a non-integer multiple of $-\pi/2$, comprised between $-\pi/2$ and $-\pi$), around the frequency, ω_u. For the nominal and non-nominal plant, this phase locking defines a *vertical template* in the Nichols plane. The change from the second to the third generation results from the generalization of the *vertical template*, in accordance with two generalization levels: in the first generalization level, and for the nominal plant, the vertical template is replaced with a straight-line segment of any direction, known as the *generalized template*; in the second generalization level, and for the nominal plant, the generalized template is replaced with a set of generalized templates, called the *multi-template*, which leads to a *curvilinear template*.

The "Solved Problems" section underlines the educational nature of this book through 13 solved problems. Each begins with a presentation that generally reflects a synthesis method, and a solution that details, literally and numerically, the responses to the questions set out in the presentation. Problems 1–3 relate to Chapters 1 and 2, Problems 4 and 5 relate to Chapter 3, Problems 6–9 relate to Chapter 4 and finally, Problems 10–13 relate to Chapter 5.

1

The Control Loop: Characterization and Behavior in Open Loop and Closed Loop

1.1. Introduction

After demonstrating that the plant output disturbance is none other than the effect of the set of so-called "original" disturbances on its output, this chapter develops *tracking-regulation duality* in the operational and frequency domains, applying the most dual approach possible in order to best structure the presentation of the control loop.

After a very detailed and broad definition of transmittance in open loop, which responds to the linear systems coming under engineering, the open loop is used to physically present the stable, unstable or barely oscillating nature of the control loop, as well as the left-hand criterion, which governs this character for a stable open loop with minimum phase shifting.

Transmittances in closed loop, which are the tracking and regulation transmittances, as well as the resulting tracking and regulation functions, which introduce *reference tracking* and *disturbance rejection*, constitute the substrate of tracking-regulation duality.

The structural nature, dictated by such a duality, is then used to study, in turn, the input sensitivity (of the plant), which results in replacing the academic Proportional, Integral and Derivative controller (PID) with the non-academic PID, the behavior and performances of the control loop, as well as its dynamics, before ending with the charts in tracking and regulation, whose duality speaks for itself.

1.2. Definition and terminology

Referring to the dictionary definition of the verb "to track", a *servo-control* can be defined by its function, which consists of subjecting (or servo-controlling) the output of a plant to a reference, independently of the presence of disturbances.

More precisely, this definition introduces both the idea of a *reference tracking* (up to a factor, which can be a single unit) and the idea of a *disturbance rejection* (consistent with independence from disturbances): these two ideas, which relate to the reference and disturbances, respectively, indeed contribute in a complementary manner to *global reference tracking*, which can be defined as the joint response to the reference and disturbances. Moreover, the nature of a tracking such as this justifies the designation, "servo-control" that has long been, and continues to be, attributed in this field.

A servo-control can also be referred to as a "servo-system", "corrected-error control" or "control loop" (or, more simply, "control"). Within the framework of our developments, we have a preference for the designation *control loop* because in the genericity of the term "control", it provides a true reflection of the specific structural nature of a servo-control.

1.3. The plant

1.3.1. *Definition*

The plant is the system to be servo-controlled (see Figure 1.1). It is controlled by the *input* (or *control*) $u(t)$ and disturbed by *disturbances* (said to be *original*) $d'_1(t)$, ... , $d'_q(t)$. The *output*, $y(t)$, depends on the input and disturbances: in fact, the input and disturbances are the *stresses*, and the output, $y(t)$, is *the response to the stresses*.

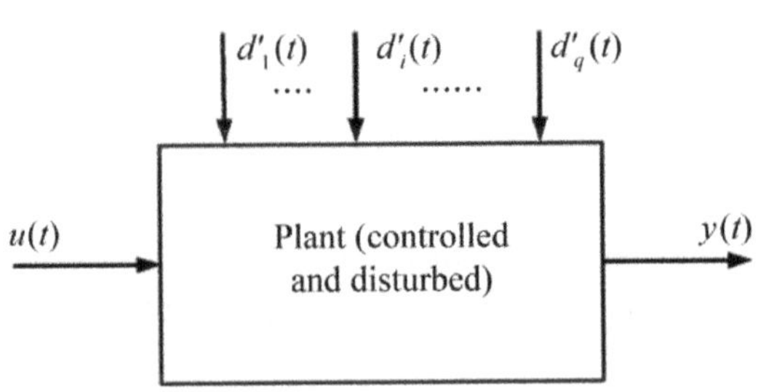

Figure 1.1. *The plant, with output, $y(t)$, is controlled by $u(t)$ and disturbed by $d'_1(t)$, ..., $d'_q(t)$*

1.3.2. *From a modeling difficulty to a pragmatic approach to disturbances*

The relationship between the stresses and output is a *cause-and-effect relationship*, with the stresses as the cause of the effect on the output.

In accordance with a declination of the stresses, the relationship between the input and output (or *input–output relationship*) is a cause-and-effect relationship that *can be modeled*, definable using a differential equation (which may or not be linear, according to the nature of the plant). While the relationship between a disturbance and the output (or *disturbance–output relationship*) is also a cause-and-effect relationship, it is, however, *difficult (or even impossible) to model.*

It is precisely this *modeling difficulty* that is at the origin of the *pragmatic approach to disturbances*, as presented here.

The idea is to relate the set of original disturbances, $d'_1(t), \ldots, d'_q(t)$, processed by the plant, to the plant output, in the form of an *output disturbance*, $d(t)$.

To this end, let us consider the block diagram of the plant (see Figure 1.2), in which $G(s)$ represents the transmittance relative to the input, $u(t)$, (or *plant transmittance*) and $G'_i(s)$ the transmittance relating to the original disturbance, $d'_i(t)$, that is,

$$G(s) = \frac{Y(s)}{U(s)} \text{ and } G'_i(s) = \frac{Y(s)}{D'_i(s)}, \qquad [1.1]$$

with each of the transmittances defined for a stress, and the others considered to be zero.

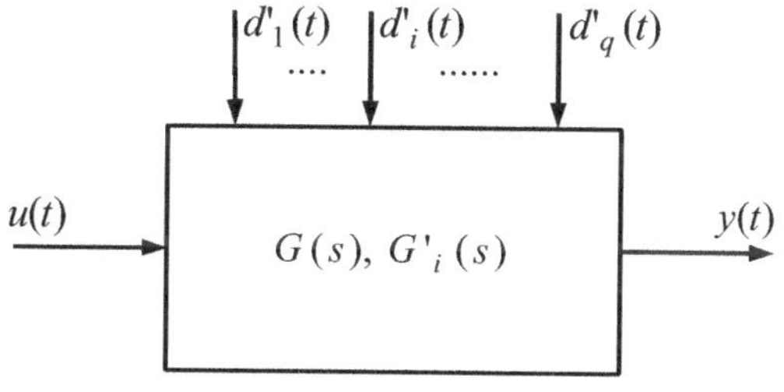

Figure 1.2. *Functional representation of the plant*

The output Laplace transform, $y(t)$, admits an expression of the form:

$$Y(s) = G(s)U(s) + G'_1(s)D'_1(s) + ... + G'_i(s)D'_i(s) + ... + G'_q(s)D'_q(s), \qquad [1.2]$$

or:

$$Y(s) = G(s)U(s) + \sum_i G'_i(s)D'_i(s). \qquad [1.3]$$

That is, by writing

$$D(s) = \sum_i G'_i(s)D'_i(s): \qquad [1.4]$$

$$Y(s) = G(s)U(s) + D(s), \qquad [1.5]$$

with this equation reflecting the block diagram in Figure 1.3, in which $d(t) = \mathcal{L}^{-1}[D(s)]$.

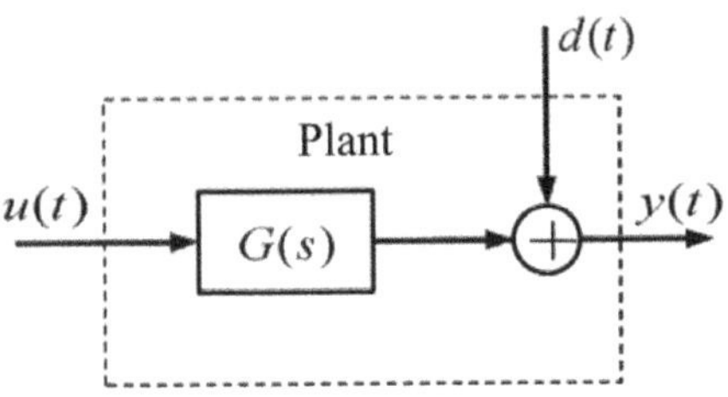

Figure 1.3. *Plant controlled by $u(t)$ and disturbed at the output by $d(t)$*

The disturbance, $d(t)$, known as the *plant output disturbance* or, more simply, *disturbance* where there is no ambiguity, therefore represents *the effect on the plant output of the set of original disturbances.*

In conclusion, when the plant is integrated into a control loop, the *control system* (which generates the control, $u(t)$, of the plant) processes, not the original disturbances, but *the effect of these disturbances on the plant output.*

As a warning, it should be specified that if the control output is none other than the plant output, it is nonetheless important (as we will see below) not to confuse *plant output disturbance*, $D(s)$, and *control output disturbance*, $D(s)/(1+\beta(s))$, with $\beta(s)$ designating the open-loop transmittance of the control. Moreover, $1+\beta(s)$ or $1+\beta(j\omega)$ defines the (complex) *efficiency factor*, which *divides* the disturbance, $D(s)$ or $D(j\omega)$, existing at the plant output in the absence of a control system.

1.4. Functional representation of the control loop

1.4.1. *Block diagram and corresponding transfers*

The block diagram in Figure 1.4 provides a functional representation of a corrected error control of a disturbed plant.

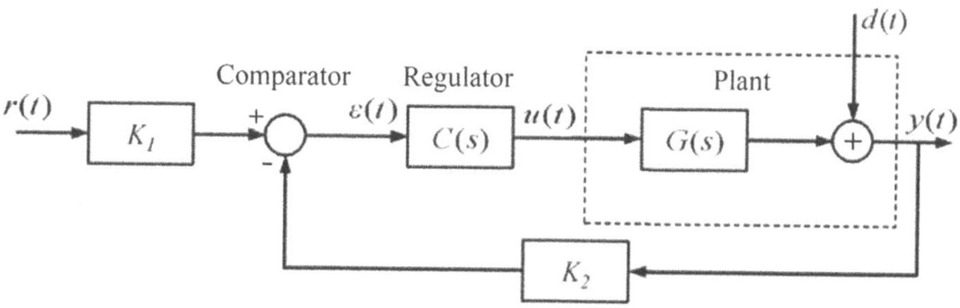

Figure 1.4. *Block diagram of a control loop: the dotted line framing the plant (controlled and disturbed) in compliance with the previous figure, is not indispensable and does not generally feature in the functional representation*

There are two different types of variable involved:

– the *external variables*, which, where stresses are concerned, constitute the reference input (or reference), $r(t)$, and the output disturbance of the plant (or disturbance), $d(t)$, and where the response to these stresses is concerned, the output, $y(t)$, of the plant and the control loop;

– the *internal variables*, which are the error (or gap) signal $\varepsilon(t)$ and the input (or control) $u(t)$ of the plant.

With regard to the symbolic operators of the *direct (or action) chain*, $G(s)$ represents the transmittance of the *system to be servo-controlled (or plant)* and $C(s)$ the transmittance of the *controller (or regulator)*. K_1 and K_2 designate the transfer factors of the *display system* and the *return (or reaction) chain*, generally constituted by the sensor of the output variable: these transfer factors result from the reduction of the transfer functions, $K_1(s)$ and $K_2(s)$, dictated by the significance of their bandwidth with respect to the control loop. The sub-assembly of the control loop, which delivers (or generates) the control law, $u(t)$, from the reference, $r(t)$, and the output, $y(t)$, is called the *system*. In the specific case where K_1 and K_2 are respectively reduced to become one, the control loop is said to have a *unit display and reaction* or, more simply, an elementary form.

1.4.2. *Regarding the gap detector*

The gap detector delivers the error (or gap) signal, $\varepsilon(t)$, from the reference, $r(t)$, and the output, $y(t)$. It comprises three sub-systems: the output comparator, $\varepsilon(t)$, the input display system, $r(t)$, and the input sensor, $y(t)$.

1.4.2.1. *The comparator*

The *comparator*, which delivers the error signal through a *subtraction operation*, is generally electronic in nature, and aims to compare physical values of the same type. In analog, it can be achieved by a differential amplifier with operational amplifier.

1.4.2.2. *The display system*

The *display system* can be a sensor, which transforms the reference value, $r(t)$, into an electronic value, notably a voltage, which is indeed compatible with the nature of the comparator, as defined technologically. If the reference value, $r(t)$, is an electrical voltage to begin with, it can be applied to the non-inverting input of the comparator, either directly or via a potentiometer or a low-power amplifier.

1.4.2.3. *The sensor (of the output variable)*

The *sensor* consists of the sensor itself and its environment. An illustrative example is that of a *position sensor*, which can be achieved by a potentiometer (the sensor itself) supplied by two opposite direct voltages via two identical resistors located on either side of the resistive track of the potentiometer. In the case of a rectilinear position, the track is straight; for an angular position, the track is circular for a rotation of less than one turn, or helical for a rotation of more than one turn.

In light of the elements presented in the above section and generalized here, a *sensor transforms (or converts) one physical value into another physical value.* For example, it transforms a mechanical value into an electrical value in a position or speed control.

In the angular position control of a motor shaft, the sensor constituting the reaction chain converts the shaft's angular position into an electrical voltage. If the sensor converts 1 radian into 5 V, the transfer factor, K_2, is equal to 5 V/rd, that is, $K_2 = 5$ (unitless). If the same sensor is modified with respect to its environment (via the supply voltages or the series resistors) to ensure conversion from 1 radian to 1 V, K_2 is then reduced to 1 V/rd, that is, $K_2 = 1$ (unitless): thus, the sensor output gives (up to 1 *measurement noise* aside) the measurement of the output, $y(t)$, that is, the *output measured* $y_m(t)$, such that (unitless) $y_m(t) = y(t)$.

If the angular position of the motor shaft is servo-controlled to that of another motor shaft (in a motor-shaft synchronization scenario), the display system of the control can be composed of a sensor similar to that constituting the reaction chain, with the display system having to convert the angular position of the other shaft into an electrical voltage. If the conversion is brought to 1 V for 1 radian, the transfer factor, K_1, is reduced to 1 V/rd, that is, $K_1 = 1$ (unitless), the output of the sensor then giving (up to 1 *measurement noise* aside) the measurement of the reference, $r(t)$, that is, the *measured reference*, $r_m(t)$, such that (unitless) $r_m(t) = r(t)$.

1.4.2.4. *Benefit of transfer functions* $K_1(s)$ *and* $K_2(s)$*, reduced to transfer factors*

The block diagram in Figure 1.4 enables the error signal, $\varepsilon(t)$, to be expressed in the form:

$$\varepsilon(t) = K_1 r(t) - K_2 y(t) \,. \qquad [1.6]$$

In the frequency band in which the control loop is *effecient (good reference tracking and good disturbance rejection)*, the open-loop gain is high, as is, consequently, the action-chain gain, given that the reaction chain is not supposed to provide any gain (as a sensor is not, in essence, an amplifier).

In other words, for a finite output, $y(t)$, the error signal, $\varepsilon(t)$, is then close to zero, thus enabling us to write:

$$K_1 r(t) - K_2 y(t) = 0 \,, \qquad [1.7]$$

that is:

$$y(t) = \frac{K_1}{K_2} r(t) \cdot \quad [1.8]$$

This relationship demonstrates that the value and variations of $y(t)$ are (as is appropriate for reference tracking and disturbance rejection) like those of $r(t)$, *because the ratio* K_1/K_2 *is a proportionality coefficient, given that* K_1 *and* K_2 *are transfer factors*. Hence, the benefit of such factors, which moreover merely confirm that a comparator is, essentially, supposed to compare values of the same type, which reflect the reference and the output (therefore proportional to $r(t)$ and $y(t)$).

The transfer factors K_1 and K_2, whose ratio, K_1/K_2, determines the proportionality coefficient between $r(t)$ and $y(t)$, thus need to be precise and reliable in order to meet the precision and reliability requirements of the output, $y(t)$. To this end, their achievement will rely more greatly on passive components, which are more precise and reliable than active components.

1.4.2.5. *An absolute reference*

For $K_1 = K_2$ or $K_1 = K_2 = 1$ (in the elementary control loop scenario), the relationship

$$y(t) = (K_1/K_2) r(t) \quad [1.9]$$

is reduced to the new relationship:

$$y(t) = r(t), \quad [1.10]$$

two relationships that call for a clear distinction to be made.

The first relationship expresses that the *desired output*, $(K_1/K_2) r(t)$, reflects (in the proportionality sense) the reference, $r(t)$. The second expresses that the *desired output*, $r(t)$, is the reference itself: the reference, $r(t)$, then appears as an *absolute reference* for the output, $y(t)$.

1.5. Open-loop transmittance

1.5.1. *Definition*

The transmittance in open loop, $\beta(s)$, of a control loop is the opposite of the ratio of the Laplace transforms of the output and input of the open loop, with its output then charged by its input impedance, and the reference, $R(s)$, and the disturbance, $D(s)$, being considered nil (see Figure 1.5), that is:

$$\beta(s) = \left[-\frac{V_s(s)}{V_e(s)} \right]_{R(s)=D(s)=0} \quad [1.11]$$

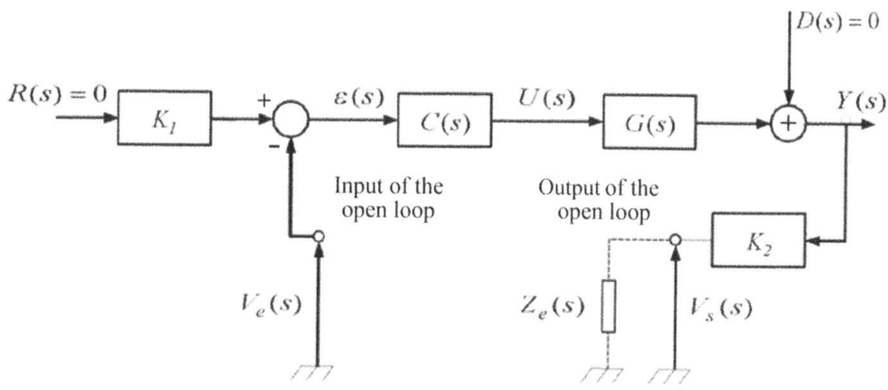

Figure 1.5. *Block diagram, from which the transmittance in open loop, $\beta(s)$, is determined: $Z_e(s)$ designates the impedance presented by the inverting input of the comparator*

1.5.2. *General expression*

Observing Figure 1.5 enables us to express the open-loop output symbolically:

$$V_s(s) = K_2 G(s) C(s)(-V_e(s)), \quad [1.12]$$

from which we obtain:

$$\frac{V_s(s)}{V_e(s)} = -K_2 G(s) C(s), \quad [1.13]$$

or, given the definition of $\beta(s)$ given by the relationship [1.11]:

$$\beta(s) = C(s)G(s)K_2, \qquad [1.14]$$

a result that reflects the fact that *the open-loop transmittance is none other than the product of the transmittances of the action and reaction chains.*

1.5.3. *Regarding the input impedance of the open loop*

Taking the inverting input impedance of the comparator in the definition of open-loop transmittance into account enables any looped system that is electrical, electronic, mechanical, thermal, etc., in nature to be presented in the form of a control loop block diagram.

One characteristic example is that of the COLPITTS oscillator (see Figure 1.6), which uses:

– an amplifier stage with a transistor connected in common base;

– loaded by a resonant circuit (with resonance frequency $\omega_0 = 1/\sqrt{LC}$);

– looped by a capacitive voltage divider.

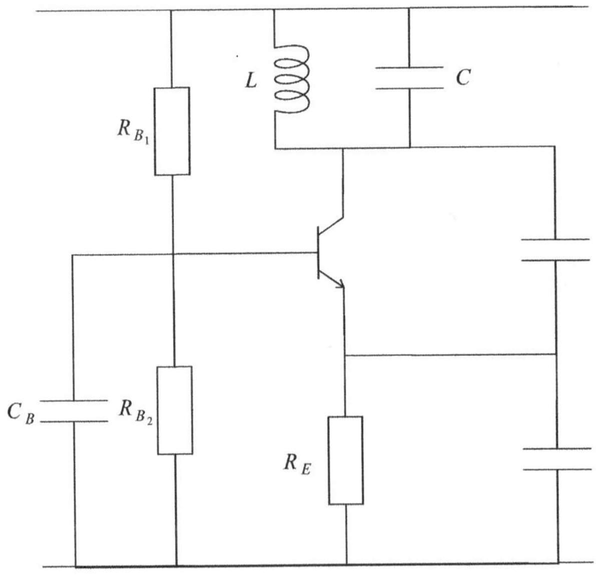

Figure 1.6. *COLPITTS oscillator*

Without taking the input resistance, R_e, of the amplifier stage (see Figure 1.7), into account, the transfer function, $K_2(s)$, of the reaction chain is reduced to the transfer factor, K_2, such that:

$$K_2 = \frac{\frac{1}{C_2 s}}{\frac{1}{C_1 s} + \frac{1}{C_2 s}} = \frac{\frac{1}{C_2}}{\frac{1}{C_1} + \frac{1}{C_2}} = \frac{C_1}{C_1 + C_2}. \quad [1.15]$$

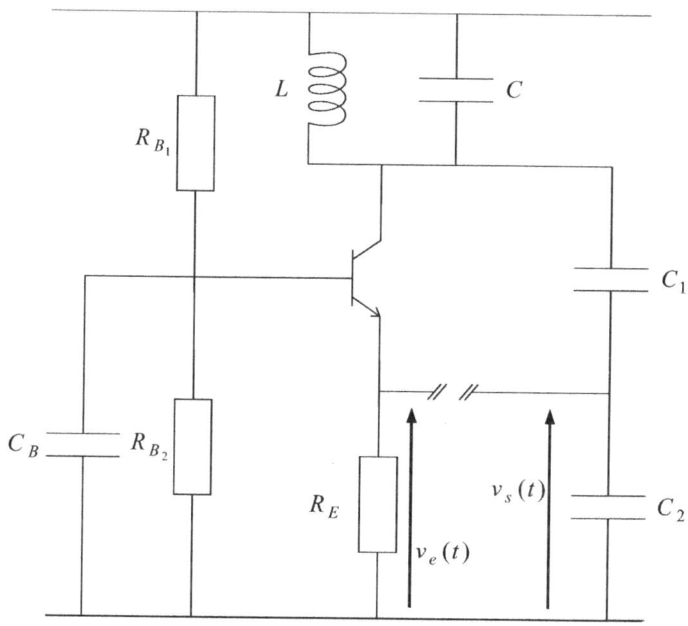

Figure 1.7. *Loop opening with the input resistance of the amplifier stage not taken into account*

Taking the input resistance, R_e, of the amplifier stage (see Figure 1.8) into account enables the transfer, $K_2(s)$, to be expressed in the form:

$$K_2(s) = \frac{R_e \,//\, \frac{1}{C_2 s}}{\frac{1}{C_1 s} + R_e \,//\, \frac{1}{C_2 s}} = \frac{R_e C_1 s}{1 + R_e(C_1 + C_2)s}, \quad [1.16]$$

with R_e of the order of several dozen Ohms, an expression that complies with that relating to the closed loop, the loop opening thus not modifying the operating conditions.

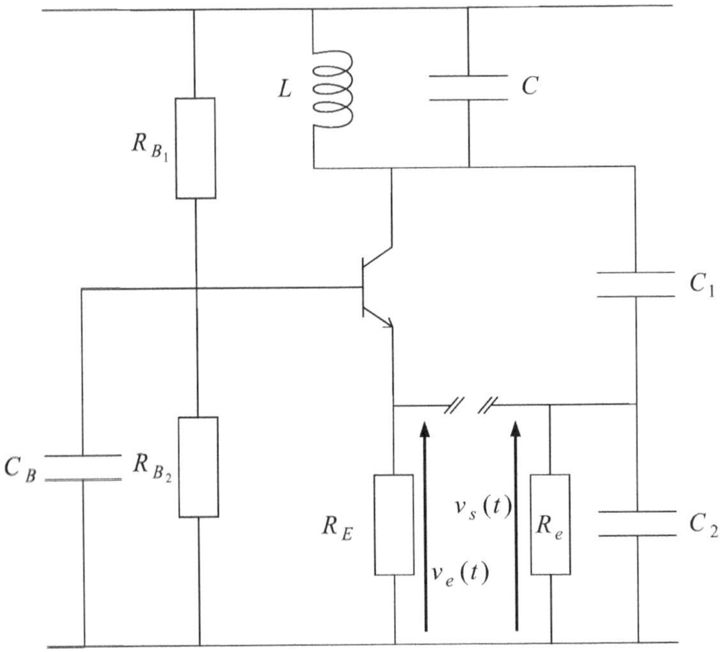

Figure 1.8. *Loop opening with the input resistance of the amplifier stage taken into account: $R_e = R_E\,//\,(r_{BE}/\beta)$ with the natural parameters; $R_e = R_E\,//\,(h_{11}/h_{21})$ with the hybrid parameters; here β designates the transistor current gain with small variations ($\beta = h_{21}$ and $r_{BE} = h_{11}$)*

1.5.4. *Open-loop Nichols locus: elementary form and characteristic quantities*

Within the framework of a general graphical representation aiming to illustrate the explanations relating to control in a straightforward manner, an *elementary-form open-loop Nichols locus* may be used. It is represented in Figure 1.9 in a graphically truncated form at low and high frequencies. Thus, we attribute to the open-loop Nichols locus, a monotonic and decreasing *"elementary" form (or shape)*, which underlines the *low-pass character* of the corrected plant, $C(j\omega)G(j\omega)$.

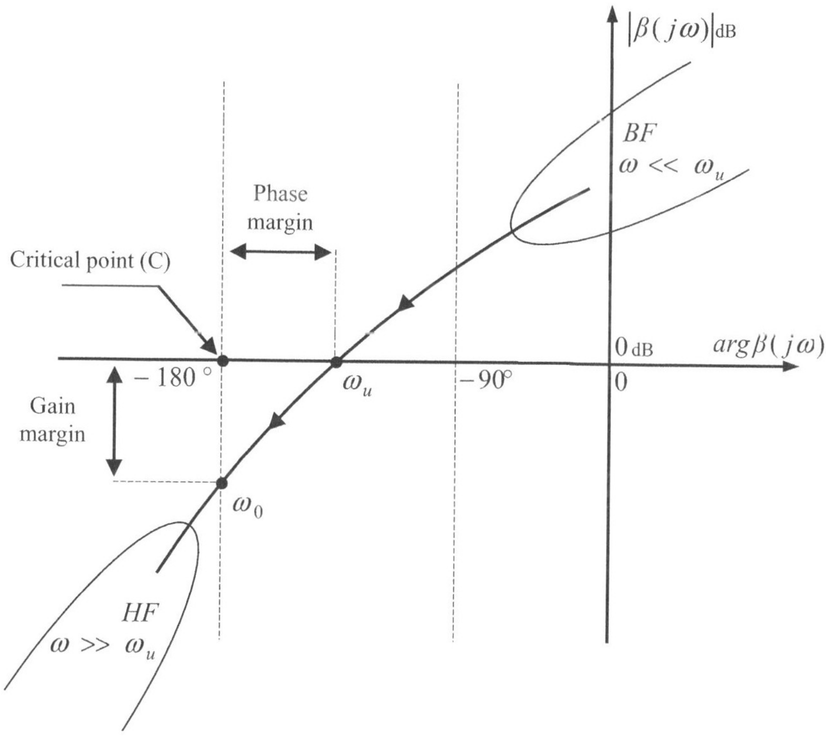

Figure 1.9. *Open-loop Nichols locus said to be of elementary form and associated characteristic quantities*

The form thus defined is widely used in qualitative analysis and reasoning in this field, enabling simple illustration, in open loop, of a certain number of *quantities (including characteristic frequencies)*:

– the *frequency at a phase of –180°*, ω_0, such that $\arg\beta(j\omega_0) = -180°$;

– the *unit gain frequency* (or *transition frequency*), ω_u, such that $|\beta(j\omega_u)| = 1$, that is, $|\beta(j\omega_u)|\text{dB} = 0$;

– the *low frequency area* ($\omega << \omega_u$), for which $|\beta(j\omega)| >> 1$;

– the *high frequency area* ($\omega >> \omega_u$), for which $|\beta(j\omega)| << 1$;

– the *critical point* (0dB, –180°), which corresponds to the *stability limit* of the control (the control barely oscillates if the locus of $\beta(j\omega)$ passes through this point);

– the gain and phase distances to the critical point, which define the *stability margins (gain and phase margins)*, with their sense conditioned by the monotony of the Nichols locus around the critical point (this monotony characterizing the elementary form Nichols locus); indeed, the stability margins, Φ_m and $G_m\mathrm{dB}$, are no longer significant to the distance to the critical point of the Nichols locus of $\beta(j\omega)$, when the latter presents one or more local resonances around frequencies ω_u and ω_0 (see Figure 1.10).

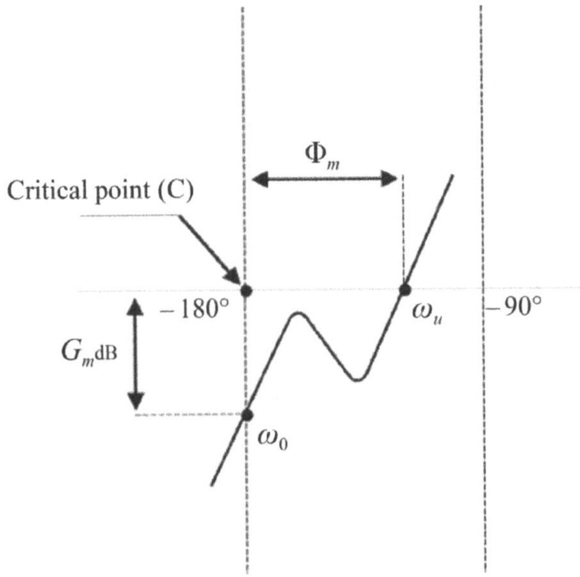

Figure 1.10. *An example of a frequency configuration in open loop, for which the stability margins no longer have a sense*

1.5.5. *Left-hand criterion in the Nichols plane*

1.5.5.1. *Presentation*

The left-hand *criterion* is a *stability criterion* that, in the Nichols plane, is presented as follows: if, when sweeping the open-loop Nichols locus in the increasing ω direction, we keep the critical point, C (0dB, – 180°), on the right, the control loop is stable; it is unstable if point C is kept on the left; and finally, it barely oscillates if the Nichols locus of $\beta(j\omega)$ passes through point C.

1.5.5.2. *Demonstration*

The demonstration of the left-hand criterion, as performed here, results from a *physical study* of the stability.

According to the definition of the frequency response in open loop, that is (for $R(j\omega) = D(j\omega) = 0$)

$$\beta(j\omega) = -\frac{V_s(j\omega)}{V_e(j\omega)}, \tag{1.17}$$

at frequency ω_0, we can write:

$$\beta(j\omega_0) = -\frac{V_s(j\omega_0)}{V_e(j\omega_0)} = -1 \times \frac{V_s(j\omega_0)}{V_e(j\omega_0)}, \tag{1.18}$$

hence:

$$\arg \beta(j\omega_0) = \arg(-1) + \arg \frac{V_s(j\omega_0)}{V_e(j\omega_0)}, \tag{1.19}$$

that is:

$$\arg \beta(j\omega_0) = -180° + \arg \frac{V_s(j\omega_0)}{V_e(j\omega_0)}. \tag{1.20}$$

Yet, $\arg \beta(j\omega_0) = -180°$ given the definition of the frequency, ω_0, therefore:

$$\arg \frac{V_s(j\omega_0)}{V_e(j\omega_0)} = 0, \tag{1.21}$$

which expresses the fact that the input and output sinusoidal signals of the open loop, $v_e(t)$ and $v_s(t)$, vibrate in phase at frequency ω_0, that is $v_e(t) = v_{e_m} \cos(\omega_0 t + \phi)$ and $v_s(t) = v_{s_m} \cos(\omega_0 t + \phi)$; in other words, the output, $v_s(t)$, of the open loop then vibrates in phase with its input, $v_e(t)$; in the declination of the following three scenarios, the input, $v_e(t)$, is assumed to be externally generated (before closing the loop).

With regard to the ratio of amplitudes v_{s_m} and v_{e_m} of $v_s(t)$ and $v_e(t)$, given by the gain

$$\left|\frac{V_s(j\omega_0)}{V_e(j\omega_0)}\right| = |\beta(j\omega_0)|, \qquad [1.22]$$

let us consider the following three scenarios.

Note that $|\beta(j\omega_0)| < 1$, that is, $|\beta(j\omega_0)|\text{dB} < 0$, therefore $v_{s_m} < v_{e_m}$; hence, after closing the loop, there is a *convergent oscillation phenomenon* (converging toward zero) after passing through the loop several times, thus demonstrating the control *stability*. Also, $|\beta(j\omega_0)| = 1$, that is, $|\beta(j\omega_0)|\text{dB} = 0$, therefore $v_{s_m} = v_{e_m}$; hence, after closing the loop, there is a *self-sustained oscillation phenomenon*, after the first passage through the loop, thus demonstrating the *barely oscillating nature* of the control (the output amplitude of which remains within the linear domain). Finally, $|\beta(j\omega_0)| > 1$, that is, $|\beta(j\omega_0)|\text{dB} > 0$, therefore $v_{s_m} > v_{e_m}$; hence, after closing the loop, there is a *divergent oscillation phenomenon* (diverging linearly to infinity) after passing through the loop several times, thus demonstrating the *instability* of the control.

First remark: concerning the three scenarios considered, passing from open loop to closed loop by closing the loop is an operation stemming from thought that through this operation, the output oscillation of the open loop in response to the input oscillation can replace the input oscillation initially applied and then be processed by the loop.

Second remark: concerning the second scenario, it should be emphasized that the *oscillation condition*

$$|\beta(j\omega_0)| = 1 \text{ and } \arg \beta(j\omega_0) = -180°, \qquad [1.23]$$

that is,

$$\beta(j\omega_0) = -1, \qquad [1.24]$$

is in accordance with the oscillation condition defined by the *Barkhausen criterion* (which is best known to electronics engineers), that is,

$$\frac{V_s(j\omega_0)}{V_e(j\omega_0)} = 1, \qquad [1.25]$$

this compliance originating from the relationship:

$$\beta(j\omega_0) = -\frac{V_s(j\omega_0)}{V_e(j\omega_0)}. \quad [1.26]$$

1.5.5.3. *Regarding the spontaneous generation of a self-sustained oscillation*

1.5.5.3.1. Discussion of recorded phenomena

The explanation of the three oscillating phenomena as defined is based:

– on a sinusoidal oscillation, $v_e(t) = v_{e_m} \cos(\omega_0 t + \phi)$, applied at the input of the open loop;

– then on the comparison, at amplitude v_{e_m}, of amplitude v_{s_m} of the sinusoidal oscillation, $v_s(t) = v_{s_m} \cos(\omega_0 t + \phi)$, collected at the output of the open loop;

– with the result of this comparison ($v_{s_m} < v_{e_m}$, $v_{s_m} = v_{e_m}$ or $v_{s_m} > v_{e_m}$) at the origin of the behavior of the control after closing the loop.

While this explanation, which falls within the *context of the loop being opened then closed*, certainly presents the benefit of setting out how the oscillatory phenomena in question manifest themselves in simple terms, this should not lead to the conclusion that these phenomena only arise in such a context (or, more simply, that these phenomena are conditioned by this context).

It is true to say that in reality, we are aware that these phenomena manifest themselves spontaneously, without the loop being opened, therefore within an *exclusively closed-loop context* where no sinusoidal oscillation of frequency ω_0 is applied.

1.5.5.3.2. A legitimate question

So what is the reality, and in particular, what can we say about the spontaneous generation of a self-sustained oscillation (or “self-oscillation”)?

More precisely, how can an oscillator that itself generates a sinusoidal oscillation, of frequency ω_0, become excited? This is certainly a legitimate question, which raises the well-known problem of *oscillator excitation*.

1.5.5.3.3. The answer to this question

The answer to the problem of oscillator excitation indisputably lies in the *noise*, which comprises all of the spectral components, and in particular, that of frequency

ω_0, which, being alone in responding to the oscillation condition, will be alone in giving rise to a self-sustained oscillation.

1.5.5.3.4. Stabilization of the self-sustained oscillation amplitude

To set the amplitude, y_m, of the output oscillation, $y(t) = y_m \cos(\omega_0 t + \phi)$, to a desired value, y_{m_d}, it is necessary for the gain in open loop at frequency ω_0, $|\beta(j\omega_0)|$, to be a function of y_m, such that (see Figure 1.11):

$$|\beta(j\omega_0)| > 1 \text{ for } y_m < y_{m_d}, \qquad [1.27]$$

divergent oscillation phenomenon in transient state;

$$|\beta(j\omega_0)| = 1 \text{ for } y_m = y_{m_d}, \qquad [1.28]$$

self-sustained oscillation phenomenon (or barely oscillating phenomenon) in steady state;

$$|\beta(j\omega_0)| < 1 \text{ for } y_m > y_{m_d}, \qquad [1.29]$$

convergent oscillation phenomenon that gives rise to nothing in steady state.

To this end, it is appropriate to introduce a *nonlinear amplification* into the loop, notably via a nonlinear amplifier incorporated into the controller. Although this amplifier is *linear* for a given amplitude of the input or output, with the output indeed responding sinusoidally to a sinusoidal input, it is *nonlinear* for a variation of the amplitude: indeed, while the input amplitude varies by a given factor, the output amplitude does not vary in the same ratio, even if the output still varies sinusoidally.

An amplifier of this kind serves as an example of an achievement, equipping a *sinusoidal oscillator of order 5/2*, as mentioned in Oustaloup [OUS 83b]. This example is based on an operational amplifier and a *field-effect transistor*, used in variable resistance between drain and source, and controlled by the gate by a negative direct voltage, proportional to the amplitude of the output signal.

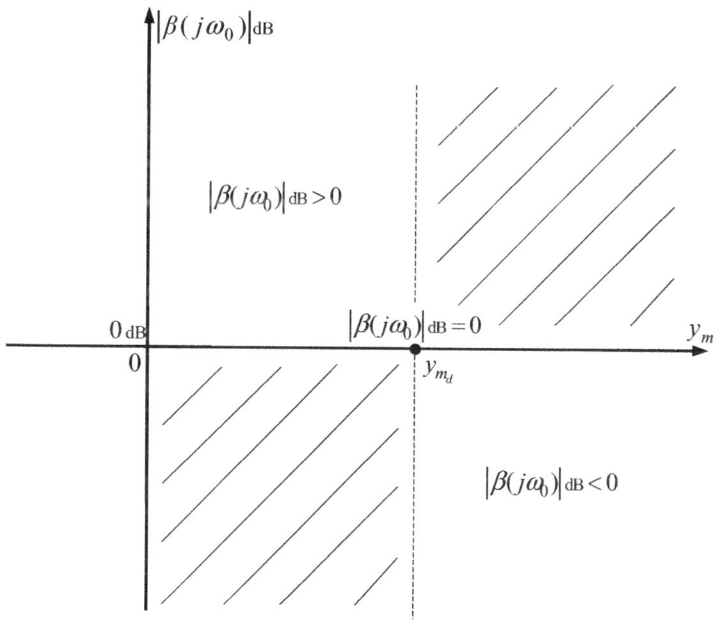

Figure 1.11. *The non-hatched areas define those areas where $|\beta(j\omega_0)|_{dB}$ belongs, enabling y_m to be set at y_{m_d}*

1.6. Closed-loop transmittances

The block diagram in Figure 1.4 enables the Laplace transform of the output, $y(t)$, to be expressed, that is:

$$Y(s) = D(s) + G(s)C(s)\left[K_1R(s) - K_2Y(s)\right], \qquad [1.30]$$

or, by gathering the terms in $Y(s)$:

$$Y(s)\left[1 + C(s)G(s)K_2\right] = K_1C(s)G(s)R(s) + D(s), \qquad [1.31]$$

from which we draw the general expression of the output, $Y(s)$, as a function of stresses $R(s)$ and $D(s)$, that is:

$$Y(s) = K\frac{\beta(s)}{1+\beta(s)}R(s) + \frac{D(s)}{1+\beta(s)}, \qquad [1.32]$$

forming the ratio

$$K = \frac{K_1}{K_2} \qquad [1.33]$$

and using the relationship established above,

$$\beta(s) = C(s)G(s)K_2. \qquad [1.34]$$

The additive nature of the general expression of $Y(s)$ defined by the relationship [1.32] leads to the definition of two transmittances characterizing the influence of the stresses on the output, one relative to $R(s)$ and the other relative to $D(s)$.

1.6.1. *Transmittance in tracking*

1.6.1.1. *Definition and general form*

The *transmittance relating to the reference* (or *transmittance in tracking*) is defined by the ratio:

$$F_r(s) = \left[\frac{Y(s)}{R(s)}\right]_{D(s)=0}, \qquad [1.35]$$

that is, taking [1.32] into account:

$$F_r(s) = K\frac{\beta(s)}{1+\beta(s)}. \qquad [1.36]$$

The general form of expression of $F_r(s)$ thus obtained suggests the definition of a normalized transmittance, that is,

$$\mathbb{F}_r(s) = \frac{F_r(s)}{K} = \frac{\beta(s)}{1+\beta(s)}, \qquad [1.37]$$

known as *reduced transmittance in tracking* or *complementary sensitivity* (noted T or $\overline{S}$ in contemporary literature).

1.6.1.2. *Tracking function*

Transmittance in tracking, $F_r(s)$, characterizes the *tracking function*, which reflects the control loop (output)'s ability to *track* the reference (of course up to a factor K).

The performances of the tracking function can be simulated from the transmittance in open loop, $\beta(s)$, in accordance with the block diagram in Figure 1.12, which represents the cascaded transfer, K, with a *unit-reaction loop*, the action chain of which is formed by $\beta(s)$.

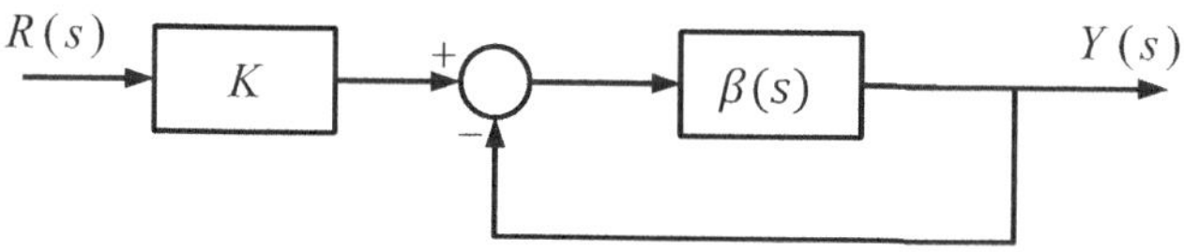

Figure 1.12. *Functional representation of* $F_r(s) = K\beta(s)(1+\beta(s))^{-1}$

1.6.2. *Transmittance in regulation*

1.6.2.1. *Definition and general form*

The transmittance relating to the disturbance (or *transmittance in regulation)* is defined by the ratio:

$$F_d(s) = \left[\frac{Y(s)}{D(s)}\right]_{R(s)=0}, \qquad [1.38]$$

that is, taking [1.32] into account:

$$F_d(s) = \frac{1}{1+\beta(s)}. \qquad [1.39]$$

The general form of expression of $F_d(s)$ thus obtained is well known under the designation, that is, *sensitivity* (noted S in contemporary literature).

Combining relationships [1.37] and [1.39] gives:

$$F_r(s) + F_d(s) = 1. \qquad [1.40]$$

That is, given the existing notations ($F_r(s) = T = \overline{S}$ and $F_d(s) = S$):

$$T + S = 1, \qquad [1.41]$$

or

$$\overline{S} + S = 1, \qquad [1.42]$$

with these expressions reflecting that $T = \overline{S}$ is the complement to 1 of sensitivity, S, thus justifying the designation, "complementary sensitivity", attributed to T.

1.6.2.2. *Regulation function*

Transmittance in regulation, $F_d(s)$, characterizes the *regulation function*, which reflects the control loop (output)'s ability to *not track* (or *reject*) the disturbance.

The performances of the regulation function can be simulated from the transmittance in open loop, $\beta(s)$, in accordance with the block diagram in Figure 1.13, which directly represents a *unit-action loop*, the reaction chain of which is formed by $\beta(s)$.

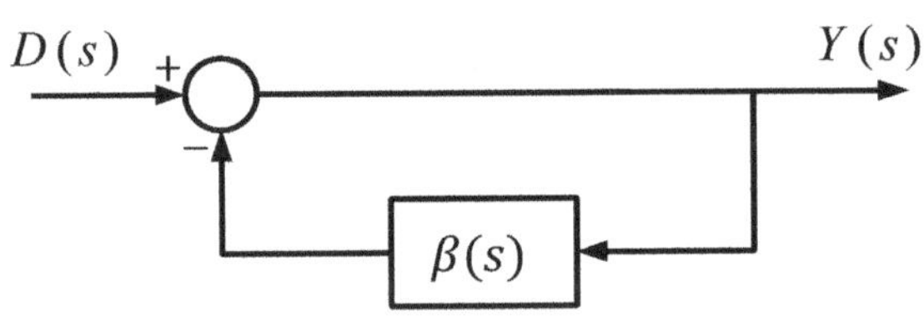

Figure 1.13. *Functional representation of* $F_d(s) = (1 + \beta(s))^{-1}$

1.6.3. *A tracking-regulation dilemma?*

1.6.3.1. *One question*

Given the interest of identifying dilemmas to best manage performance compromises, in the study of closed-loop behavior, is *tracking-regulation duality* not likely to raise questions as to whether or not a tracking-regulation dilemma exists? It is precisely this question of circumstance that forms the basis of the following section.

1.6.3.2. *A well-reasoned response*

At frequencies below ω_u, for which the open-loop gain is greater than one, that is, $|\beta(j\omega)| > 1$, any increase in this gain jointly ensures both:

– a variation toward K of $F_r(j\omega) = K\dfrac{\beta(j\omega)}{1+\beta(j\omega)}$;

– a variation toward zero of $F_d(j\omega) = \dfrac{1}{1+\beta(j\omega)}$.

These two concomittant effects indeed express the joint improvement of:

– the *reference tracking*, via the variation toward K of the frequency response in tracking, $F_r(j\omega)$;

– the *disturbance rejection*, via the variation toward zero of the frequency response in regulation, $F_d(j\omega)$.

Thus, there is an improvement to the reference tracking, accompanied by an improvement to the disturbance rejection, and, conversely, *tracking and regulation are not contradictory requirements*, which excludes the existence of any dilemma formulated in terms of tracking and regulation.

1.7. Input sensitivity

The *(plant) input sensitivity* characterizes the effect of stresses on the plant input.

In tracking, it is defined by the gain:

$$\mathcal{S}_r(\omega) = \left|\frac{U(j\omega)}{R(j\omega)}\right|_{D(j\omega)=0}, \qquad [1.43]$$

and, in regulation, by the gain:

$$\mathcal{S}_d(\omega) = \left|\frac{U(j\omega)}{D(j\omega)}\right|_{R(j\omega)=0}. \qquad [1.44]$$

The gains formulated are indeed significant of the plant input sensitivity to the stresses that are the reference and the disturbance.

1.7.1. *Input sensitivity in tracking*

For $D(j\omega) = 0,$ the block diagram in Figure 1.4 enables us to write:

$$U(j\omega) = C(j\omega)\left[K_1 R(j\omega) - K_2 G(j\omega)U(j\omega)\right], \tag{1.45}$$

that is:

$$U(j\omega)\left[1 + K_2 C(j\omega)G(j\omega)\right] = K_1 C(j\omega)R(j\omega), \tag{1.46}$$

from which we obtain the input sensitivity in tracking:

$$\mathcal{S}_r(\omega) = \left|\frac{K_1 C(j\omega)}{1 + \beta(j\omega)}\right|, \tag{1.47}$$

with $1/\mathcal{S}_r(\omega)$ defining the *input immunity in tracking*.

1.7.2. *Input sensitivity in regulation*

For $R(j\omega) = 0,$ the block diagram in Figure 1.4 enables us to write:

$$U(j\omega) = C(j\omega)\left(-K_2\left[D(j\omega) + G(j\omega)U(j\omega)\right]\right), \tag{1.48}$$

that is:

$$U(j\omega)\left[1 + K_2 C(j\omega)G(j\omega)\right] = -K_2 C(j\omega)D(j\omega), \tag{1.49}$$

from which we obtain the *input sensitivity in regulation*:

$$\mathcal{S}_d(\omega) = \left|\frac{K_2 C(j\omega)}{1 + \beta(j\omega)}\right|, \tag{1.50}$$

with $1/\mathcal{S}_d(\omega)$ defining the *input immunity in regulation*.

1.7.3. *Input sensitivity in tracking and regulation*

In the case of the elementary control loop ($K_1 = K_2 = 1$), the expressions [1.47] and [1.50] identify, respectively, with the same expression, which defines the *input sensitivity in tracking and regulation*, that is:

$$\mathcal{S}(\omega) = \mathcal{S}_r(\omega) = \mathcal{S}_d(\omega) = \left|\frac{C(j\omega)}{1+\beta(j\omega)}\right|, \qquad [1.51]$$

with $1/\mathcal{S}(\omega)$ defining the *input immunity in tracking and regulation.*

With ω_u designating the unit gain frequency in open loop, $(|\beta(j\omega_u)| = 1)$, with for $\omega/\omega_u << 1$, $|\beta(j\omega)|$ then being large with respect to the unit, $\mathcal{S}(\omega)$ is reduced to:

$$\mathcal{S}(\omega) = \left|\frac{C(j\omega)}{\beta(j\omega)}\right| = \left|\frac{C(j\omega)}{C(j\omega)G(j\omega)}\right| = \frac{1}{|G(j\omega)|}, \qquad [1.52]$$

that is, in decibels (see Figure 1.14):

$$\mathcal{S}(\omega)\text{dB} = -\,20\log|G(j\omega)| = -|G(j\omega)|\,\text{dB}\,. \qquad [1.53]$$

For $\omega/\omega_u >> 1$, $|\beta(j\omega)|$ then being small with respect to the unit, $\mathcal{S}(\omega)$ is reduced to:

$$\mathcal{S}(\omega) = |C(j\omega)|, \qquad [1.54]$$

that is, in decibels (see Figure 1.14):

$$\mathcal{S}(\omega)\text{dB} = 20\log|C(j\omega)| = |C(j\omega)|\,\text{dB}\,. \qquad [1.55]$$

The gain diagrams relative to [1.53] and [1.55] intersect at frequency ω_u (see Figure 1.14), because from the relationship,

$$|C(j\omega_u)||G(j\omega_u)| = 1, \qquad [1.56]$$

that is,

$$|C(j\omega_u)|\text{dB} + |G(j\omega_u)|\text{dB} = 0, \qquad [1.57]$$

we obtain:

$$|C(j\omega_u)|\text{dB} = -|G(j\omega_u)|\text{dB}. \qquad [1.58]$$

The shape of the curve of $-|G(j\omega)|\text{dB}$ respects the low-pass character of the plant, $G(j\omega)$. With regard to the shape of the curve of $|C(j\omega)|\text{dB}$, it observes the gain increase with the frequency that accompanies the phase lead of the regulator, $C(j\omega)$, around the frequency, ω_u.

When the frequency, ω_u, increases, the gain diagram of the regulator rises in accordance with the displacement of the intersection point, a rise which, beyond ω_u, is accentuated by the joint growth of the gain slope of the regulator at frequency, ω_u, thus reflecting that the high-frequency input sensitivity increases with ω_u. The growth (with ω_u) of the gain slope of the regulator is explained by the growth (with ω_u) of the regulator phase lead, which must compensate for the growth (with ω_u) of the plant phase-lag to maintain the same phase margin of the control.

Looking to Figure 1.14, it appears that, as a first approximation, the input sensitivity in tracking and regulation is:

– at low frequencies ($\omega << \omega_u$), given exclusively by the *inverse of the plant gain* (thus expressing that the low-frequency input sensitivity does not constitute a constraint for the regulator synthesis);

– at medium frequencies (around ω_u), a function of both the plant and the regulator;

– at high frequencies ($\omega >> \omega_u$), given exclusively by the *regulator gain* (thus expressing that the high-frequency input sensitivity constitutes, in this case, a constraint for the regulator synthesis).

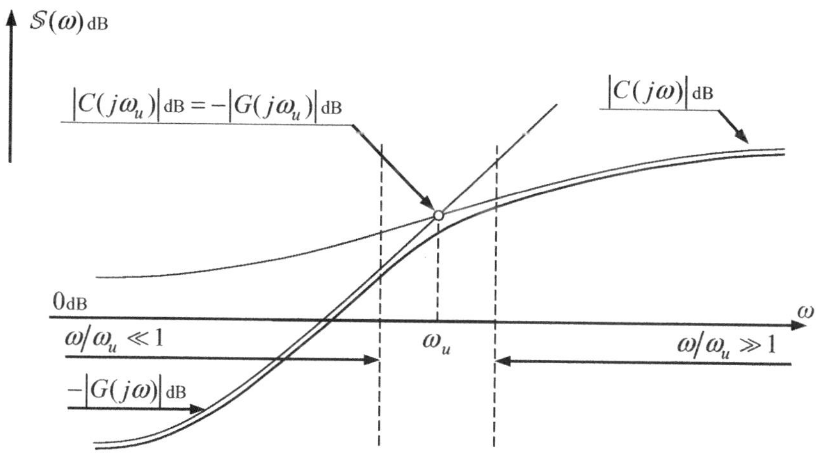

Figure 1.14. *Frequency variation (in decibels) of the input sensitivity in tracking and regulation*

In other words, in the regulator synthesis, *the high-frequency regulator gain needs to be determined with utmost attention.* Notably, it must not tend toward infinity with the frequency, to ensure that the regulator does not produce a high-frequency noise amplifier. Therefore, a "red card" can be attributed to the *academic PID*, whose high-frequency gain tends toward infinity with frequency. The following presentation of this PID offers a brief, simple explanation of this problem and introduces the *non-academic PID* as a solution.

1.7.3.1. *Academic PID*

The "academic" PID controller is defined by the elementary transmittance, in additive form:

$$PID(s) = k + \frac{\omega'_i}{s} + \frac{s}{\omega'_d}, \qquad [1.59]$$

while the three parameters, k, ω'_i and ω'_d, are defined as follows:

k : transfer factor of the *proportional action*;

ω'_i : transition (or unit gain) frequency of the *integral action*, ω'_i/s;

ω'_d : transition (or unit gain) frequency of the *differential action*, s/ω'_d.

By factoring as suggested by the relationship [1.59], the controller transmittance can be rewritten in the form:

$$PID(s) = k\left(1 + \frac{\omega'_i}{ks} + \frac{s}{k\omega'_d}\right), \qquad [1.60]$$

that is, by writing $\omega_i = \omega'_i/k$ and $\omega_d = k\omega'_d$:

$$PID(s) = k\left(1 + \frac{\omega_i}{s} + \frac{s}{\omega_d}\right), \qquad [1.61]$$

from which we obtain the frequency response:

$$PID(j\omega) = k\left(1 + \frac{\omega_i}{j\omega} + j\frac{\omega}{\omega_d}\right), \qquad [1.62]$$

a form that calls for the *reduced frequency response*:

$$PID(j\omega)/k = 1 + \frac{\omega_i}{j\omega} + j\frac{\omega}{\omega_d}. \qquad [1.63]$$

The new transition frequencies, ω_i and ω_d, such that $\omega_i < \omega_d$, lead to the frequency space being divided into three parts, defined, respectively, by:

$$\omega < \omega_i,\ \omega_i < \omega < \omega_d \text{ and } \omega > \omega_d, \qquad [1.64]$$

with these three sub-spaces enabling the following reductions, which are at the origin of the Bode asymptotic diagrams of $PID(j\omega)/k$.

For $\omega << \omega_i$, and therefore for $\omega << \omega_d$, which expresses $\omega_i/\omega >> 1$ and $\omega/\omega_d << 1$, $PID(j\omega)/k$ is reduced to:

$$PID(j\omega)/k = \frac{\omega_i}{j\omega}, \qquad [1.65]$$

an integral action of transition frequency, ω_i.

For $\omega >> \omega_d$, and therefore for $\omega >> \omega_i$, which expresses $\omega/\omega_d >> 1$ and $\omega_i/\omega << 1$, $PID(j\omega)/k$ is reduced to:

$$PID(j\omega)/k = j\frac{\omega}{\omega_d}, \qquad [1.66]$$

a differential action of transition frequency, ω_d.

Finally, for $\omega_i << \omega << \omega_d$, which expresses $\omega_i/\omega << 1$ and $\omega/\omega_d << 1$, $PID(j\omega)/k$ is reduced to:

$$PID(j\omega)/k = 1, \qquad [1.67]$$

a unitary, proportional action.

These three reductions determine the Bode asymptotic diagrams of $PID(j\omega)/k$, as presented by Figure 1.15.

The Bode asymptotic diagrams thus obtained can be interpreted as resulting from the cascading of:

– a frequency-response *proportional-integral controller*

$$PI(j\omega) = 1 + \frac{\omega_i}{j\omega} \qquad [1.68]$$

– a frequency-response *proportional-differential controller*

$$PD(j\omega) = 1 + j\frac{\omega}{\omega_d}, \qquad [1.69]$$

with the characteristic frequencies ω_i and ω_d then being the transitional frequencies of the PI and PD controllers.

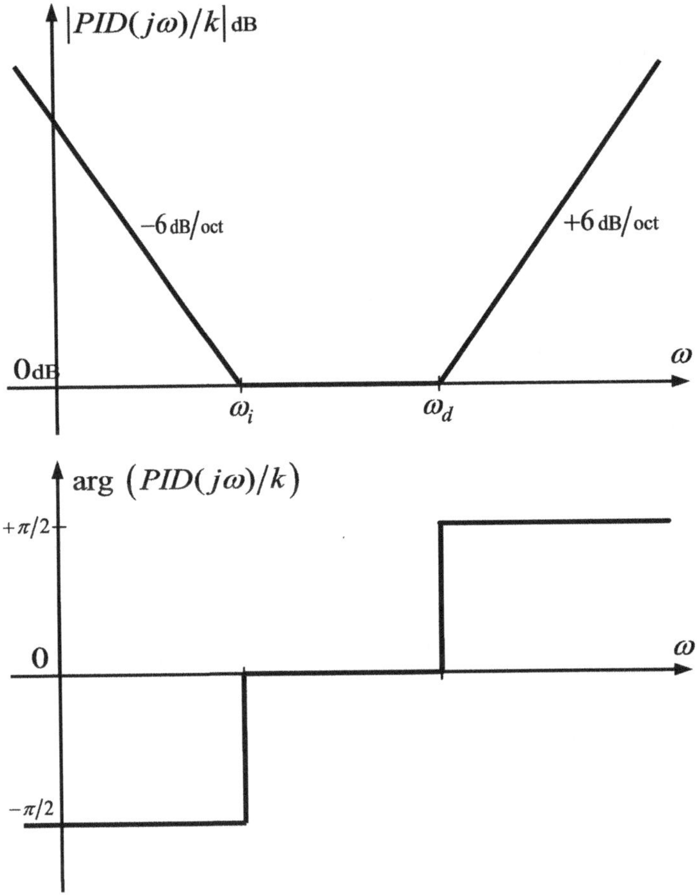

Figure 1.15. *Bode asymptotic diagrams of the academic PID*

Indeed, the product of these frequency responses, that is,

$$PI(j\omega)PD(j\omega) = \left(1 + \frac{\omega_i}{j\omega}\right)\left(1 + j\frac{\omega}{\omega_d}\right), \qquad [1.70]$$

admits, as a development:

$$PI(j\omega)PD(j\omega) = 1 + \frac{\omega_i}{j\omega} + j\frac{\omega}{\omega_d} + \frac{\omega_i}{\omega_d}, \qquad [1.71]$$

or, for $\omega_i / \omega_d << 1$:

$$PI(j\omega)PD(j\omega) = 1 + \frac{\omega_i}{j\omega} + j\frac{\omega}{\omega_d}, \qquad [1.72]$$

or, taking [1.63] into account:

$$PI(j\omega)PD(j\omega) = PID(j\omega)/k, \qquad [1.73]$$

or indeed:

$$k\,PI(j\omega)PD(j\omega) = PID(j\omega). \qquad [1.74]$$

Thus, for a significant ratio between the characteristic frequencies, ω_d and ω_i $(\omega_d / \omega_i >> 1)$, the *multiplicative form* of the PID controller, that is,

$$k\left(1 + \frac{\omega_i}{j\omega}\right)\left(1 + j\frac{\omega}{\omega_d}\right), \qquad [1.75]$$

is equivalent to its *additive form*, that is:

$$k\left(1 + \frac{\omega_i}{j\omega} + j\frac{\omega}{\omega_d}\right). \qquad [1.76]$$

It is then possible to make use of this *equivalence* to retain and make the best use of the multiplicative form, which is preferable to the additive form, owing to the computational advantages it offers and its closer correspondence to Bode asymptotic diagrams when the characteristic frequencies, ω_d and ω_i approach, or even identify with oneanother, this being confirmed by the computations conducted for $\omega_i = \omega_d = \omega_{id}$ and $\omega = \omega_{id}$.

1.7.3.2. *Non-academic PID*

In the case of the academic PID controller, the input sensitivity at high frequencies, given by the controller gain at these frequencies, admits an expression of the form:

$$\underset{\omega \gg \omega_d}{\mathcal{S}(\omega)} = \underset{\omega \gg \omega_d}{|PID(j\omega)|} = k\frac{\omega}{\omega_d}, \qquad [1.77]$$

a result that is obtained from the multiplicative or additive form of the controller, and that clearly demonstrates that through the controller gain for $\omega >> \omega_d$, the high-frequency input sensitivity tends to infinity with the frequency.

The solution to this problem thus consists of *truncating the differential action of the academic PID* using a first-order low-pass filter, the transitional frequency of which, ω_t, determines the *truncation frequency of the differential action*. The cascading of this low-pass filter with the multiplicative form of the academic PID is then reflected by the following frequency response, which defines the non-academic PID controller:

$$PID(j\omega) = k\left(1 + \frac{\omega_i}{j\omega}\right)\frac{1 + j\dfrac{\omega}{\omega_d}}{1 + j\dfrac{\omega}{\omega_t}}, \qquad [1.78]$$

with $\omega_i < \omega_d < \omega_t$, the Bode asymptotic diagrams of $PID(j\omega)/k$ being presented in Figure 1.16.

The frequency response thus obtained can be interpreted as resulting from the cascading of:

– a *proportional controller* of gain k;

– a frequency-response *proportional-integral controller*

$$1 + \frac{\omega_i}{j\omega}; \qquad [1.79]$$

– a frequency-response *phase-lead controller* (see Chapter 2)

$$\frac{1 + j\dfrac{\omega}{\omega_d}}{1 + j\dfrac{\omega}{\omega_t}}. \qquad [1.80]$$

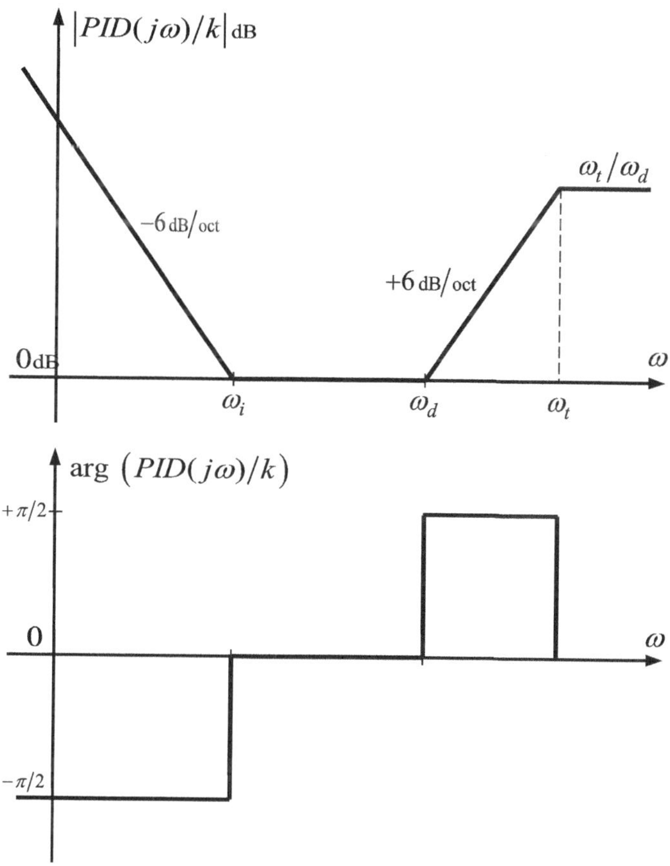

Figure 1.16. *Bode asymptotic diagrams of the non-academic PID*

The input sensitivity at high frequencies, given by the controller gain at these frequencies, then admits a new expression of the form:

$$\underset{\omega \gg \omega_t}{\mathcal{S}(\omega)} = \underset{\omega \gg \omega_t}{|PID(j\omega)|} = k\frac{\omega_t}{\omega_d}, \qquad [1.81]$$

this result demonstrating that the high-frequency input sensitivity is now reduced to a constant (in this case $k\,\omega_t/\omega_d$).

1.8. Behavior and frequency performances in tracking and regulation

By way of introduction, in order to fully understand the beginning of the next section, it is important to present the stability degree–precision dilemma, and then specify the principle of bringing it into question, which ensures the robustness of the control when the parametric variations of the plant lead only to open-loop gain variations around the frequency, ω_u.

The stability degree–precision dilemma expresses that any increase in the open-loop gain, aiming to improve precision, leads to a decrease in the stability degree by decreasing the stability margins (significant of the distance to the critical point).

Its calling into question is ensured by an open-loop Nichols locus, which forms, around frequency ω_u, a vertical-line segment, known as the template, which is sufficiently long, with an abscissa between $-\pi/2$ and $-\pi$ (see Figure 1.17): indeed, this straight-line segment, which slides on itself in the event of any gain variation in open loop, ensures the invariance of the phase margin, which alone signifies the distance to the critical point (since it is the smallest stability margin owing to the sufficient length of the template).

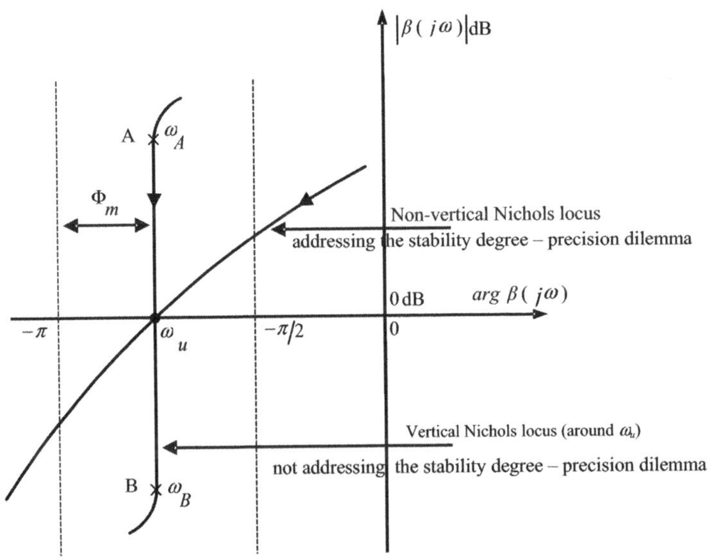

Figure 1.17. *Calling into question of the stability degree–precision dilemma by the (vertical) template of the second-generation CRONE control, and this template being described by:* $\beta(j\omega) = \left(\frac{\omega_u}{j\omega}\right)^n$ *with* $1 < n < 2$ *and* $\omega_A \le \omega \le \omega_B$

1.8.1. *Frequency and time behavior*

If in the second-generation CRONE control (see Chapter 5), we consider a template extending across the entire frequency domain, we obtain a frequency response in open loop, $\beta(j\omega)$, which identifies a non-integer integrator, that is:

$$\beta(j\omega) = \left(\frac{\omega_u}{j\omega}\right)^n \text{ with } 1 < n < 2 \qquad [1.82]$$

The general expressions of the closed-loop frequency responses, directly deduced from [1.36] and [1.39], that is:

$$F_r(j\omega) = K\frac{\beta(j\omega)}{1+\beta(j\omega)} \qquad [1.83]$$

and

$$F_d(j\omega) = \frac{1}{1+\beta(j\omega)}, \qquad [1.84]$$

can then be particularized according to the relationships:

$$F_r(j\omega) = \frac{K}{1+\left(j\frac{\omega}{\omega_u}\right)^n} \qquad [1.85]$$

and

$$F_d(j\omega) = \frac{\left(j\frac{\omega}{\omega_u}\right)^n}{1+\left(j\frac{\omega}{\omega_u}\right)^n}. \qquad [1.86]$$

The corresponding gain asymptotic diagrams, represented in Figure 1.18, reflect that:

– first, *the control loop behaves like a low-pass filter in tracking and a high-pass filter in regulation*;

– second, $F_r(j\omega)$ *and* $F_d(j\omega)$ *have the same transitional frequency*, in this case ω_u, expressing that (temporally) *the dynamics rapidity is the same in tracking and regulation*.

The *open-loop transition frequency*, ω_u, therefore determines the *closed-loop transitional frequency*, that is, the *transitional frequency in tracking and regulation.* The existence of the same transitional frequency in both tracking and regulation expresses (frequentially) the identity of the *transmission band in tracking* and the *rejection band in regulation*; these two bands can be referred to using a common designation, namely the *(control) transmission and rejection band*, because, for the control, this frequency band is both a *transmission band* (implied to be in tracking) and a *rejection band* (implied to be in regulation).

The gain asymptotic diagrams in Figure 1.18 suggest a comparative analysis of the corresponding temporal behaviors to be conducted via the unit-step responses in tracking and regulation, $y_r(t)$ and $y_d(t)$.

Concerning the reference or disturbance unit step:

– the *instantaneous transition* of the step generates infinite frequencies at instant zero; as Figure 1.18 indicates, these frequencies are not transmitted in tracking (therefore $y_r(0^+) = 0$), but are entirely transmitted in regulation (therefore $y_d(0^+) = 1$);

– the step *level*, which becomes a continuous signal long after the transition (which can then be forgotten), generates, at infinite instants, a zero frequency; as shown in Figure 1.18, this frequency is transmitted with the factor K in tracking (therefore $y_r(\infty) = K$), but is not transmitted in regulation (therefore $y_d(\infty) = 0$).

Thus, for $t > 0$ (with t varying from zero to infinity), $y_r(t)$ evolves from 0 to K, while $y_d(t)$ evolves from 1 to 0, these evolutions being compliant with the general shape of the step responses of a low-pass filter and a high-pass filter.

In fact, these results are obtained without computation from the principle drawn from our analysis, according to which: “when the time domain is swept to the right (with time varying from zero to infinity), the frequency domain is swept to the left (with frequency indeed varying from infinity to zero)”.

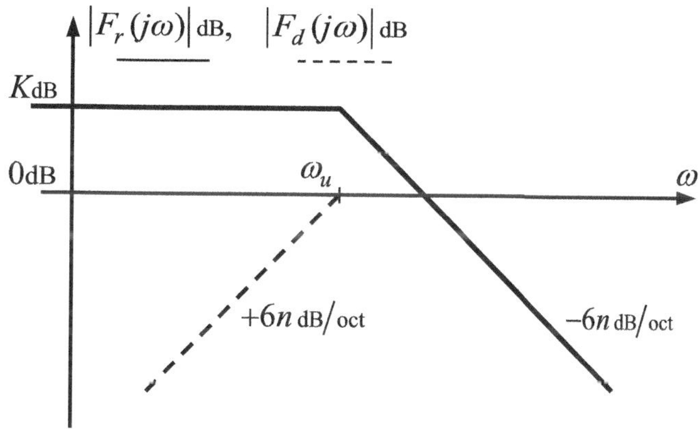

Figure 1.18. *Gain asymptotic diagrams in tracking and regulation*

1.8.2. *Frequency performances*

Given that the control loop presents two very distinct behaviors in tracking and regulation (those of low-pass and high-pass filters), it is useful to distinguish between the frequency performances of $F_r(j\omega)$ and $F_d(j\omega)$, notably the cut-off frequency and the resonance frequency and ratio.

To define these performances in accordance with the dual context of low-pass and high-pass filters, it is appropriate to define beforehand, for each type of filter, the *reference gain*, which is the gain at *a particular frequency, characteristic of the frequencies that the filter is supposed to transmit.* Thus, for a low-pass filter (which is supposed to transmit low frequencies), the reference gain is the *zero-frequency gain*, and for a high-pass filter (which is supposed to transmit high frequencies), the reference gain is the *infinite-frequency gain*: these two gains serve as a reference in the following dual definitions.

1.8.2.1. *Cut-off frequency*

In tracking, the cut-off frequency at –6dB, ω_c, satisfies the equation:

$$|F_r(j\omega_c)| = \frac{1}{2}|F_r(j0)| \quad [1.87]$$

In regulation, it satisfies the equation:

$$\left|F_d(j\omega_c)\right| = \frac{1}{2}\left|F_d(j\infty)\right| \qquad [1.88]$$

In both cases, the frequency gain, ω_c, is equal to half of the reference gain.

1.8.2.2. *Resonance frequency*

In tracking, the resonance frequency, ω_r , is such that:

$$\left|F_r(j\omega_r)\right| = \left|F_r(j\omega)\right|_{\max}. \qquad [1.89]$$

In regulation, it is such that:

$$\left|F_d(j\omega_r)\right| = \left|F_d(j\omega)\right|_{\max}. \qquad [1.90]$$

In both cases, the frequency gain, ω_r , is equal to the maximum gain.

1.8.2.3. *Resonance ratio*

In tracking, the resonance ratio, Q, is defined by the ratio:

$$Q = \frac{\left|F_r(j\omega_r)\right|}{\left|F_r(j0)\right|} = Q_r. \qquad [1.91]$$

In regulation, it is given by the ratio:

$$Q = \frac{\left|F_d(j\omega_r)\right|}{\left|F_d(j\infty)\right|} = Q_d. \qquad [1.92]$$

In both cases, the resonance ratio is the ratio between the maximum gain and the reference gain.

1.8.3. *Regarding the effect of the increase in frequency ω_u*

When the frequency ω_u increases, with the transmission and rejection band also increasing (see section 1.8.1), the output of the control loop (and therefore of the plant) then varies more rapidly, although the plant bandwidth (low-pass system) has not for its part increased. So what about the plant input?

The answer to this question indisputably lies in the following property (see section 1.8.3.1), a property according to which *the plant input increases with the output rapidity*, this input increase moreover being in accordance with the rise in input sensitivity, as shown in Figure 1.14 when ω_u increases.

1.8.3.1. *A property of low-pass systems*

The faster the output of a low-pass system varies, the more the input is subject to significant stresses. In other words, for a low-pass system (therefore for the plant), *the more rapid the output, the more significant the input.*

1.8.3.2. *An illustration of this property*

One notable illustration of this property is undoubtedly a gamma RC circuit, whose output varies according to a step.

For an elementary plant, let us therefore consider a gamma RC circuit (see Figure 1.19), the input–output model of which is obtained from the electrical equation:

$$u(t) = R\,i(t) + y(t), \quad [1.93]$$

that is, given that

$$i(t) = C\frac{dy(t)}{dt} \quad [1.94]$$

$$u(t) = RC\frac{dy(t)}{dt} + y(t), \quad [1.95]$$

a well-known differential equation describing the dynamic behavior between $u(t)$ and $y(t)$.

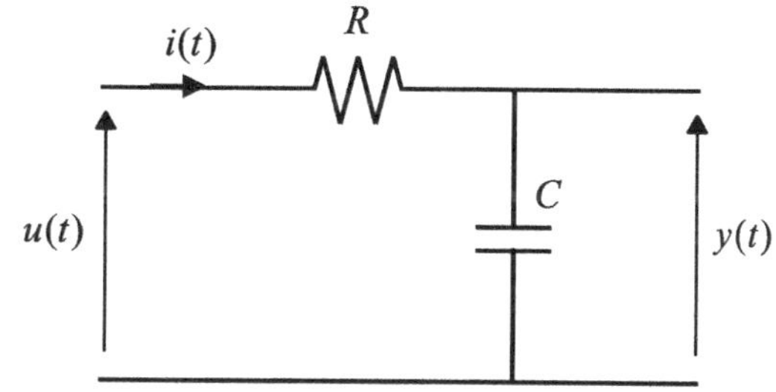

Figure 1.19. *Gamma RC circuit*

The goal is to determine the input, $u(t)$, which ensures a step variation at the output, $y(t)$, with $y(t)$ then presenting a *discontinuity of finite amplitude*, notably unitary for a unit step, that is, $y(t) = 0$ for $t < 0$ and $y(t) = 1$ for $t > 0$.

With $y(t)$ thus being particularized, the differential equation of the RC circuit becomes:

$$u(t) = RC\,\delta(t) + y(t)\,, \qquad [1.96]$$

the presence of the Dirac impulse, $\delta(t)$, expressing that $u(t)$ thus presents a *discontinuity of infinite amplitude.*

1.8.3.3. *A significant conclusion*

For the output of a low-pass system to present a finite amplitude discontinuity, the input must present a discontinuity of infinite amplitude: in other words, owing to the infinitely rapid character of a discontinuity, *an infinitely rapid output of finite amplitude requires an infinitely rapid input of infinite amplitude.*

1.9. Dynamics in tracking and regulation

The general form of transmittances $F_r(s)$ and $F_d(s)$, given by the relationships [1.36] and [1.39], enables us to conduct an immediate comparison of the dynamics in tracking and regulation. Indeed, the general relationship between the step responses in tracking and regulation simply needs to be established, notably by expressing the step response in regulation as a function of the step response in tracking (the latter being better known).

The Laplace transform of the unit step response in regulation, $y_d(t)$, is written as:

$$Y_d(s) = F_d(s)\frac{1}{s} = \frac{1}{1+\beta(s)}\frac{1}{s}\,, \qquad [1.97]$$

that is:

$$Y_d(s) = \left[1 - \frac{\beta(s)}{1+\beta(s)}\right]\frac{1}{s} = \frac{1}{s} - \frac{F_r(s)}{K}\frac{1}{s}\,, \qquad [1.98]$$

or:

$$Y_d(s) = \frac{1}{s} - \frac{1}{K} Y_r(s), \quad [1.99]$$

with $Y_r(s)$ designating the Laplace transform of the unit step response in tracking, $y_r(t)$.

In the time domain, the symbolic equation [1.99] admits, as an original equation:

$$y_d(t) = u(t) - \frac{1}{K} y_r(t), \quad [1.100]$$

that is, for $t > 0$ (given that $u(t)$ is then unitary):

$$y_d(t) = 1 - \frac{1}{K} y_r(t). \quad [1.101]$$

This result expresses that $y_d(t)$ is a *linear function* of $y_r(t)$ and vice versa. These responses therefore have the same nature, aperiodic or oscillatory, and their transience is characterized by the *same dynamic performances*, same reduced first overshoot, same damping ratio and same natural frequency, thus revealing the *same dynamics* in tracking and regulation.

In the case of the elementary control loop ($K = 1$), the relationship [1.101] becomes:

$$y_r(t) + y_d(t) = 1, \quad [1.102]$$

with the sum of the responses then reduced to a simple constant; in this case: one. Figure 1.20 illustrates this scenario with the graphical construction of $y_d(t)$, starting from $y_r(t)$. The initial values, $y_r(0^+) = 0$ and $y_d(0^+) = 1$, corresponding to high-frequency behaviors, as well as the end values, $y_r(\infty) = 1$ and $y_d(\infty) = 0$, corresponding to low-frequency behaviors, show that the control loop behaves like a low-pass filter in tracking and like a high-pass filter in regulation.

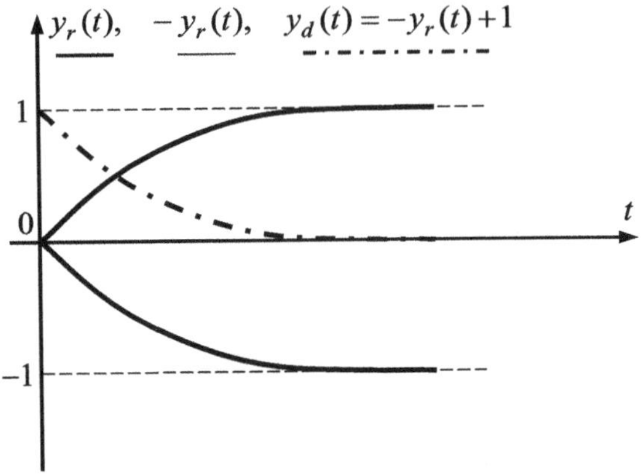

Figure 1.20. *Passage from $y_r(t)$ to $y_d(t)$*

1.10. Charts in tracking and regulation

1.10.1. *Chart in tracking: Nichols chart*

1.10.1.1. *Definition*

Starting from the open-loop frequency-response Nichols locus $\beta(j\omega)$, the Nichols chart (plotted in the Nichols plane) enables us to determine, for each frequency ω, the modulus (in decibels) and the argument (in degrees) of the reduced frequency response in tracking, $\mathbb{F}_r(j\omega)$.

To obtain the modulus and argument of the frequency response in tracking, $F_r(j\omega)$, starting from those of $\mathbb{F}_r(j\omega)$, we simply need to use the passage relationship:

$$F_r(j\omega) = K\,\mathbb{F}_r(j\omega)\,, \qquad [1.103]$$

from which the gain and phase are obtained:

$$\left|F_r(j\omega)\right| \text{dB} = \left|\mathbb{F}_r(j\omega)\right| \text{dB} + K \text{ dB} \qquad [1.104]$$

and

$$\arg F_r(j\omega) = \arg \mathbb{F}_r(j\omega)\,. \qquad [1.105]$$

1.10.1.2. *Construction*

Let us write:

$$\beta(j\omega) = A(\omega)\, e^{j\psi(\omega)} \qquad [1.106]$$

and

$$\mathbb{F}_r(j\omega) = B(\omega)\, e^{j\varphi(\omega)}. \qquad [1.107]$$

The general form of $\mathbb{F}_r(j\omega)$ expressed as a function of $\beta(j\omega)$ then becomes:

$$B(\omega)e^{j\varphi(\omega)} = \frac{A(\omega)\, e^{j\psi(\omega)}}{1 + A(\omega)\, e^{j\psi(\omega)}}, \qquad [1.108]$$

from which we deduce the modulus and argument of $\mathbb{F}_r(j\omega)$ as a function of $A(\omega)$ and $\psi(\omega)$:

$$B(\omega) = \frac{A(\omega)}{\left[1 + A^2(\omega) + 2A(\omega)\cos\psi(\omega)\right]^{1/2}} \qquad [1.109]$$

and

$$\varphi(\omega) = \arctan\frac{\sin\psi(\omega)}{A(\omega) + \cos\psi(\omega)} \qquad [1.110]$$

The Nichols chart is obtained by plotting in the Nichols plane:

– the loci of the coordinate points, $A\text{dB}$, $\psi°$, for which

$$B\text{dB} = 20\log\frac{A}{\left(1 + A^2 + 2A\cos\psi\right)^{1/2}} = \text{cte}, \qquad [1.111]$$

with these loci known as *amplitude (or gain) contours*;

– the loci of the coordinate points, AdB , $\psi°$, for which

$$\varphi° = \arctan \frac{\sin\psi}{A + \cos\psi} = \text{cte} , \qquad [1.112]$$

with these loci known as *phase contours*.

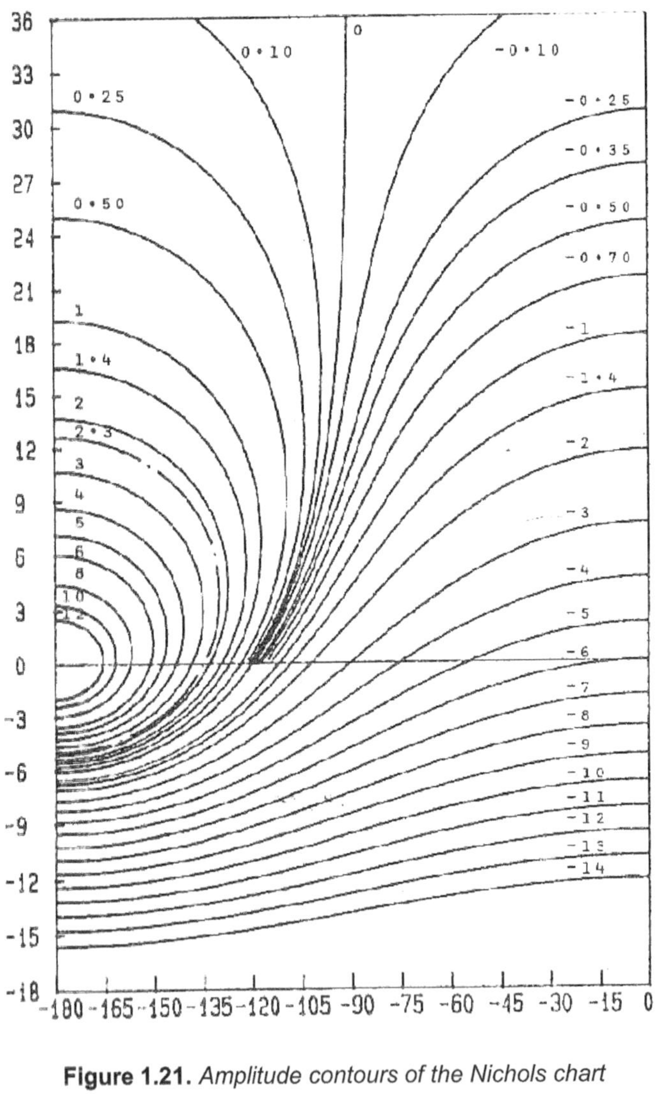

Figure 1.21. *Amplitude contours of the Nichols chart*

These loci are composed of two families of curves presenting a periodicity of 360°. The amplitude contours are represented in Figure 1.21, limiting ψ to the domain most widely used in practice, that is, [0, –180°], with the domain [–180°, –360°] for its part symmetrical about the vertical axis of abscissa –180°.

1.10.1.3. *Use*

Let there be an open-loop frequency-response Nichols locus plotted on the Nichols chart. If a point, M, of this locus corresponding to the frequency, ω_0, can be found, for example, at the intersection of contours $\lambda = -9\text{dB}$ and $\varphi = -60°$, of the Nichols chart, -9dB and $-60°$ constitute, respectively, the modulus and argument of $\mathbb{F}_r(j\omega)$ at frequency ω_0.

1.10.2. *Chart in regulation: dual of the Nichols chart*

1.10.2.1. *Definition*

From the open-loop frequency-response Nichols locus $\beta(j\omega)$, the dual of the Nichols chart (plotted in the Nichols plane) enables us to deduce, for each frequency ω, the modulus (in decibels) and the argument (in degrees) of the frequency response in regulation, $F_d(j\omega)$.

1.10.2.2. *Construction*

By writing

$$\beta(j\omega) = A(\omega)\, e^{j\psi(\omega)} \qquad [1.113]$$

and

$$F_d(j\omega) = B(\omega)\, e^{j\varphi(\omega)}, \qquad [1.114]$$

the general expression of $F_d(j\omega)$ given as a function of $\beta(j\omega)$ is particularized according to the equation:

$$B(\omega)e^{j\varphi(\omega)} = \frac{1}{1 + A(\omega)\, e^{j\psi(\omega)}}, \qquad [1.115]$$

from which we deduce the modulus and argument of $F_d(j\omega)$ as a function of $A(\omega)$ and $\psi(\omega)$:

$$B(\omega) = \frac{1}{\left[1 + A^2(\omega) + 2A(\omega)\cos\psi(\omega)\right]^{1/2}} \qquad [1.116]$$

and

$$\varphi(\omega) = -\arctan\frac{A(\omega)\sin\psi(\omega)}{1 + A(\omega)\cos\psi(\omega)}. \qquad [1.117]$$

The dual of the Nichols chart is obtained by plotting in the Nichols plane:

– the loci of the coordinate points, AdB , $\psi°$, for which:

$$B\text{dB} = 20\log\frac{1}{\left(1 + A^2 + 2A\cos\psi\right)^{1/2}} = \text{cte}, \qquad [1.118]$$

with these loci known as *amplitude (or gain) contours*;

– the loci of the coordinate points, AdB, $\psi°$, for which:

$$\varphi° = -\arctan\frac{A\sin\psi}{1 + A\cos\psi} = \text{cte}, \qquad [1.119]$$

with these loci known as *phase contours*.

As with the Nichols chart, these loci are composed of two families of curves having a periodicity of 360°. The amplitude contours are represented in Figure 1.22, limiting ψ to the domain [0, –180°], with the domain [–180°, –360°] for its part being symmetrical about the vertical axis of abscissa –180°.

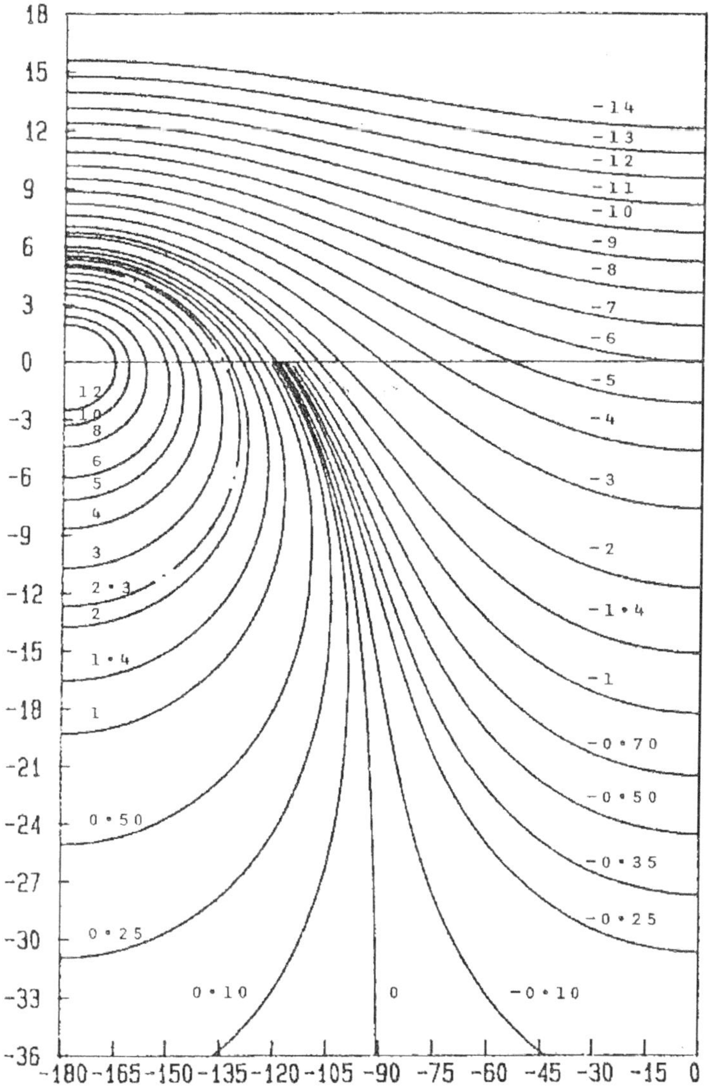

Figure 1.22. *Amplitude contours of the dual of the Nichols chart*

1.10.2.3. *Use*

This type of chart is used in the same way as the Nichols chart. Given an open-loop frequency-response Nichols locus plotted on this chart, if a point, M, of this locus corresponding to the frequency, ω_0, can be found, for example, at the intersection of contours $\lambda = -0.5\text{dB}$ and $\varphi = 10°$, -0.5dB and $10°$ constitute, respectively, the modulus and argument of $F_d(j\omega)$ at frequency ω_0.

1.10.3. *Passage from the Nichols chart to its dual*

According to the relationships [1.111] and [1.112], the amplitude and phase contours of the Nichols chart are such that:

$$B_r(A,\psi)\text{dB} = 20\log\frac{A}{\left(1 + A^2 + 2A\cos\psi\right)^{1/2}} = \text{cte} \qquad [1.120]$$

and

$$\varphi°_r(A,\psi) = \arctan\frac{\sin\psi}{A + \cos\psi} = \text{cte}\,. \qquad [1.121]$$

Likewise, according to relationships [1.118] and [1.119], the amplitude and phase contours of the dual chart are such that:

$$B_d(A,\psi)\text{dB} = 20\log\frac{1}{\left(1 + A^2 + 2A\cos\psi\right)^{1/2}} = \text{cte} \qquad [1.122]$$

and

$$\varphi°_d(A,\psi) = -\arctan\frac{A\sin\psi}{1 + A\cos\psi} = \text{cte}\,. \qquad [1.123]$$

For a given value of ψ, let us change A to $1/A$ in relationships [1.120] and [1.121], which, in the Nichols plane, comes down to passing from the point $(A\text{dB}, \psi)$ to its symmetry around the axis of arguments $(-A\text{dB}, \psi)$. The set of expressions [1.120] to [1.123] thus enables the passage relationships to be established:

$$B_d(A,\psi)\text{dB} = B_r(1/A,\psi)\text{dB} \qquad [1.124]$$

and

$$\varphi°_d(A,\psi) = -\varphi°_r(1/A,\psi). \qquad [1.125]$$

These relationships reveal that between the Nichols chart and its dual, there is a symmetry with respect to the axis of arguments, on the one hand, between the amplitude contours of the same parameter and, on the other hand, between the phase contours of opposite parameters. In addition, the dual chart can be simply deduced from the Nichols chart: starting from the latter, the sign of the scales on the ordinate and the parameters of the phase contours simply need to be changed.

The dual chart (or the dual of the Nichols chart) can therefore be interpreted as the *inverse of the Nichols chart* (or *inverted Nichols chart*), the origin of which is given in the following section.

1.10.4. *A little background*

The use of the dual chart as a means of studying the regulation function was published in 1978 by Oustaloup [OUS 78a].

A recent bibliography search reveals that while the use of this chart in regulation is not called into question, in itself, the chart's "new" nature dates back beyond 1978, because we can find it in the book [GIB 63].

In this book, this chart, referred to as the "inverted Nichols chart", or "Lohcin chart", enables the determination, from the Nichols locus of $\beta(j\omega)$, for each frequency ω, of the modulus (in decibels) and the argument (in degrees) of the frequency response between the error signal, $\varepsilon(t)$, and the reference, $r(t)$, of an elementary control loop of zero disturbance, $d(t)$.

Again, in his book, Gibson refers to an article by Hill [HIL 62]. In this article, Hill points out that similar ideas were expressed in an article by Balchen [BAL 56].

In this context, it is worth remembering that the Nichols plane (decibel gain on the ordinate, degree phase on the abscissa) and the Nichols chart plotted in this plane date back to 1947, with the Black plane (gain on the abscissa, degree phase on the ordinate) and the Black chart plotted in this plane dating back to 1934. In fact, in the Black plane, from which the Nichols plane originates, the Cartesian coordinates correspond directly to the polar coordinates of a point on the Nyquist locus dating back to 1932. The Bode diagrams that separate the gain and phase frequency plots date back to 1938.

2

The Control Loop: Stability and Stability Degree, Precision, Dynamic Performances and Controller Synthesis

2.1. Introduction

This chapter begins by defining stability, and presenting the fundamental stability condition, which can be verified using algebraic or graphical criteria.

Next, the frequency stability margins, that is, the gain and phase margins, are defined. The phase margin is examined in more detail via the determination of an optimum phase margin in the sense of the minimization of a quadratic criterion based on the steepness time and the first overshoot of the step response in tracking or regulation.

The stability degree, an essential concept in system dynamics, perfectly exemplified by the control loop owing to the coupling between open loop and closed loop, is presented in time domain, then in frequency domain, where particular use is made of this coupling by means of the amplitude contours presented in Chapter 1. This approach demonstrates, *inter alia*, that while the circle centered at (–1) in the complex plane offers (in the mathematical sense) a true measurement of the distance to the critical point, it is nevertheless not significant of a time dynamic performance able to quantify the stability degree in time domain. When compared with pole-placement control, the frequency approach stands out by satisfying both a reduced first overshoot specification and a damping ratio specification.

After defining the precision based on the difference between the sought and actual values of the control loop output, precision is presented according to a structure that observes tracking-regulation duality and the steady states corresponding to step and ramp tests.

Once the stability degree–precision dilemma has been defined, it is highlighted in the Nichols plane, then solved by an open-loop gain, which provides a compromise between stability degree and precision. Next, this compromise is improved based on one of three possibilities: increasing the stability degree without decreasing precision, increasing precision without decreasing the stability degree, or jointly increasing both stability degree and precision. Finally, the dilemma is brought into question by an open loop Nichols locus described by the frequency response of a non-integer order integrator.

A pragmatic definition of dynamics is provided, after which the time dynamic performances that quantify the characteristics of dynamics are defined, and the frequency dynamic performances are identified to create a pool from which those used for the controller synthesis are drawn. Section 2.10 identifies the frequency dynamic performances significant of time dynamic performances, which thus determine dynamics.

In accordance with one study consideration for the controller synthesis and with the determination of its phase contribution at the open-loop unit gain frequency, the nature of the controller is imposed and its synthesis forms part of a straightforward, systematic methodological approach that offers the user efficient methods in a user-guide form.

2.2. Stability

2.2.1. *Definition*

A control loop is stable if the free response, expressed by the solution to the free differential equation, $(r(t) = d(t) = 0)$, converges toward its resting value (in this case, $y(t)$ tending toward zero when t tends toward infinity).

2.2.2. *Fundamental stability condition*

The necessary, sufficient condition for a control loop to be stable is for the roots of its *characteristic equation*, that is,

$$1 + \beta(s) = 0, \qquad [2.1]$$

to have their negative real part (or, in other words, for the zeros of the *characteristic polynomial*, $1+\beta(s)$, to be with negative real part).

Bearing in mind that the characteristic equation is none other than the denominator equaling zero of the transmittances in tracking and regulation; in other words, the necessary, sufficient stability condition (of a control loop) is for the closed-loop transmittance poles to be with negative real part.

2.2.3. *Stability algebraic criteria*

The stability algebraic criteria are criteria (or rules) that enable us to determine whether the roots of the characteristic equation, written in the form of an algebraic equation

$$B_n s^n + B_{n-1} s^{n-1} + \cdots + B_1 s + B_0 = 0, \qquad [2.2]$$

have their negative real part *without the need to solve the equation*. The Routh criterion is the most widely used in practice.

2.2.3.1. *Necessary condition*

A necessary stability condition is that all of the coefficients, B_i, of the denominator of the transmittances in tracking and regulation must be *present and have the same sign.*

This condition is only sufficient in the case of first- and second-order control loops.

2.2.3.2. *Necessary, sufficient condition: Routh criterion*

With all of the coefficients, B_i, present and having the same sign (positive, for example), we can build the table presented in Figure 2.1, known as the *Routh table.*

s^n	B_n	B_{n-2}	B_{n-4}	……
s^{n-1}	B_{n-1}	B_{n-3}	B_{n-5}	……
s^{n-2}	$\dfrac{B_{n-1}B_{n-2} - B_n B_{n-3}}{B_{n-1}}$	$\dfrac{B_{n-1}B_{n-4} - B_n B_{n-5}}{B_{n-1}}$	$\dfrac{B_{n-1}B_{n-6} - B_n B_{n-7}}{B_{n-1}}$	……
…	………………	………………	………………	……

Figure 2.1. *Construction of the Routh table*

The coefficient, B_{n-1}, is the *pivot* for the third row. The fourth row, like the third, is obtained by diagonally multiplying the terms of the second and third rows, with the terms obtained all being divided by the pivot of the fourth row (the term in the first column of the previous row), and so on.

If all of the terms in the first column of the Routh table are positive, the characteristic equation only possesses roots with negative real part and the loop is *stable*. If there are η sign changes in this column, the characteristic equation possesses η roots with positive real parts.

When the characteristic equation possesses two pure conjugate imaginary roots, the Routh table contains a row where all coefficients are nil: if some of the other roots have positive real parts, the loop is *unstable*, whereas if all other roots have negative real parts, the loop is *barely oscillating*. In order to verify whether the loop is unstable or barely oscillating, it is necessary to continue the Routh table beyond the row where all coefficients are nil. To this end, it is necessary to write in the place of the row in question the coefficients of the derived polynomial, with respect to s, of an *auxiliary polynomial*; the coefficients of which are the terms of the last non-nil row. The roots with nil real parts are the zeros of the auxiliary polynomial.

2.2.4. *Stability graphic criteria*

The stability graphic criteria enable us to conclude on the closed-loop stability based on the frequency response loci in open loop.

Two different graphic criteria are to be distinguished between:

– the *left-hand criterion*, demonstrated physically in Chapter 1, which can be applied to a stable open loop with minimal phase shifting (poles and zeros with negative real parts);

– the *Nyquist criterion*, which can be applied to all open loops.

2.2.4.1. *Left-hand criterion in the Nyquist plane*

When the open loop Nyquist locus is swept in the increasing ω direction, the control loop is stable if the critical point, *C(-1)*, is kept on the left (Figure 2.2(a)); it is unstable if point C is kept on the right (Figure 2.2(b)); and finally, it is barely oscillating if the Nyquist locus of $\beta(j\omega)$ passes through point C (Figure 2.2(c)).

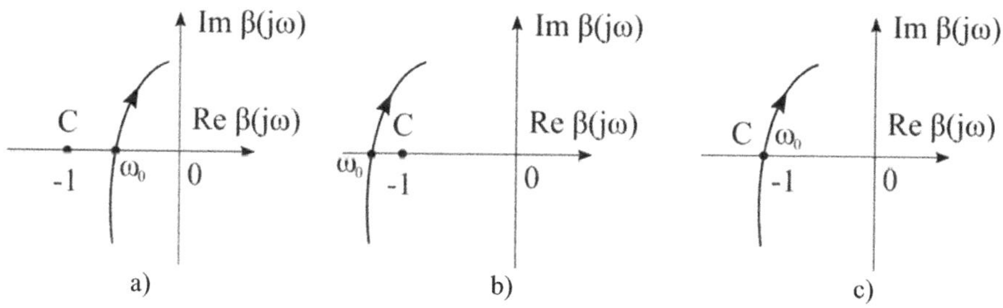

Figure 2.2. *Illustration of the left-hand criterion in the Nyquist plane: the control loop is stable (a), unstable (b) and barely oscillating (c)*

2.2.4.2. *Left-hand criterion in the Nichols plane*

The stability condition in the Nyquist plane (Figure 2.2(a)), that is,

$$|\beta(j\omega)| < 1 \quad [2.3]$$

at the frequency, ω_0, such that

$$\arg \beta(j\omega_0) = -180°, \quad [2.4]$$

is reflected in the Nichols plane (Figure 2.3(a)) by

$$|\beta(j\omega)|\mathrm{dB} < 0 \quad [2.5]$$

at the frequency, ω_0, such that

$$\arg \beta(j\omega_0) = -180° \quad [2.6]$$

The left-hand criterion in the Nichols plane can thus be expressed as follows: when the open loop Nichols locus is swept in the increasing ω direction, the control loop is stable if the critical point, C(0dB, –180°), is kept on the right (Figure 2.3(a)); it is unstable if point C is kept on the left (Figure 2.3(b)); and finally, it is barely oscillating if the Nichols locus of $\beta(j\omega)$ passes through point C (Figure 2.3(c)).

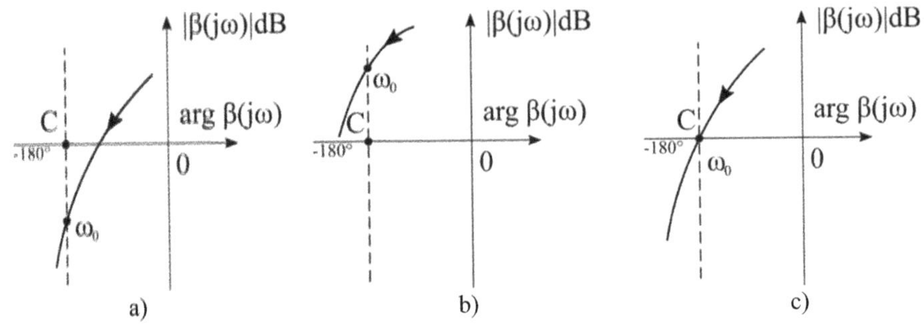

Figure 2.3. *Illustration of the left-hand criterion in the Nichols plane: the control loop is stable (a), unstable (b) and barely oscillating (c)*

2.2.4.3. *Nyquist criterion*

The Nyquist criterion is an application of a Cauchy theorem: for this reason, it is widely referred to by numerous authors as the "Nyquist–Cauchy criterion".

This criterion is applicable regardless of the form of the transmittance in open loop. It can therefore be used to study the stability of a control loop *with unstable open loop*: the transmittance in open loop, $\beta(s)$, and the transmittance of the open loop, $\alpha(s) = -\beta(s)$, which of course have the same poles, then possess the same poles with positive real parts.

Regarding the Cauchy theorem: let $F(s)$ be a function of the complex variable, s. If the point, M, of affix, s, describes a closed curve, (C), in the complex plane, the point, M', of affix, $F(s)$, describes a curve, (Γ), such that (C) and (Γ) correspond point by point. This being so, Cauchy demonstrates that when the point of affix, s, fully describes the curve, (C), in the clockwise direction, the locus, (Γ), of $F(s)$ performs a number of rotations, N, around the origin in the trigonometric direction, such that:

$$N = P - Z, \qquad [2.7]$$

where P and Z are the numbers of poles and zeros, respectively, of the function, $F(s)$, that are located inside the closed contour, (C).

From the Cauchy theorem to the Nyquist criterion: let us recall that a transmittance control loop in open loop $\beta(s)$ is *stable*, if the denominator, $1+\beta(s)$, of the transmittances in closed loop *does not possess any zero with positive real part.*

Let us therefore apply the Cauchy theorem to the function, $F(s) = 1 + \beta(s)$, considering a closed contour, (C_N), known as the *Nyquist contour*, containing all of the poles and zeros of $F(s)$ and consisting of:

– one parallel to the imaginary axis of infinitely small positive abscissa;

– one half-circumference located in the right half-plane with infinitely large radius.

When the point, M of affix, s, describes this contour in the clockwise direction, the point, M', of affix, $1 + \beta(s)$, rotates in the trigonometric direction *around the origin* of a number, N, of rotations equaling $P - Z$, where P and Z represent, respectively, the number of poles and the number of zeros of $1 + \beta(s)$ located inside (C_N), that is, given the form of (C_N), the numbers of poles and zeros of $1 + \beta(s)$ with positive real parts.

With $\beta(s)$ written in the form $P(s) / Q(s)$, $\beta(s)$ and $1 + \beta(s) = (Q(s) + P(s)) / Q(s)$ have the same denominator and therefore the same poles. This being so, if we note that when the point of affix, $1 + \beta(s)$, performs N rotations around the origin, the point of affix, $\beta(s)$, performs the same number of rotations around -1, the above statement is equivalent to the following: when the point, M, of affix, s, describes the Nyquist contour in the clockwise direction, the image of $\beta(s)$ rotates in the trigonometric direction *around the critical point* of a number of rotations, $N = P - Z$, where P is the number of poles of $\beta(s)$ with positive real parts and Z the number of zeros of $1 + \beta(s)$ with positive real parts.

With the stability condition (no zero of $1 + \beta(s)$ with positive real part) imposing:

$$Z = 0, \qquad [2.8]$$

N is then reduced to P, that is,

$$N = P \qquad [2.9]$$

The stability condition can therefore be presented as follows: a control loop is stable on the necessary, sufficient condition that when the point, $M(s)$, describes the Nyquist contour in the clockwise direction, the locus described by the image of

$\beta(s)$ performs a number of rotations around the critical point in the trigonometric direction, this number being equal to the number of poles of $\beta(s)$ with positive real parts; in the particular scenario where $P = 0$ (control loop *stable in open loop*), the necessary, sufficient stability condition is for the locus described by the image of $\beta(s)$ not to surround the critical point.

Complete Nyquist locus of $\beta(j\omega)$: the complex variable, $s = \sigma + j\omega$, is reduced to $s = j\omega$ on the straight line, $(-\infty,+\infty)$, of the Nyquist contour, (C_N). The straight-line transform, $(0,+\infty)$, is the locus of the image of $\beta(j\omega)$ for ω varying from $\omega = 0$ to $\omega = +\infty$, that is, the *Nyquist locus* of $\beta(j\omega)$. The half-straight transform, $(-\infty,0)$, is the locus of the image of $\beta(j\omega)$ for ω varying from $\omega = -\infty$ to $\omega = 0$, that is, the locus of the image of $\beta(-j\omega)$ for ω varying from $\omega = 0$ to $\omega = +\infty$: with $\beta(j\omega)$ and $\beta(-j\omega)$ being conjugate complex numbers, this locus is symmetrical to the Nyquist locus of $\beta(j\omega)$ with respect to the real axis. This locus and the Nyquist locus of $\beta(j\omega)$ constitute the locus of the image of $\beta(j\omega)$ for ω varying from $\omega = -\infty$ to $\omega = +\infty$, the locus known as the *complete Nyquist locus* of $\beta(j\omega)$. As we need to take the transform of the entire Nyquist contour, (C_N), it should be specified that the semi-circle transform with infinitely large radius (s tending toward infinity) is none other than the origin of the plane of the complex function, $\beta(s)$ (therefore zero).

Presentation of the Nyquist criterion: a control loop is stable on the necessary, sufficient condition that the complete Nyquist locus of the frequency response in open loop $\beta(j\omega)$ sweeped from $\omega = -\infty$ to $\omega = +\infty$ surrounds the critical point, in the trigonometric direction, a number, N, of rotations equaling the number, P, of poles with positive real parts of the transmittance in open loop, $\beta(s)$: in the particular case where $P = 0$ (control loop *stable in open loop*), the necessary, sufficient stability condition is for the complete Nyquist locus of $\beta(j\omega)$ not to surround the critical point; this result confirms the left-hand criterion in the Nyquist plane (section 2.2.4.1).

2.3. Stability margins

The stability margins make it possible to quantify the distance to the critical point of the Nyquist or Nichols locus in open loop. There are two types of stability margin: the *gain margin* and the *phase margin* (Figure 2.4).

2.3.1. *Gain margin*

The gain margin (or *marginal gain*) represents the gain distance to the critical point of the Nyquist or Nichols locus of $\beta(j\omega)$ at the frequency, ω_0, such that $\arg\beta(j\omega_0) = -180°$, that is:

$$G_m = \frac{\text{gain at critical point}}{|\beta(j\omega_0)|} = \frac{1}{|\beta(j\omega_0)|}, \qquad [2.10]$$

or, in decibels:

$$G_m\text{dB} = 20\log 1 - 20\log|\beta(j\omega_0)| = 0\text{dB} - |\beta(j\omega_0)|\text{dB} \qquad [2.11]$$

Therefore, the gain margin represents the number of decibels by which the open-loop gain can increase without provoking closed-loop instability. A satisfactory gain margin has a value of the order of 10 dB.

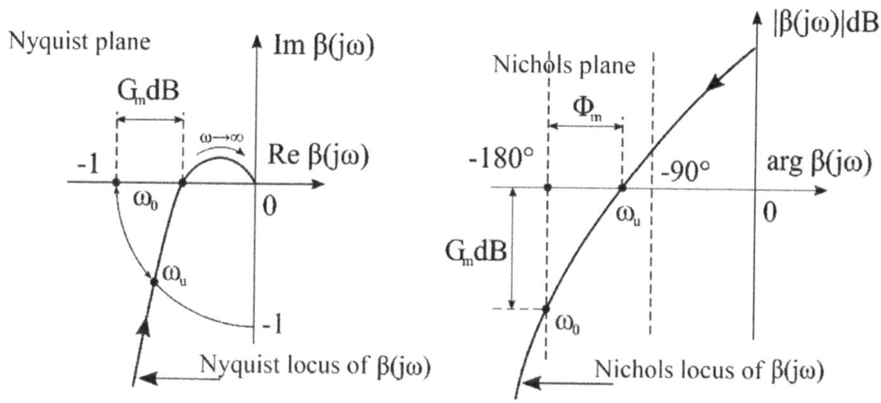

Figure 2.4. *Stability margins defined in the Nyquist and Nichols planes*

2.3.2. *Phase margin*

2.3.2.1. *Definition*

The *phase margin* (or *marginal phase shifting*) represents the angular (or phase) distance to the critical point of the Nyquist or Nichols locus of $\beta(j\omega)$ at the frequency, ω_u, such that $|\beta(j\omega_u)| = 1$, that is:

$$\begin{aligned} \Phi_m &= \arg \beta(j\omega_u) - \text{phase at critical point} \\ &= \arg \beta(j\omega_u) - (-180°) \\ &= \arg \beta(j\omega_u) + 180°. \end{aligned} \quad [2.12]$$

Therefore, the phase margin represents the number of degrees by which the open loop phase can decrease without provoking instability in closed loop.

2.3.2.2. *Satisfactory phase margin and optimum phase margin*

A satisfactory phase margin lies between 50° and 55°, an interval that is often extended, particularly by 5° on either side, therefore between 45° and 60°, in accordance with the results obtained in section 2.3.2.3. Indeed, in both tracking and regulation, any value from this interval offers a compromise between the steepness time and the first overshoot of the step response, with the steepness time defined as the instant at which the response passes through its steady-state value for the first time (Figure 2.5). Such a property can be immediately verified when the frequency response in open loop is that of a non-integer integrator, (Figure 2.6):

$$\beta(j\omega) = \left(\frac{\omega_u}{j\omega}\right)^n \quad \text{with } 1 < n < 2, \quad [2.13]$$

the phase margin, Φ_m, then being expressed very simply as a function of the order n (Figure 2.6):

$$\Phi_m = \arg \beta(j\omega_u) + 180° = 90°(2 - n) \quad [2.14]$$

Figure 2.7 gives the steepness-time curve, θ_r, as a function of the first overshoot, D_1, scaled in values of n comprised between 1.1 and 1.9. At first glance, it appears to show that below $n = 1.4$, D_1 decreases little, whereas θ_r increases significantly, while above $n = 1.4$, θ_r decreases little, whereas D_1 increases significantly; these areas not satisfying the search for a compromise: an optimal value of Φ_m corresponds to the value of n as particularized, considered here as optimum, and which thus determines the curve bend, that is,

$$\Phi_m{}° = 90(2 - 1.4) = 54. \quad [2.15]$$

Thus, contrary to a phase margin of 90° ($n = 1$), which leads to an infinite steepness time, a phase margin from 50° to 55° can be defined as ensuring *a suitable finite steepness time in return for an acceptable first overshoot* (of the order of 20%).

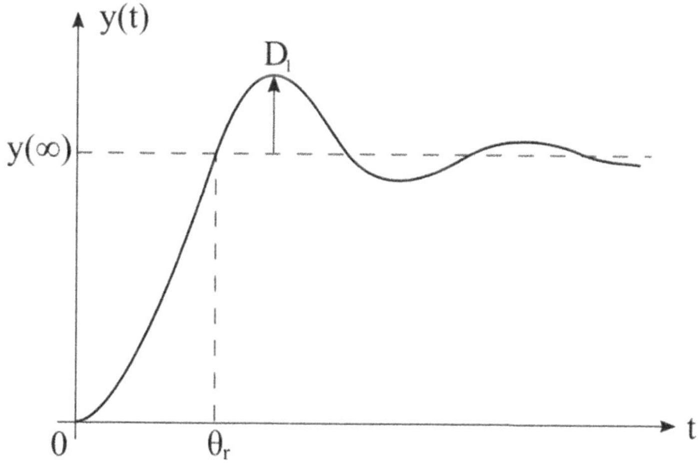

Figure 2.5. *Illustration (on a step response in tracking) of the steepness time, θ_r, and of the first overshoot, D_1*

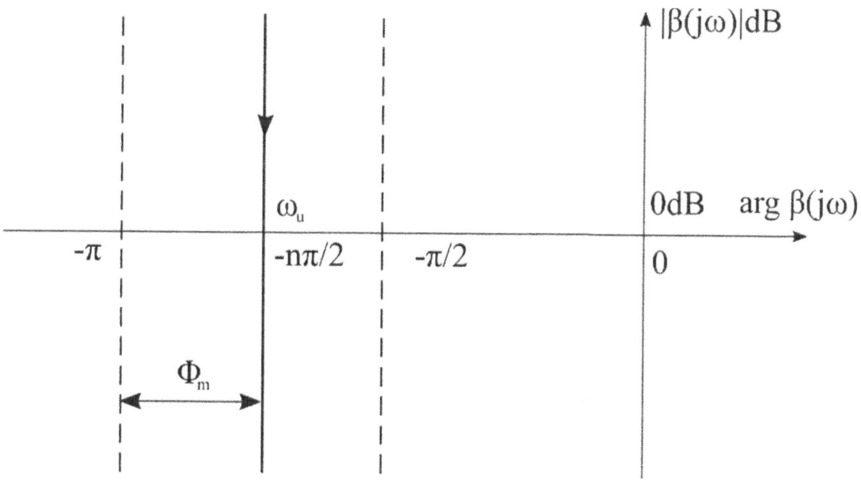

Figure 2.6. *Nichols locus of a non-integer integrator*

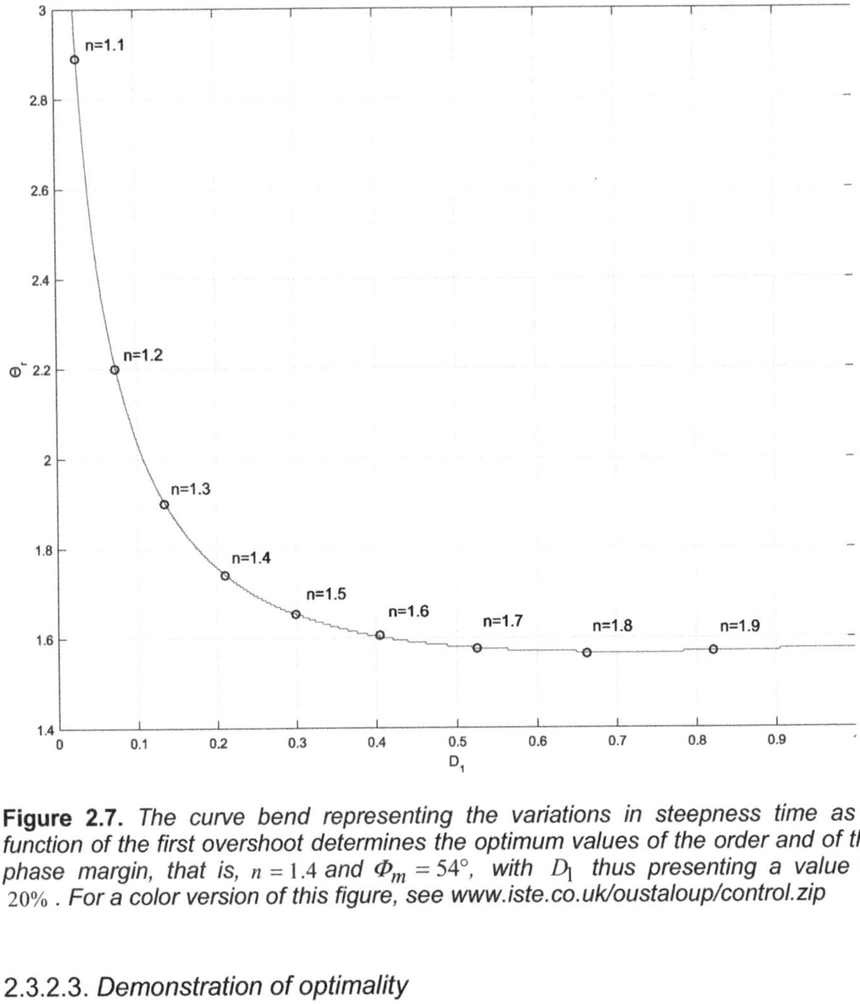

Figure 2.7. *The curve bend representing the variations in steepness time as a function of the first overshoot determines the optimum values of the order and of the phase margin, that is,* $n = 1.4$ *and* $\Phi_m = 54°$, *with* D_1 *thus presenting a value of* 20% . *For a color version of this figure, see www.iste.co.uk/oustaloup/control.zip*

2.3.2.3. *Demonstration of optimality*

The numerical values of θ_r and D_1 are obtained through simulation based on the transmittance in tracking, $F_r(s)$, of an elementary control loop, whose transmittance in open loop, $\beta(s)$, is that of an integrator of non-integer order, that is:

$$F_r(s) = \frac{\beta(s)}{1+\beta(s)} = \frac{\left(\frac{\omega_u}{s}\right)^n}{1+\left(\frac{\omega_u}{s}\right)^n}, \qquad [2.16]$$

therefore:

$$F_r(s) = \frac{1}{1+\left(\dfrac{\omega_u}{s}\right)^n} = \frac{1}{1+(\tau s)^n}, \qquad [2.17]$$

with

$$1 < n < 2 \text{ and } \tau = 1/\omega_u . \qquad [2.18]$$

The unit step response, $y_r(t)$, numerically particularized by $\tau = 1$ ($\omega_u = 1$), is then given by the original of $Y_r(s)$, that is:

$$y_r(t) = \mathcal{L}^{-1}\left[Y_r(s)\right], \qquad [2.19]$$

with:

$$Y_r(s) = F_r(s)\frac{1}{s} = \frac{1}{1+(\tau s)^n}\frac{1}{s}. \qquad [2.20]$$

Given the expression of the response natural frequency, that is,

$$\omega_p = \omega_u \sin\frac{\pi}{n}, \qquad [2.21]$$

the maximum natural frequency is given by:

$$\omega_{p_{\max}} = \left[\omega_p\right]_{n=2} = \omega_u \sin\frac{\pi}{2} = \omega_u . \qquad [2.22]$$

Yet, $\omega_{p_{\max}}$ can also be written as:

$$\omega_{p_{\max}} = 2\pi f_{p_{\max}} = \frac{2\pi}{T_{p_{\min}}}, \qquad [2.23]$$

therefore:

$$\frac{2\pi}{T_{p_{\min}}} = \omega_u , \qquad [2.24]$$

from which we obtain:

$$T_{p_{\min}} = \frac{2\pi}{\omega_u}. \qquad [2.25]$$

Given that the minimum steepness time is given by

$$\theta_{r_{\min}} = \frac{1}{4} T_{p_{\min}}, \qquad [2.26]$$

$\theta_{r_{\min}}$ admits the expression:

$$\theta_{r_{\min}} = \frac{1}{4}\frac{2\pi}{\omega_u} = \frac{\pi}{2}\frac{1}{\omega_u}, \qquad [2.27]$$

or, given that $1/\omega_u = \tau$:

$$\theta_{r_{\min}} = \frac{\pi}{2}\tau. \qquad [2.28]$$

A non-nil minimum steepness time (confirmed by Figure 2.7) results in introducing an orthonormal system (Figure 2.8), the axes of which constitute the asymptotes of the curve representing the variations of θ_r as a function of D_1 or, more precisely, of θ_r' as a function of D_1', θ_r' and D_1' being reduced (dimensionless) variables, defined from θ_r and D_1, $\theta_{r_{\min}}$, τ and $y(\infty)$, in accordance with:

$$\theta_r' = \frac{\theta_r - \theta_{r_{\min}}}{\tau} = \frac{\theta_r}{\tau} - \frac{\pi}{2} \qquad [2.29]$$

and

$$D_1' = \frac{D_1}{y(\infty)}, \qquad [2.30]$$

or, for $\tau = 1$ ($\omega_u = 1$) and $y(\infty) = 1$:

$$\theta_r' = \theta_r - \frac{\pi}{2} \qquad [2.31]$$

and

$$D_1' = D_1 . \qquad [2.32]$$

Given the curve of $\theta_r' = f(D_1')$ (Figure 2.8), the distance, *d*, from a point, P, on the curve to origin, *O*, of the plane (θ_r', D_1'), can be measured by the radius, R, of a circle centered at the origin and passing through the point, that is,

$$d = R = \sqrt{(\theta_r')^2 + (D_1')^2} = \sqrt{J} , \qquad [2.33]$$

by writing

$$J = (\theta_r')^2 + (D_1')^2 , \qquad [2.34]$$

a criterion whose minimum value, $\min J$, determines the minimum distance:

$$d_{\min} = R_{\min} = \sqrt{\min J} . \qquad [2.35]$$

This *minimum distance* (or *distance from the curve to origin*) is none other than the *minimum radius* defined by the radius of the circle tangent to the curve, the point of tangency, meanwhile, defining the *bend* of the curve (Figure 2.9).

Thus, Figure 2.9, which shows that the point of tangency corresponds to $n = 1.4$, demonstrates the optimum nature of this order (in the sense of criterion *J*) and, consequently, confirms the interpretation of Figure 2.7, as summarized in its key (and having led to $n = 1.4$, $\Phi_m = 54°$ and $D_1 = D_1' = 0.21 = 20\%$).

In order to study how the distance to origin varies around $n = 1.4$, which comes down to studying the sensitivity of this distance to the order n around $n = 1.4$, it is appropriate to represent graphically, as a function of the order, *n*, plotted on the abscissa, the distance to origin of a point on the curve of $\theta_r' = f(D_1')$ (Figure 2.10). Given that this distance is measured by the radius of the circle centered at the origin and passing through the point, the ordinate axis then plots the radii of the set of intersecting circles, one of which becomes tangent for the optimum value of *n*, in this case: $n = 1.4$.

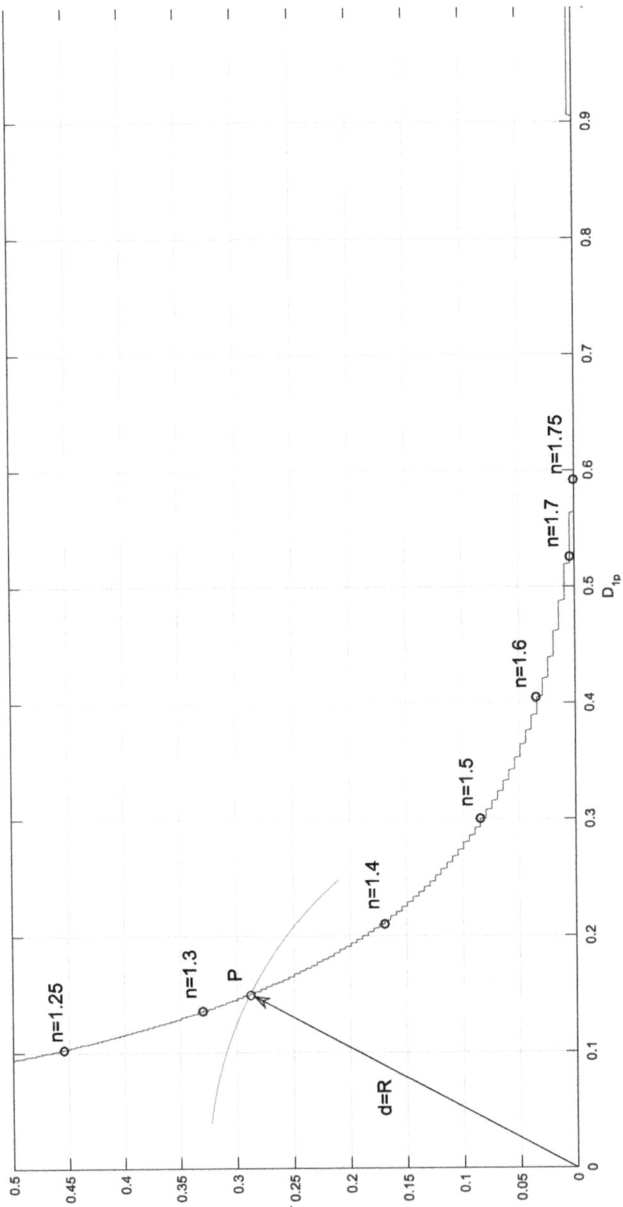

Figure 2.8. *System of axes asymptotic to the curve and distance from a point to origin. For a color version of this figure, see www.iste.co.uk/oustaloup/control.zip*

Considering that a 45° phase margin (corresponding to $n = 1.5$) is known to be acceptable and often retained owing to its value favorably lying between 0° and 90°, it is appropriate, based on the curve shown in Figure 2.10:

– first, to plot on the ordinate the distance to origin for $n - 1.5$, namely $d = 0.3122$, a value 15% greater than $d_{\min} = 0.2701$;

– second, to plot on the abscissa the order below 1.4, corresponding to the same distance, namely $n = 1.333$ which then determines a phase margin of 60°.

Thus, by admitting a difference of 15% from the minimum distance, the extreme values of the phase margin, 45 and 60°, are determined. In this interval, assistance in choosing the phase margin can be provided by the extreme values of the reduced first overshoot, D_1', namely $D_1' = 30\%$ for 45° and $D_1' = 0.158 = 15\%$ for 60°.

Given the numerical values of D_1' and, more generally, those proposed in this section for $n = 1.333$, $n = 1.4$ and $n = 1.5$, they can be confirmed using polynomial approximation:

$$D_1' = an^2 + bn + c \,, \qquad [2.36]$$

which (taking account of a correlation coefficient of 0.99995 characterizing the degree of approximation) enables the admission that D_1' constitutes a veritable parabolic function of *n*. This formulation originates directly from the book on CRONE control [OUS 91, formula 3.28], with the coefficients of this formula having the following numerical values:

$$a = 79.195;\; b = -138.507;\, c = 59.528 \,. \qquad [2.37]$$

Thus, the verification as defined leads to:

$$D_1'(\%) = 15.627 = 15 \quad \text{for} \quad n = 1.333 \;\; (\Phi_m{}^\circ = 60)\,, \qquad [2.38]$$

$$D_1'(\%) = 20.840 = 20 \quad \text{for} \quad n = 1.4 \quad (\Phi_m{}^\circ = 54) \qquad [2.39]$$

and

$$D_1'(\%) = 29.956 = 30 \quad \text{for} \quad n = 1.5 \quad (\Phi_m{}^\circ = 45)\,, \qquad [2.40]$$

with these results in accordance with those already obtained, and reiterating the set of results provided.

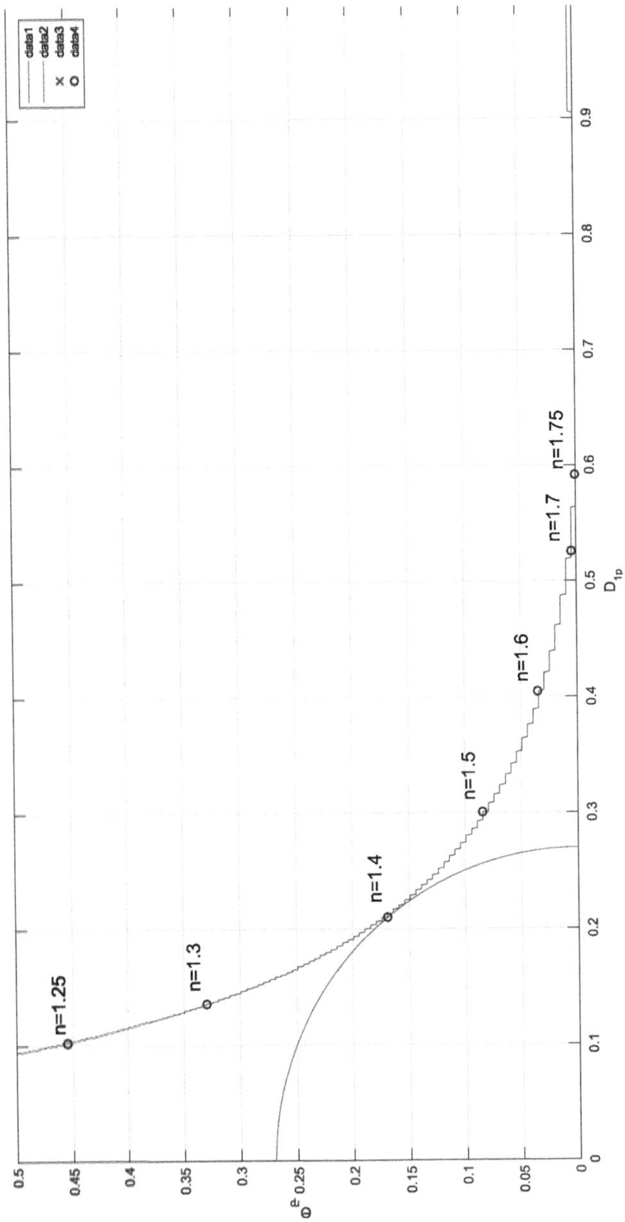

Figure 2.9. *The tangency to the curve of a circle centered at the origin is obtained for* n=1.4. *For a color version of this figure, see www.iste.co.uk/oustaloup/control.zip*

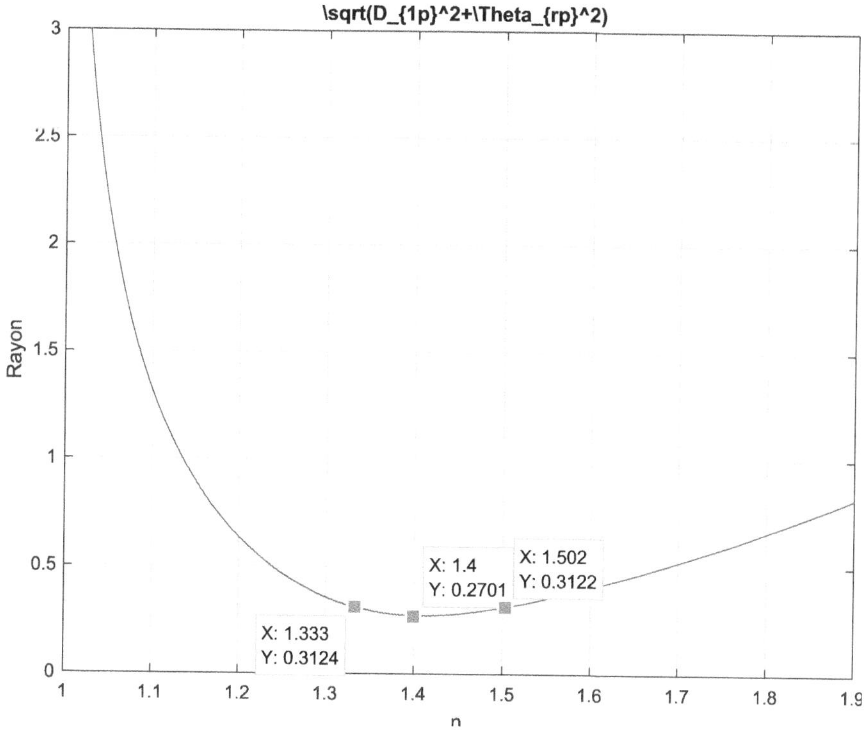

Figure 2.10. *Sensitivity to the order, n, of the distance to the origin. For a color version of this figure, see www.iste.co.uk/oustaloup/control.zip*

2.4. Stability degree

2.4.1. *Time domain*

In the *time domain*, the stability degree of a control loop is quantified (or measured):

– either by the *reduced first overshoot* of the step response in tracking or in regulation, $D_1' = D_1 / y(\infty)$, D_1 being the first overshoot of the response and $y(\infty)$ the steady-state value in tracking;

– or by the *damping ratio* in tracking and regulation, ζ, characterizing the decrease rate of the overshoots of oscillatory mode (the dominant one if there are several) of the step response in tracking or in regulation, and defined as the *reduced damping coefficient* of this mode, namely $\zeta = \alpha / \omega_n$, the ratio between the *damping coefficient*, α, which defines the decrease in $e^{-\alpha t}$ of the oscillating mode, and the *natural frequency without damping*, ω_n, such that $\omega_p = \omega_n\sqrt{1-\zeta^2}$, ω_p designating the oscillating-mode natural frequency (or pseudo-frequency).

2.4.2. *Frequency domain*

In the *frequency domain*, the stability degree can be characterized by the "*distance*" *to the critical point*. Such a distance is measured:

– either by the *stability margins* constituted by the gain and phase margins;

– or by a *Nichols amplitude contour* (the one that is tangent to the open loop Nichols locus), parameterization of which gives the *resonance ratio in tracking*, Q_r (dB); this contour corresponds to a Hall λ-circle in the Nyquist plane (Figure 2.11); it creates an *iso-overshoot contour* bearing in mind that Q_r is, significant of the reduced first overshoot, D_1';

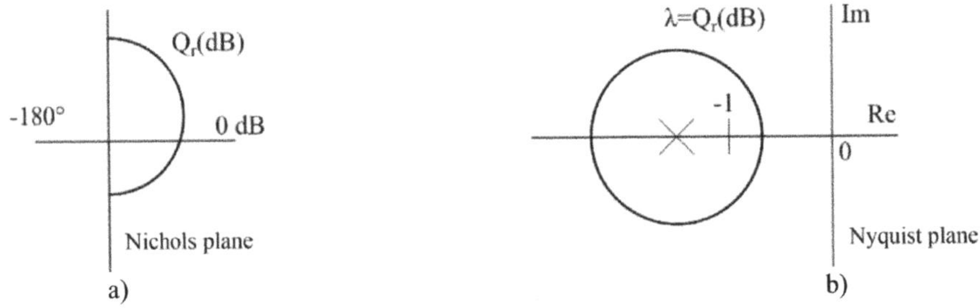

Figure 2.11. *Nichols amplitude contour (a) and its transposition into the Nyquist plane (b), namely a Hall λ-circle eccentric with respect to the point,* (–1)

– or by an *inverted Nichols amplitude contour* (tangent to the open loop Nichols locus), whose parametrization gives the *resonance ratio in regulation*, Q_d (dB); this contour corresponds to a circle centered at the critical point in the Nyquist plane (Figure 2.12); *this circle centered on* (–1) in the complex plane, which constitutes (in the mathematical sense) a true measure of the distance to the critical point, makes an inverse contour a veritable measure of this distance; nevertheless, this circle and the corresponding inverse contour are not significant for any time dynamic performance, because Q_d is not significant for any performance of this type;

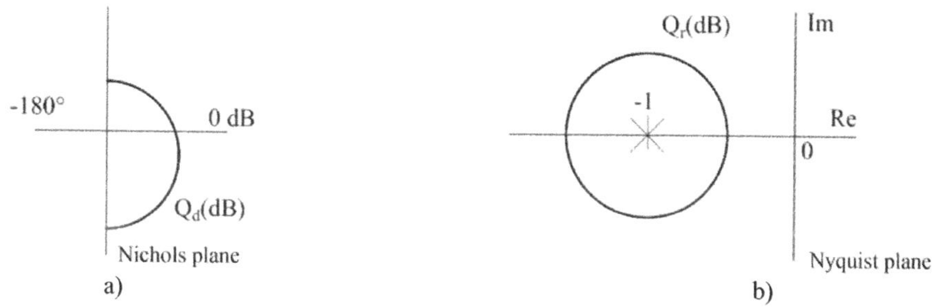

Figure 2.12. *Inverse Nichols amplitude contour (a) and its transposition into the Nyquist plane (b), namely a circle centered at the point* (-1)

– or by an *isodamping contour* (tangent to the open loop Nichols locus), the parameterization of which gives the damping ratio in tracking and regulation, ζ; this contour, symmetrical with respect to the axis 0 dB, results from the arithmetic mean of a Nichols amplitude contour and the dual amplitude contour (with the same parameter) (Figure 2.13).

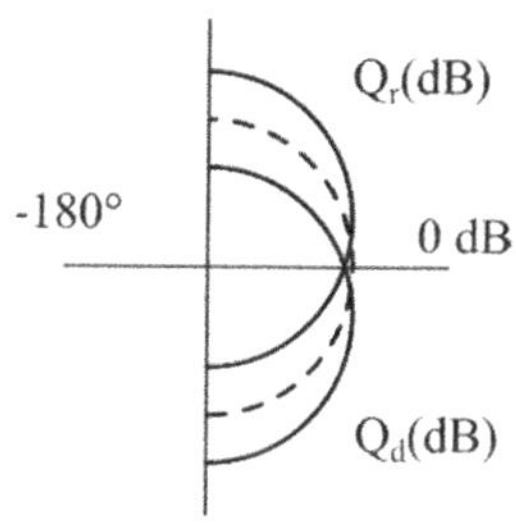

Figure 2.13. *Construction of an isodamping contour based on amplitude contours with the same parameter*

2.4.3. *Comparison with pole-placement control: an observation confirmed by section 2.4.2*

Within the context of a dynamics specification (section 2.7), which consists of specifying the reduced first overshoot together with the damping ratio and the natural frequency (whose order of magnitude is determined by the frequency, ω_u, then set in consequence), it is appropriate to compare how the frequency approach and pole-placement control satisfy a reduced first overshoot specification and a damping ratio specification.

To conduct this comparison in a straightforward manner, the idea is to express $\beta(s)$ in the form of a rational fraction, whereby $N(s)$ and $D(s)$ represent the numerator and the denominator, that is, $\beta(s) = N(s) / D(s)$, then analyze how $N(s)$ and $D(s)$ are taken into consideration in the frequency approach and pole-placement control.

In the case of the frequency approach, $N(s)$ and $D(s)$ are taken into account, through the (tangent) Nichols locus of $\beta(j\omega) = N(j\omega) / D(j\omega)$, on the one hand, and the amplitude contour tangential to this locus, on the other hand, determined from

$$F_r(j\omega) = K\frac{\beta(j\omega)}{1+\beta(j\omega)} = \frac{KN(j\omega)}{D(j\omega)+N(j\omega)} \quad \text{(for the Nichols contour)} \qquad [2.41]$$

or

$$F_d(j\omega) = \frac{1}{1+\beta(j\omega)} = \frac{D(j\omega)}{D(j\omega)+N(j\omega)} \quad \text{(for the inverted Nichols contour).} \qquad [2.42]$$

That is to say, in this scenario, $N(s)$ and $D(s)$ are always taken into account, through both the Nichols locus and an amplitude contour, or, more generally, a performance contour (an isodamping contour, for example).

In the case of the pole-placement control, the way in which $N(s)$ and $D(s)$ are taken into account can be studied by rapidly analyzing the general form of closed loop transmittances expressed as a function of $N(s)$ and $D(s)$:

$$F_r(s) = K\frac{\beta(s)}{1+\beta(s)} = \frac{KN(s)}{D(s)+N(s)} \qquad [2.43]$$

and

$$F_d(s) = \frac{1}{1+\beta(s)} = \frac{D(s)}{D(s)+N(s)}. \qquad [2.44]$$

By placing the poles of $F_r(s)$, which are the zeros of $D(s)+N(s)$, only the zeros of the denominator of $F_r(s)$ are taken into account, which means overlooking both the zeros of the numerator and the numerator itself, $KN(s)$. Likewise, by placing the poles of $F_d(s)$, which are the zeros of $D(s)+N(s)$, only the zeros of the denominator of $F_d(s)$ are taken into account, which means overlooking both the zeros of the numerator and the numerator itself, $D(s)$.

When we know that the zeros both of the numerator and the denominator of $F_r(s)$ or $F_d(s)$ characterize these transmittances and thus determine the form of their step response, we can understand why this partial taking into account of the zeros does not comply with this form of response, notably the reduced first overshoot.

Moreover, in a dynamics specification context, while the pole-placement control enables a damping ratio specification to be satisfied, it cannot satisfy a reduced first overshoot specification, because unlike the damping ratio, the reduced first overshoot depends on the full set of parameters of the corresponding transmittance.

In conclusion, in meeting the specifications for both the reduced first overshoot and the damping ratio, the frequency approach stands out as favorable in terms of performance, but at the cost, it must be said, of a sound knowledge of frequency techniques. While comparative analysis as conducted via $N(s)$ and $D(s)$ can offer a means of arriving at this observation, what is more important is to explain this advantage of the frequency approach by using the content of section 2.4.2: if the controller, $C(j\omega)$, is such that the Nichols locus of $\beta(j\omega)=C(j\omega)G(j\omega)$ is tangential to both an iso-overshoot contour (or Nichols amplitude contour) and an isodamping contour with parameters corresponding to the specifications of the reduced first overshoot and of the damping ratio, the control, through its step response in tracking or in regulation, satisfies both these performance specifications.

2.5. Precision

2.5.1. *Definition*

A control loop is all the more precise the smaller the difference, $\varepsilon_y(t)$, between the *desired* value, $y_d(t)$, of the output and the *true* value, $y(t)$. Its precision can therefore be quantified by the difference, known as the *error* or *gap*.

$$\varepsilon_y(t)=y_d(t)-y(t), \qquad [2.45]$$

that is, in Laplace transforms:

$$\varepsilon_y(s)=Y_d(s)-Y(s). \qquad [2.46]$$

2.5.2. *Precision in tracking and regulation*

The output desired values in tracking and regulation correspond to an infinitely large open-loop gain in the entire frequency domain. Thus, by tending $\beta(s)$ toward infinity in the general expressions of closed loop transmittances (relationships [1.36] and [1.39]), it is possible to deduce therefrom:

$$Y_d(s) = KR(s) \qquad [2.47]$$

in tracking, and

$$Y_d(s) = 0 \qquad [2.48]$$

in regulation.

Equation [2.46] can then be particularized according to the relationships:

$$\varepsilon_y(s) = KR(s) - K\frac{\beta(s)}{1+\beta(s)}R(s) = K\frac{R(s)}{1+\beta(s)} \qquad [2.49]$$

in tracking, and

$$\varepsilon_y(s) = 0 - \frac{D(s)}{1+\beta(s)} = -\frac{D(s)}{1+\beta(s)} \qquad [2.50]$$

in regulation.

Thus, both in tracking and in regulation, *the precision is all the better, the greater the efficiency factor*, and therefore the open-loop gain is higher, considering that *the efficiency factor is defined by the quantity,* $|1+\beta(j\omega)|$.

2.5.3. *Precision in steady state*

Precision in steady state is characterized by the *error in steady state*, that is, according to the end-value theorem:

$$\lim_{t\to\infty} \varepsilon_y(t) = \lim_{s\to 0} s\varepsilon_y(s). \qquad [2.51]$$

Given the relationships [2.49] and [2.50], the error thus defined is expressed in tracking and regulation:

$$\lim_{t\to\infty} \varepsilon_y(t) = \lim_{s\to 0} Ks \frac{R(s)}{1+\beta(s)} \tag{2.52}$$

in tracking, and

$$\lim_{t\to\infty} \varepsilon_y(t) = \lim_{s\to 0} -s \frac{D(s)}{1+\beta(s)} \tag{2.53}$$

in regulation.

2.5.3.1. *Step-test scenario*

In the specific case of a step test, steady state is none other than static (or continuous) state. We then speak of *static precision* and *position error* (*or step error*).

In tracking, that is, $r(t) = u(t)$ and $R(s) = 1/s$, the relationship [2.52] becomes:

$$\lim_{t\to\infty} \varepsilon_y(t) = \frac{K}{1+\beta(0)}. \tag{2.54}$$

In regulation, that is, $d(t) = u(t)$ and $D(s) = 1/s$, the relationship [2.53] becomes:

$$\lim_{t\to\infty} \varepsilon_y(t) = -\frac{1}{1+\beta(0)}. \tag{2.55}$$

Both in tracking and in regulation, the position error is canceled out for $\beta(0) = \infty$, that is, when the open loop transfer comprises at least one low-frequency integration.

2.5.3.2. *Ramp-test scenario*

In the particular scenario of a ramp test, the error in steady state is known as the *velocity error* (*or ramp error*).

In tracking, that is, $r(t) = tu(t)$ and $R(s) = 1/s^2$, the relationship [2.52] becomes:

$$\lim_{t \to \infty} \varepsilon_y(t) = \lim_{s \to 0} \frac{K}{s} \frac{1}{1+\beta(s)}. \qquad [2.56]$$

In regulation, that is, $d(t) = tu(t)$ and $D(s) = 1/s^2$, the relationship [2.53] becomes:

$$\lim_{t \to \infty} \varepsilon_y(t) = \lim_{s \to 0} -\frac{1}{s} \frac{1}{1+\beta(s)}. \qquad [2.57]$$

Both in tracking and in regulation, the velocity error is canceled out when:

$$\lim_{s \to 0} [s + s\beta(s)] = \infty, \qquad [2.58]$$

or

$$\lim_{s \to 0} s\beta(s) = \infty, \qquad [2.59]$$

that is, if $\beta(s)$ comprises, *inter alia*, a generalized integration defined by a term in $1/s^n$ with $n \in \mathbb{R}^+$:

$$\lim_{s \to 0} s\frac{\text{cte}^+}{s^n} = \lim_{s \to 0} \frac{\text{cte}^+}{s^{n-1}} = \infty, \qquad [2.60]$$

a condition that is satisfied for:

$$n > 1. \qquad [2.61]$$

The velocity error is therefore canceled out when the low-frequency open loop transfer comprises at least:

– two integrations (namely, $n = 2$) if n is an integer;

– or one integration, its order exceeding 1 if n is a non-integer. In other words, extending the second-generation CRONE control template to low frequencies (Figure 2.14) cancels out both the position error and the velocity error.

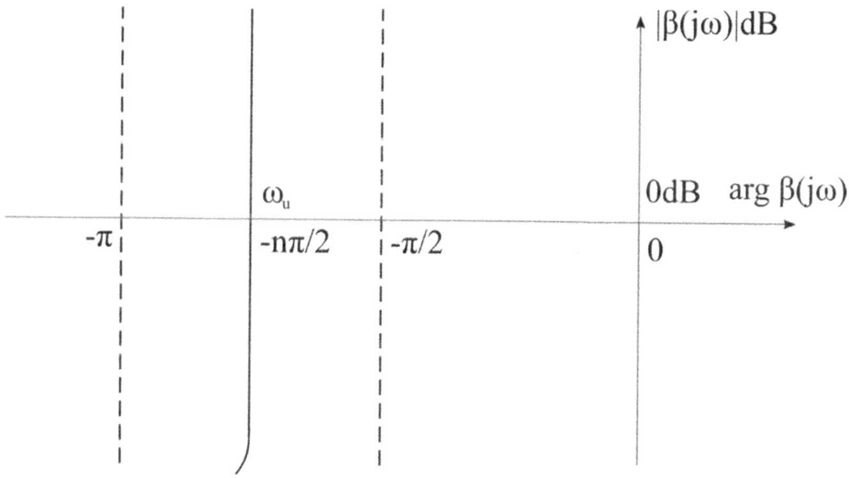

Figure 2.14. *Extending the CRONE control vertical template to low frequencies cancels out the position and velocity errors*

2.6. Stability degree–precision dilemma

2.6.1. *Definition*

As with any dilemma, the stability degree–precision dilemma reflects the existence of two contradictory requirements: stability degree and precision; an increase in one of these performances leads to a decrease in the other and vice versa.

2.6.2. *Highlighting the dilemma in the Nichols plane*

Let there be an open-loop frequency-response locus, $\beta(j\omega)$, plotted in the Nichols plane (Figure 2.15).

An increase in the gain of $\beta(j\omega)$ aiming to increase precision in accordance with relationships [2.49] and [2.50] is reflected by a decrease in the distance to the critical point of the Nichols locus of $\beta(j\omega)$, which results in a deterioration in the stability degree (section 2.4).

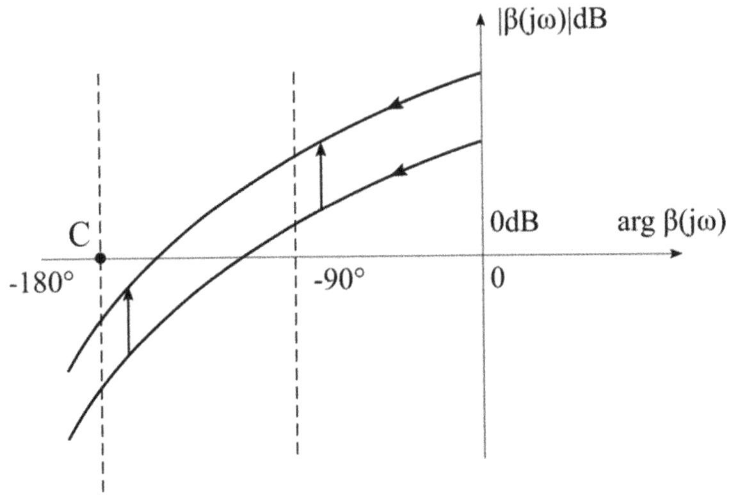

Figure 2.15. *Graphical illustration of the stability degree–precision dilemma: sliding to the top of the Nichols locus of $\beta(j\omega)$, consecutive to a gain increase, reduces its distance to the critical point*

2.6.3. *Compromise between stability degree and precision, and optimum adjustment*

A way of settling the dilemma thus defined consists of choosing an open-loop gain that offers a *compromise between stability degree and precision*. The corresponding adjustment is said to be optimum.

Numerous performance criteria demonstrate that a resonance ratio in tracking of 1.3 (that is, 2.3 dB) offers such a compromise.

2.6.4. *Improving the compromise between stability degree and precision*

The compromise between stability degree and precision can be improved in three distinct ways. Indeed, the improvement is based on three solutions, with the third ensuring the best control performances overall because it combines the advantages of the first two solutions.

The first solution consists of increasing the distance to the critical point without affecting the low-frequency open-loop gain, which comes down to increasing the stability degree without decreasing the precision, this solution being ensured by a phase-lead controller, which acts around the frequency, ω_u (through its differential action).

The second solution consists of increasing the low-frequency open-loop gain without affecting the distance to the critical point, which comes down to increasing the precision without decreasing the stability degree, this solution being ensured by a proportional-integral controller, which acts at low frequency (through its integral action).

Lastly, the third solution, most certainly the best, consists of increasing both distance to the critical point and the low-frequency open-loop gain, which comes down to increasing both the stability degree and the precision, with this solution ensured by a PID controller that combines the actions of the phase-lead controller and the proportional-integral controller.

In fact, in summary, the improvement to the compromise between stability degree and precision lies in the controller synthesis, $C(s)$:

– which (to improve stability degree) ensures the best distance to the critical point, by best deforming the open loop Nichols locus in the vicinity of this point, notably through *a phase lead around the frequency,* ω_u (with the maximum lead provided at ω_u);

– which (to improve precision) ensures the best gain for the rest of the locus, notably through a *low-frequency gain increase.*

2.6.5. *Bringing the dilemma into question*

The correction that brings the stability degree–precision dilemma into question reduces the Nichols locus of $\beta(j\omega)$ to a vertical straight line (Figure 2.16), described by the frequency response of a non-integer integrator, that is

$$\beta(j\omega) = \left(\frac{\omega_u}{j\omega}\right)^n \text{ with } 1 < n < 2 . \qquad [2.62]$$

In the event of a gain variation, the Nichols locus thus defined indeed slides on itself, thus ensuring the invariance of its distance to the critical point. This locus results from extending the template, which characterizes the second-generation CRONE control, to the entire frequency domain (see the introduction to Chapter 1 as well as sections 5.2 and 5.3 in Chapter 5 concerning the CRONE control).

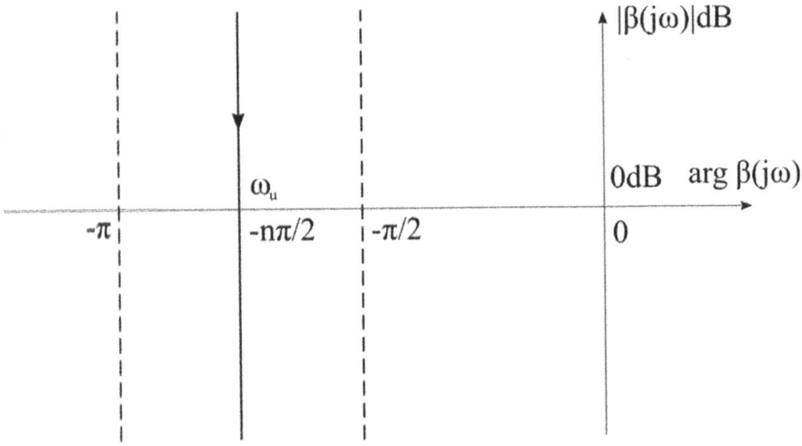

Figure 2.16. *Open loop Nichols locus calling the stability degree–precision dilemma into question*

2.7. Dynamics

The dynamics of a control loop is a conceptual reflection of the transient of the step response in tracking or in regulation. It can be characterized by:

– the overshoot (of the steady state) measured by the reduced first overshoot;

– the damping measured by the damping ratio;

– the rapidity measured by the natural frequency.

In fact, as with any concept, dynamics are quantified (or measured) by performances, or is characterized by concepts (overshoot, damping and rapidity), themselves quantifiable (or measurable) by performances (reduced first overshoot, damping ratio and natural frequency), with this assertion offering a direct introduction to the section below.

2.8. Time dynamic performances

These performances quantify (the characteristics of) the dynamics, namely:

– the reduced first overshoot, D_1', defined as the absolute value of the ratio between the first overshoot and the difference between the initial and end values of the step response in tracking or in regulation;

– the damping ratio, ζ, defined by the reduced damping coefficient of the oscillatory mode (the dominant one if there are several) of the step response in tracking or in regulation;

– the natural frequency, ω_p, defined by the pseudo-frequency of this mode.

2.9. Frequency dynamic performances

These performances, resulting from a frequency representation (therefore coming under frequency domain), make up the pool from which those used for the controller synthesis are drawn. There are two categories involved:

– those relating to the closed loop, namely:

- resonance frequency in tracking,
- resonance ratio in tracking,
- resonance frequency in regulation,
- resonance ratio in regulation;

– those relating to the open loop, namely:

- unit gain frequency,
- gain margin,
- phase margin.

2.10. Determination of dynamics

This section identifies the frequency dynamic performances, which are indicative of the time dynamic performances and thus determine the dynamics.

2.10.1. *Overshoot*

The resonance ratio in tracking, Q_r, which parameterizes (in dB) a Nichols amplitude contour, determines the reduced first overshoot, and therefore the dynamics overshoot.

2.10.2. *Damping*

The stability margins (gain and phase margins) as well as an isodamping contour (as defined by the arithmetic mean of symmetrical amplitude contours) determine the damping ratio, ζ, and therefore the dynamic damping.

Although the stability margins only quantify the distance to the critical point for an elementary form Nichols locus (that is, uniformly decreasing with frequency), it is these margins that we will retain for the controller synthesis within the framework of basic automation teaching.

In the general scenario where the gain margin is sufficiently large to ensure, to some extent, robustness of the control by a "pseudo-verticality" of the Nichols locus in the vicinity of the frequency, ω_u, the phase margin proves to be the smallest stability margin, thus significant (in its own right) of the distance to the critical point in the sense of an isodamping contour, with the phase margin, Φ_m, alone then determining the dynamics damping (Figure 2.17).

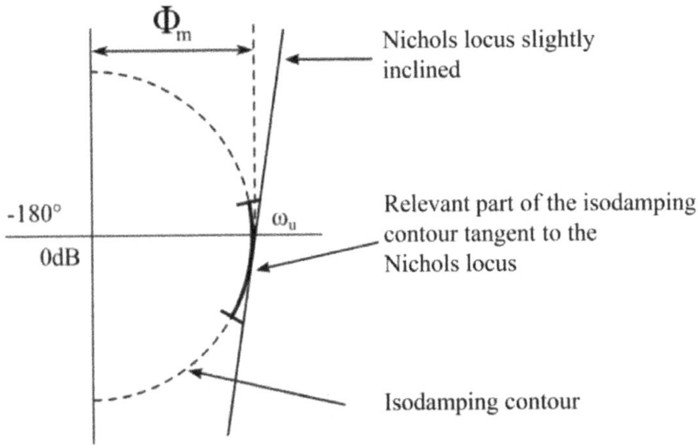

Figure 2.17. *In the case of a Nichols locus slightly inclined around the frequency, ω_u, the relevant part of an isodamping contour is defined by the phase margin in a first approximation*

2.10.3. *Rapidity*

The resonance frequencies in tracking and regulation, as well as the unit gain frequency in open loop, determine the order of magnitude of the natural frequency, ω_p, and therefore the rapidity of the dynamics.

For the controller synthesis, we will retain the unit gain frequency in open loop, ω_u, which determines the transitional frequency in closed loop (or transitional frequency in tracking and regulation).

The choice of frequency, ω_u, results from a compromise between the two contradictory requirements presented by input immunity, on the one hand, and reference tracking and disturbance rejection, on the other hand.

In order for the control to be rapid enough without the plant input being subject to significant stresses, it is appropriate to choose a frequency, ω_u, above a given factor (determined by the designer) at the plant cut-off frequency, ω_c (defined at -3dB or -6dB). A factor between 5 and 20 is generally a suitable order of magnitude.

2.11. Study consideration for the controller synthesis

As a study hypothesis for the controller synthesis, the gain margin is considered large enough to ensure a frequency configuration in open loop:

– heading in the control robustness direction;

– by a tendency toward verticality of the Nichols locus in the vicinity of the frequency, ω_u.

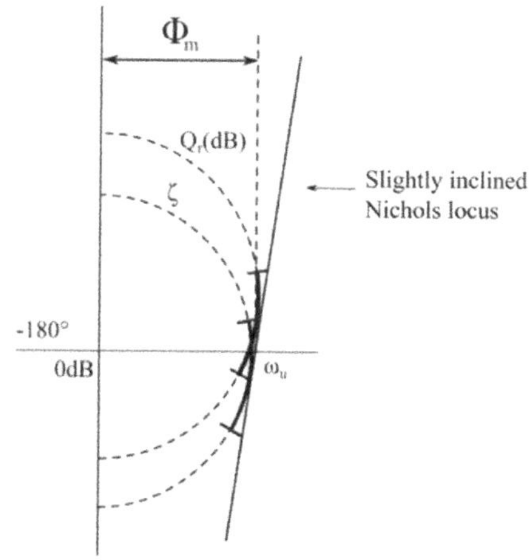

Figure 2.18. *In the case of a Nichols locus slightly inclined around the frequency, ω_u, the relevant parts of a Nichols amplitude contour and an isodamping contour are defined, as a first approximation, by the phase margin, which then determines the overshoot and the damping of the dynamics*

In this simplifying hypothesis, the phase margin can, as a first approximation, act as a substitute for (Figure 2.18):

– a Nichols amplitude contour;

– an isodamping contour.

The phase margin then becomes significant for both:

– the reduced first overshoot D_1';

– the damping ratio, ζ.

2.12. Controller phase at frequency ω_u

Given its definition, the phase margin is written as:

$$\Phi_m = \arg \beta(j\omega_u) - (-\pi) = \arg \beta(j\omega_u) + \pi , \qquad [2.63]$$

or, given that

$$\beta(j\omega_u) = C(j\omega_u)G(j\omega_u)K_2 : \qquad [2.64]$$

$$\Phi_m = \arg C(j\omega_u) + \arg G(j\omega_u) + \pi , \qquad [2.65]$$

from which we obtain:

$$\arg C(j\omega_u) = \Phi_m - \arg G(j\omega_u) - \pi \qquad [2.66]$$

This relationship expresses the phase that the controller, $C(s)$, must bring for the frequency, ω_u, determined by the designer, in accordance with:

$$5\omega_c \le \omega_u \le 20\omega_c \qquad [2.67]$$

ω_c designating the plant cut-off frequency, Φ_m needing to satisfy the recommendation (section 2.3.2.2)

$$50 \le \Phi_m{}^\circ \le 55 , \qquad [2.68]$$

or at least satisfy the least restrictive recommendation (section 2.3.2.3),

$$45 \le \Phi_m{}^\circ \le 60 . \qquad [2.69]$$

The extremal frequencies ($5\omega_c$ and $20\omega_c$) defining the interval corresponding to [2.67] are geometrically distributed with respect to the frequency, $\omega_u = 10\omega_c$, which provides the control loop with a transmission and rejection band that is sufficiently (but not excessively) large in relation to the plant bandwidth. This condition, fulfilled because of a suitable value of ω_u, is indeed necessary to respond favorably to a satisfactory reference tracking and a satisfactory disturbance rejection (each compatible with a satisfactory input immunity).

The interest of the framework of the frequency, ω_u, is due, first, to a low bound, which is supposed to ensure a minimum level of reference tracking and disturbance rejection and, second, to a high bound, which is supposed to ensure a minimum level of input immunity. Between these bounds, a higher frequency, ω_u, favors tracking and rejection at the expense of immunity and, conversely, a lower frequency favors immunity at the expense of tracking and rejection (these all being elements that may guide the designer in the choice of frequency, ω_u).

As recommended, the interval defined by [2.67] is not absolute. It is in fact given as an indication to satisfactorily initialize the regulator synthesis operation. Within this context, pragmatic determination of a frequency, ω_u, compatible with the reality (which does not escape an unknown output disturbance), can be conducted according to the following procedure.

Let there be a regulator synthesized from an initial value of ω_u belonging to the recommended interval: if, for tests reflecting real conditions of use, the synthesized regulator leads to a plant input considerably lower than the saturation thresholds, the frequency, ω_u, can be increased (then tested as to the validity of the plant's new input); it will of course need to be decreased in the event that the plant input proves to be truncated by the saturation thresholds.

2.13. Type of controller

2.13.1. *Phase-lead correction*

If $\arg C(j\omega_u)$ is *positive*, the controller, $C(s)$, must be *phase-lead* and satisfy the following two scenarios.

For

$$0 < \arg C(j\omega_u) < \pi / 2, \qquad [2.70]$$

the controller, $C(s)$, can be limited to the *single phase-lead controller* (Figure 2.19) and thus present a transmittance of the form:

$$C(s) = C_0 \frac{1 + s / \omega_1}{1 + s / \omega_2}, \qquad [2.71]$$

with

$$\omega_1 < \omega_2 \text{ and } (\omega_1 \omega_2)^{1/2} = \omega_u, \qquad [2.72]$$

the maximum phase lead, brought at frequency ω_u, being defined by (section 2.14):

$$\varphi_m = \arcsin \frac{\alpha - 1}{\alpha + 1} \text{ with } \alpha = \frac{\omega_2}{\omega_1} > 1. \qquad [2.73]$$

For

$$\arg C(j\omega_u) > \pi / 2, \qquad [2.74]$$

the single phase-lead controller is no longer sufficient, with the corrector, $C(s)$, now needing to be a *double or triple phase-lead controller* (if not more), according to the phase lead needing to be brought, and thus present a transmittance of the form:

$$C(s) = C_0 \left(\frac{1 + s / \omega_1}{1 + s / \omega_2} \right)^2 \text{ or } C(s) = C_0 \left(\frac{1 + s / \omega_1}{1 + s / \omega_2} \right)^3, \qquad [2.75]$$

with

$$\omega_1 < \omega_2 \text{ and } (\omega_1 \omega_2)^{1/2} = \omega_u, \qquad [2.76]$$

the maximum phase lead, brought at frequency ω_u, being defined by:

$$\varphi_m = 2 \arcsin \frac{\alpha - 1}{\alpha + 1} \text{ or } \varphi_m = 3 \arcsin \frac{\alpha - 1}{\alpha + 1} \text{ with } \alpha = \frac{\omega_2}{\omega_1} > 1. \qquad [2.77]$$

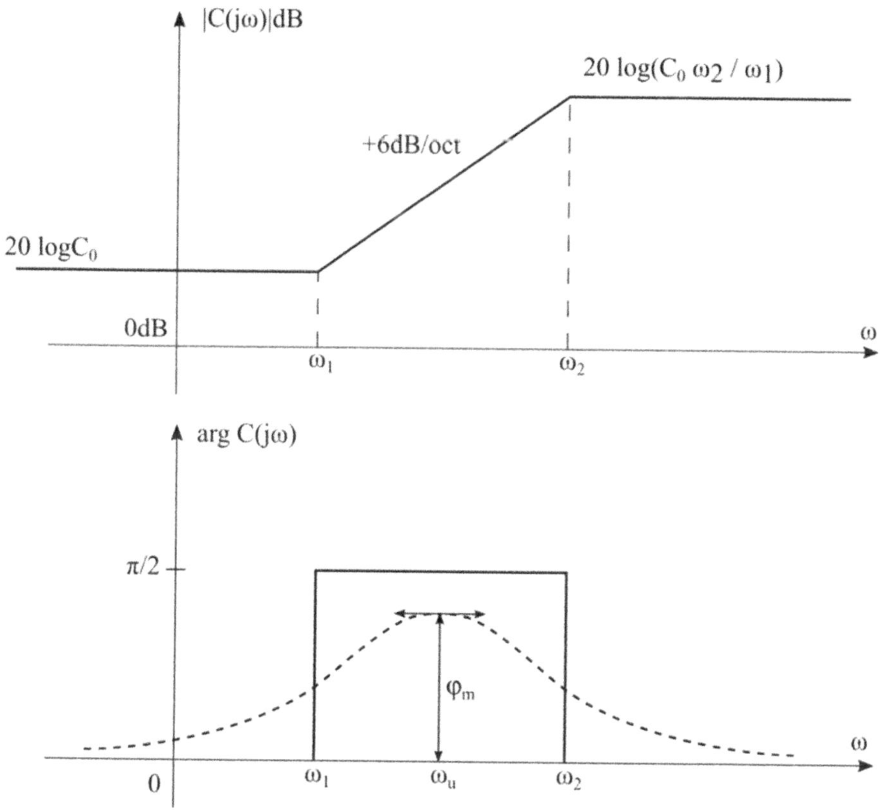

Figure 2.19. *Bode asymptotic diagrams of the single phase-lead controller, whose frequency placement is defined (or determined) by the relationship,* $(\omega_1\omega_2)^{1/2} = \omega_u$

2.13.2. *Phase-lag correction*

If $\arg C(j\omega_u)$ is *negative*, the controller, $C(s)$, must provide *phase lag* and comply with the following two scenarios.

For

$$-\pi / 2 < \arg C(j\omega_u) < 0, \tag{2.78}$$

the controller, $C(s)$, can be limited *to the single phase-lag controller* (Figure 2.20) and thus present a transmittance of the form:

$$C(s) = C_0 \frac{1 + s / \omega_1}{1 + s / \omega_2}, \qquad [2.79]$$

with

$$\omega_2 < \omega_1 \text{ and } (\omega_1 \omega_2)^{1/2} = \omega_u, \qquad [2.80]$$

the maximum (positive) phase lag, brought at frequency ω_u, being defined by:

$$\varphi_m = \arcsin \frac{\alpha - 1}{\alpha + 1} \text{ with } \alpha = \frac{\omega_1}{\omega_2} > 1. \qquad [2.81]$$

For

$$\arg C(j\omega_u) < -\pi / 2, \qquad [2.82]$$

the controller, $C(s)$, must be a *double or triple phase-lag controller* (if not more) and thus present a transmittance of the form:

$$C(s) = C_0 \left(\frac{1 + s / \omega_1}{1 + s / \omega_2} \right)^2 \text{ or } C(s) = C_0 \left(\frac{1 + s / \omega_1}{1 + s / \omega_2} \right)^3, \qquad [2.83]$$

with

$$\omega_2 < \omega_1 \text{ and } (\omega_1 \omega_2)^{1/2} = \omega_u, \qquad [2.84]$$

the maximum (positive) phase lag, brought at frequency ω_u, being defined by:

$$\varphi_m = 2 \arcsin \frac{\alpha - 1}{\alpha + 1} \text{ or } \varphi_m = 3 \arcsin \frac{\alpha - 1}{\alpha + 1} \text{ with } \alpha = \frac{\omega_1}{\omega_2} > 1. \qquad [2.85]$$

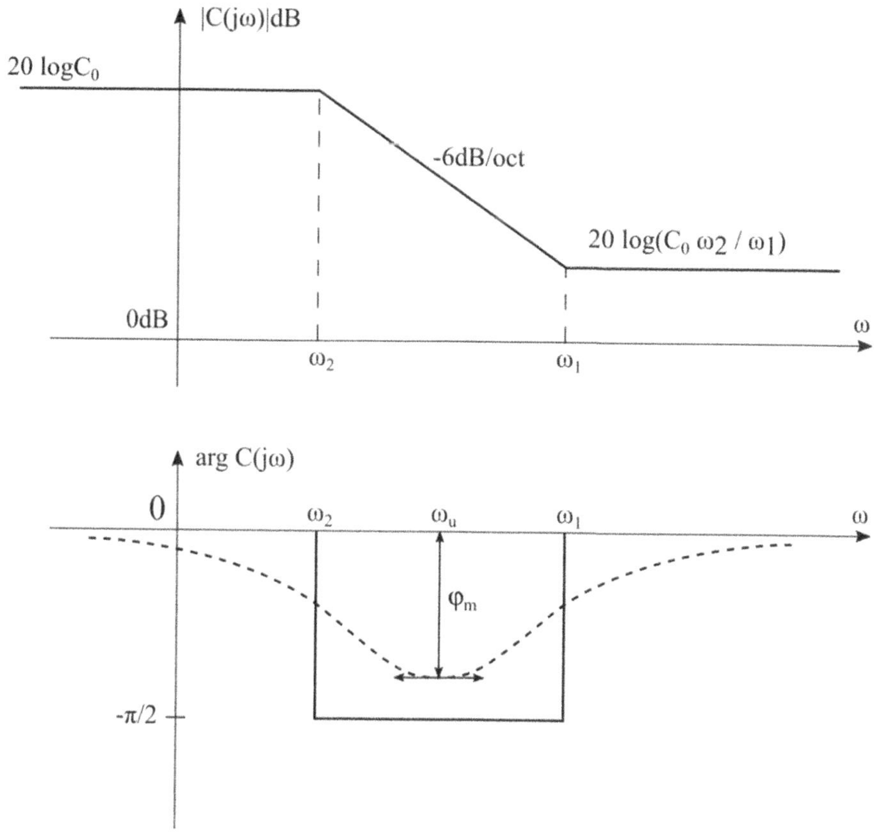

Figure 2.20. *Bode asymptotic diagrams of the single phase-lag controller, whose frequency placement is defined (or determined) by the relationship,* $(\omega_1\omega_2)^{1/2} = \omega_u$

2.13.3. *An additional correction to obtain a PID regulator*

To transform the single phase-lead or phase-lag controller into a non-academic PID-type regulator, it can be completed by cascading a *proportional-integral controller* aimed at increasing the low-frequency open-loop gain to improve the precision of the control at these frequencies.

Such a controller is defined by a transmittance of the form:

$$PI(s) = 1 + \frac{\omega_0}{s}, \text{ with } \omega_0 >> \omega_u; \qquad [2.86]$$

while the condition on ω_0 ensures the transparency of the controller, $PI(s)$ at frequency ω_u, it is not necessary if the designer is not seeking the simplifying nature of this transparency.

2.14. Example of a practical task: detailed study of the single phase-lead controller

The most elementary phase-lead controller, known as the *single phase-lead controller* (or, commonly, phase-lead controller), is characterized by the frequency response:

$$C(j\omega) = C_0 \frac{1 + j\dfrac{\omega}{\omega_1}}{1 + j\dfrac{\omega}{\omega_2}}, \text{ with } \omega_1 < \omega_2 . \qquad [2.87]$$

Defined by the argument of $C(j\omega)$, the phase (or in this case the phase lead) is expressed by:

$$\varphi(\omega) = \arg C(j\omega) = \arctan\frac{\omega}{\omega_1} - \arctan\frac{\omega}{\omega_2} . \qquad [2.88]$$

It passes through a maximum, φ_m, at the frequency, ω_m, such that:

$$\left[\frac{d\varphi(\omega)}{d\omega}\right]_{\omega=\omega_m} = 0 . \qquad [2.89]$$

Let us therefore express $d\varphi(\omega) / d\omega$:

$$\begin{aligned}\frac{d\varphi(\omega)}{d\omega} &= \frac{\dfrac{1}{\omega_1}}{1+\left(\dfrac{\omega}{\omega_1}\right)^2} - \frac{\dfrac{1}{\omega_2}}{1+\left(\dfrac{\omega}{\omega_2}\right)^2} \\ &= \frac{\omega_1}{\omega_1^2 + \omega^2} - \frac{\omega_2}{\omega_2^2 + \omega^2} \\ &= \frac{\omega_1(\omega_2^2 + \omega^2) - \omega_2(\omega_1^2 + \omega^2)}{(\omega_1^2 + \omega^2)(\omega_2^2 + \omega^2)} .\end{aligned} \qquad [2.90]$$

The frequency, ω_m, which cancels out this derivative, satisfies the equation:

$$\omega_1(\omega_2^2 + \omega_m^2) - \omega_2(\omega_1^2 + \omega_m^2) = 0, \qquad [2.91]$$

that is:

$$\omega_1\omega_2^2 + \omega_1\omega_m^2 - \omega_2\omega_1^2 - \omega_2\omega_m^2 = 0, \qquad [2.92]$$

or:

$$\omega_m^2(\omega_1 - \omega_2) - \omega_1\omega_2(\omega_1 - \omega_2) = 0, \qquad [2.93]$$

or (dividing by $\omega_1 - \omega_2 \neq 0$):

$$\omega_m^2 = \omega_1\omega_2, \qquad [2.94]$$

or indeed:

$$\omega_m = \sqrt{\omega_1\omega_2}. \qquad [2.95]$$

Let us recall that in logarithmic abscissas, the abscissa of the geometric mean, ω_m, of the transitional frequencies, ω_1 and ω_2, is none other than the arithmetic mean of the abscissas of these frequencies. Indeed:

$$\log(\omega_1\omega_2)^{1/2} = \frac{1}{2}\log(\omega_1\omega_2) = \frac{\log(\omega_1) + \log(\omega_2)}{2}. \qquad [2.96]$$

The maximum phase lead, φ_m, is written as:

$$\varphi_m = \varphi(\omega_m) = \arctan\frac{\omega_m}{\omega_1} - \arctan\frac{\omega_m}{\omega_2}, \qquad [2.97]$$

that is, given the trigonometric relationship

$$\arctan a - \arctan b = \arctan\frac{a-b}{1+ab}: \qquad [2.98]$$

$$\varphi_m = \arctan\frac{\dfrac{\omega_m}{\omega_1} - \dfrac{\omega_m}{\omega_2}}{1 + \dfrac{\omega_m^2}{\omega_1\omega_2}} = \arctan\frac{\omega_m}{2}\left(\frac{1}{\omega_1} - \frac{1}{\omega_2}\right)$$

$$= \arctan\frac{\omega_m}{2}\left(\frac{\omega_2 - \omega_1}{\omega_1\omega_2}\right) = \arctan\frac{\omega_m}{2}\frac{\omega_2 - \omega_1}{\omega_m^2} \qquad [2.99]$$

$$= \arctan\frac{1}{2}\frac{\omega_2 - \omega_1}{\omega_m} = \arctan\frac{1}{2}\frac{\omega_2 - \omega_1}{\sqrt{\omega_1\omega_2}}.$$

Based on the expression of $\tan\varphi_m$, that is,

$$\tan\varphi_m = \frac{1}{2}\frac{\omega_2 - \omega_1}{\omega_m} = \frac{1}{2}\frac{\omega_2 - \omega_1}{\sqrt{\omega_1\omega_2}}, \qquad [2.100]$$

which, for a given value of ω_m, expresses a proportionality between the tangent of φ_m and the difference between the transitional frequencies ω_2 and ω_1, $\sin\varphi_m$ can be expressed using the following computation. $\tan\varphi_m$ is indeed written as:

$$\tan\varphi_m = \frac{\sin\varphi_m}{\cos\varphi_m} = \frac{\sin\varphi_m}{\sqrt{1 - \sin^2\varphi_m}}, \qquad [2.101]$$

from which we obtain, after squaring the product of the extremes and the means:

$$\sin^2\varphi_m = \frac{\tan^2\varphi_m}{1 + \tan^2\varphi_m}, \qquad [2.102]$$

that is, given the expression of $\tan\varphi_m$:

$$\sin^2\varphi_m = \frac{\dfrac{1}{4}\dfrac{(\omega_2 - \omega_1)^2}{\omega_1\omega_2}}{1 + \dfrac{1}{4}\dfrac{(\omega_2 - \omega_1)^2}{\omega_1\omega_2}}$$

$$= \frac{(\omega_2 - \omega_1)^2}{4\omega_1\omega_2 + (\omega_2 - \omega_1)^2}, \qquad [2.103]$$

$$= \left(\frac{\omega_2 - \omega_1}{\omega_2 + \omega_1}\right)^2,$$

$\sin\varphi_m$ being expressed as follows:

$$\sin\varphi_m = \frac{\omega_2 - \omega_1}{\omega_2 + \omega_1}, \quad [2.104]$$

or:

$$\sin\varphi_m = \frac{\frac{\omega_2}{\omega_1} - 1}{\frac{\omega_2}{\omega_1} + 1}, \quad [2.105]$$

or, by writing,

$$\alpha = \frac{\omega_2}{\omega_1}, \quad [2.106]$$

the ratio between the transitional frequencies, ω_2 and ω_1 $(\alpha > 1)$:

$$\sin\varphi_m = \frac{\alpha - 1}{\alpha + 1}, \quad [2.107]$$

that is, ultimately:

$$\varphi_m = \arcsin\frac{\alpha - 1}{\alpha + 1}. \quad [2.108]$$

3

An Overview of Linearizing Approaches

3.1. Introduction

This chapter results from an educational exercise aiming to expand upon the idea of different linearizing approaches by using strategies that are proposed thanks to their simplicity. Being part of the engineering sciences, these strategies, which enrich basic automation and can be applied to the majority of plants, are conducive to favoring the application of linearizing approaches in the most diverse sectors, notably in industrial sectors.

More precisely, the presentation of each approach is structured by a principle reduced to its simplest expression and by the application of this principle to one or more nonlinear plants, one of which is used for all approaches; this nonlinear plant, common to all monovariable applications, is a motor shaft that rotates a small mass secured to the shaft by means of a straight arm.

In linearization by immersion, extended to a multivariable plant, the principle consists of linearizing the plant with respect to new inputs. To this end, the strategy consists of putting the nonlinear terms into the second member of each differential equation, with each second member thus becoming a new input.

In linearization by high gain, the principle is based on a high-gain control used for its disturbance rejection. To this end, the nonlinear plant is interpreted as a linear plant disturbed at the input by the nonlinearity (or more precisely, the nonlinear part of the plant). The strategy consists of integrating a cascaded high gain with the plant, then looping by means of a linear transfer arising from the linear part of the plant.

The basic principle of linearization by high gain expresses the idea of a total disturbance rejection, whereas linearization by disturbance rejection expresses the idea of a partial rejection, which is more realistic in terms of the plant input sensitivity.

Indeed, its principle is based on an elementary control loop whose gain in open loop remains compatible with satisfactory plant input immunity.

To this end, the nonlinear plant is interpreted as a disturbed linear plant at its output through the nonlinearity effect, so the strategy consists of synthesizing the regulator, which provides a regulation function (and therefore a disturbance rejection), compatible with an acceptable plant input (the disturbance rejection then being partial). In fact, through its ability to reject disturbance, which in this case represents the plant's nonlinearity, the control loop has a linearizing character that has come to be frequently used in the industry to overcome nonlinearities.

Whilst the linearization of the plant around a nominal trajectory constitutes a well-known, relatively widely used linearizing approach, its use presented here stands out with the introduction of uncertainty domains. Indeed, in linearization by high gain or disturbance rejection, the nonlinearity is represented temporally in the form of a disturbance; in our approach of linearization around a nominal trajectory, the nonlinearity is represented frequentially in the form of uncertainty domains: each uncertainty domain is built, for a given frequency, from the tangents linearized and determined for a set of operating points compliant with the trajectory. Displacing uncertainty domains to minimize their penetration into areas with a low degree of stability is what allows one to take into account nonlinearities in third-generation CRONE control (Chapter 5).

Finally, given that the simplicity of the strategies proposed is really a specific aspect of this chapter, we have asked Brigitte d'Andréa-Novel to complete it with a final section that provides expertise of different levels of sensitivity on these strategies and notably on potential limits inherent to their simplicity.

3.2. Linearization by immersion

3.2.1. *Principle*

Linearization by "immersion" consists of linearizing the plant with respect to a new input. More precisely, it consists of "immersing" the nonlinear plant G, of input, $u(t)$, into a linear plant, G', of input, $u'(t)$, known as the *precorrected plant*. The solution to this problem consists of obtaining the (nonlinear) precontroller that uses the precorrected plant to this effect.

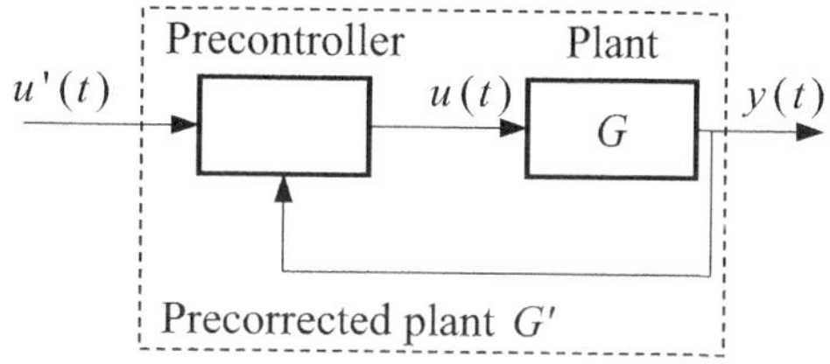

Figure 3.1. *Immersion of the plant into another plant whose dynamic behavior is linear with respect to a new input*

3.2.2. *Application*

3.2.2.1. *Monovariable (or scalar) plant*

As an example, let us consider a shaft of torque, Γ, and moment of inertia, J, which rotates a small mass, m, at a distance, r, from the longitudinal axis, Δ, of the shaft. The small mass is secured to the shaft by means of a stiff, straight arm of length OG and what is assumed to be negligible mass. $q(t)$ designates the angular position of the shaft and the small mass (with respect to the vertical). k and f represent the stiffness and the viscous friction coefficient associated with elasticity and damping, respectively.

Using I to designate the moment of inertia of the small mass with respect to axis Δ, that is, $I = mr^2$, in applying the fundamental law of dynamics for rotating bodies, we can write:

$$(J+I)\ddot{q}(t) = \Gamma(t) + mgr\sin q(t) - kq(t) - f\dot{q}(t), \tag{3.1}$$

that is:

$$(J+I)\ddot{q}(t) + f\dot{q}(t) + kq(t) - mgr\sin q(t) = \Gamma(t). \tag{3.2}$$

The nonlinear differential equation thus obtained may be rewritten in the form:

$$(J+I)\ddot{q}(t) + f\dot{q}(t) + kq(t) = \Gamma(t) + mgr\sin q(t), \tag{3.3}$$

that is:

$$(J+I)\ddot{q}(t) + f\dot{q}(t) + kq(t) = \Gamma^{'}(t), \tag{3.4}$$

by writing:

$$\Gamma'(t) = \Gamma(t) + mgr \sin q(t), \qquad [3.5]$$

from which we obtain:

$$\Gamma(t) = \Gamma'(t) - mgr \sin q(t), \qquad [3.6]$$

with this equation reflected by the functional diagram in Figure 3.2, which displays the nonlinear character of precorrection.

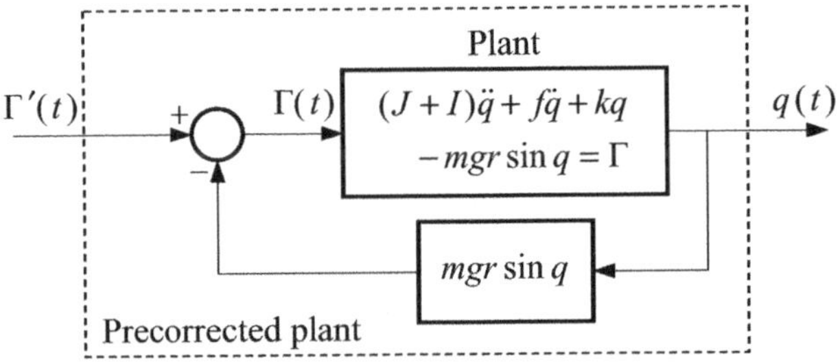

Figure 3.2. *Functional diagram of the precorrected plant*

The transmittance $G'(s)$ of the precorrected plant gives an expression of the form:

$$G'(s) = \frac{Q(s)}{\Gamma'(s)} = \frac{1}{k + fs + (J + I)s^2}. \qquad [3.7]$$

3.2.2.2. *Multivariable plant*

As an example of a nonlinear multivariable plant, a polar table is considered because of the many nonlinearities it is subject to. This table consists of a rule driven in rotation by a motor, M_1, on which a guided small mass, m_2, travels in a rectilinear manner, driven by a motor, M_2, by means of a belt (linked to m_2) and two pulleys of radius, R (one free, the other secured to the motor shaft, M_2). $q_1(t)$ designates the angular position of the rule and $q_2(t)$ the straight position of the small mass with respect to the rotation axis of the rule.

The dynamic behavior of the table with respect to the degrees of freedom, $q_1(t)$ and $q_2(t)$, is described by a system of two second-order nonlinear differential equations, that is:

$$\left(I_1 + m_2 q_2^2\right)\ddot{q}_1 + \left(f_1 + \frac{\alpha_1 \beta_1}{R_1} + 2m_2 q_2 \dot{q}_2\right)\dot{q}_1 + k_1 q_1 = \frac{\alpha_1}{R_1} u_1 \qquad [3.8]$$

and

$$\frac{I_2}{R^2}\ddot{q}_2 + \left(f_2 + \frac{\alpha_2 \beta_2}{R^2 R_2}\right)\dot{q}_2 + \left(k_2 - m_2 \dot{q}_1^2\right) q_2 = \frac{\alpha_2}{R R_2} u_2, \qquad [3.9]$$

with the different quantities featured defined on page 52 of the book on CRONE control [OUS 91]:

- I_i: moment of inertia with respect to the axis of rotation of the motor M_i (irrespective of the small mass, m_2, for I_1);
- f_i: viscous friction coefficient associated with q_i;
- k_i: stiffness (or stiffness constant) linked to q_i;
- α_i: proportionality coefficient between torque and armature current of motor, M_i;
- β_i: coefficient of proportionality between counter-electromotive force and motor shaft angular speed, M_i;
- R_i: armature resistance of motor, M_i;
- u_i: armature voltage of motor, M_i (plant input variable).

The nonlinear differential system thus obtained can be rewritten by putting the nonlinear terms into the second member of each of the equations, that is:

$$I_1 \ddot{q}_1 + \left(f_1 + \frac{\alpha_1 \beta_1}{R_1}\right)\dot{q}_1 + k_1 q_1 = \frac{\alpha_1}{R_1} u_1 - \left(m_2 q_2^2 \ddot{q}_1 + 2m_2 q_2 \dot{q}_2 \dot{q}_1\right) \qquad [3.10]$$

and

$$\frac{I_2}{R^2}\ddot{q}_2 + \left(f_2 + \frac{\alpha_2 \beta_2}{R^2 R_2}\right)\dot{q}_2 + k_2 q_2 = \frac{\alpha_2}{R R_2} u_2 - \left(-m_2 \dot{q}_1^2 q_2\right), \qquad [3.11]$$

or indeed:

$$\frac{R_1 I_1}{\alpha_1}\ddot{q}_1 + \frac{R_1}{\alpha_1}\left(f_1 + \frac{\alpha_1\beta_1}{R_1}\right)\dot{q}_1 + \left(\frac{R_1}{\alpha_1}k_1\right)q_1$$
$$= u_1 - \frac{R_1}{\alpha_1}\left(m_2 q_2^2 \ddot{q}_1 + 2m_2 q_2 \dot{q}_2 \dot{q}_1\right) \tag{3.12}$$

and

$$\left(\frac{RR_2}{\alpha_2}\frac{I_2}{R^2}\right)\ddot{q}_2 + \frac{RR_2}{\alpha_2}\left(f_2 + \frac{\alpha_2\beta_2}{R^2 R_2}\right)\dot{q}_2 + \left(\frac{RR_2}{\alpha_2}k_2\right)q_2$$
$$= u_2 - \frac{RR_2}{\alpha_2}\left(-m_2 \dot{q}_1^2 q_2\right), \tag{3.13}$$

or, in a more condensed form:

$$a_1\ddot{q}_1 + b_1\dot{q}_1 + c_1 q_1 = u_1 - F_1 = u_1' \tag{3.14}$$

and

$$a_2\ddot{q}_2 + b_2\dot{q}_2 + c_2 q_2 = u_2 - F_2 = u_2', \tag{3.15}$$

by writing:

$$a_1 = \frac{R_1 I_1}{\alpha_1} \quad a_2 = \frac{R_2 I_2}{R\alpha_2} \tag{3.16}$$

$$b_1 = \frac{R_1}{\alpha_1}\left(f_1 + \frac{\alpha_1\beta_1}{R_1}\right) \quad b_2 = \frac{RR_2}{\alpha_2}\left(f_2 + \frac{\alpha_2\beta_2}{R^2 R_2}\right) \tag{3.17}$$

$$c_1 = \frac{R_1}{\alpha_1}k_1 \quad c_2 = \frac{R_2}{\alpha_2}k_2; \tag{3.18}$$

$$F_1 = \frac{R_1}{\alpha_1}\left(m_2 q_2^2 \ddot{q}_1 + 2m_2 q_2 \dot{q}_2 \dot{q}_1\right) \tag{3.19}$$

and

$$F_2 = -\frac{RR_2}{\alpha_2} m_2 \dot{q}_1^2 q_2 \qquad [3.20]$$

nonlinear functions of the positions, speeds and accelerations, that is,

$$F_1 = F_1(q, \dot{q}, \ddot{q}) \text{ and } F_2 = F_2(q, \dot{q}, \ddot{q}) \cdot \qquad [3.21]$$

The inputs, u_1 and u_2, of the plant are expressed, respectively, as a function of the new inputs, $u_1^{'}$ and $u_2^{'}$, of the precorrected plant, that is:

$$u_1 = u_1^{'} + F_1 \qquad [3.22]$$

and

$$u_2 = u_2^{'} + F_2, \qquad [3.23]$$

with these equations reflected by the functional diagram in Figure 3.3.

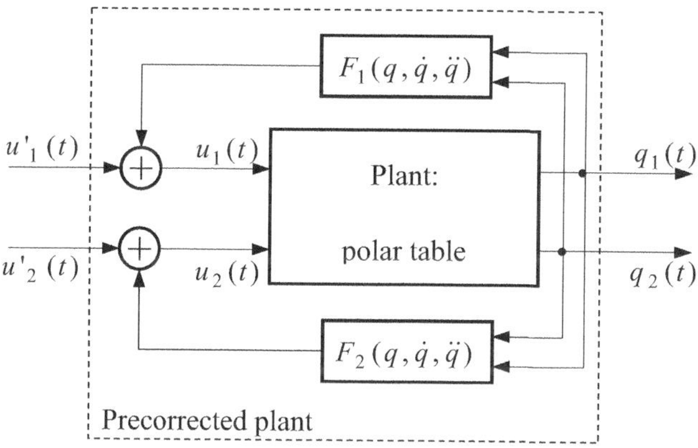

Figure 3.3. *Functional representation of the precorrected plant*

Transmittances $G_1^{'}(s)$ and $G_2^{'}(s)$ of the precorrected plant give expressions of the form:

$$G_1^{'}(s) = \frac{Q_1(s)}{U_1^{'}(s)} = \frac{1}{a_1 s^2 + b_1 s + c_1} \qquad [3.24]$$

and

$$G_2'(s) = \frac{Q_2(s)}{U_2'(s)} = \frac{1}{a_2 s^2 + b_2 s + c_2}. \qquad [3.25]$$

The precorrected plant thus obtained is linear and decoupled. Indeed, it constitutes a set of two second-order, linear, scalar plants. We then talk of a *linearizing decoupling*.

With regard to the nonlinear functions, F_1 and F_2, they are represented functionally in accordance with the block diagrams of Figure 3.4.

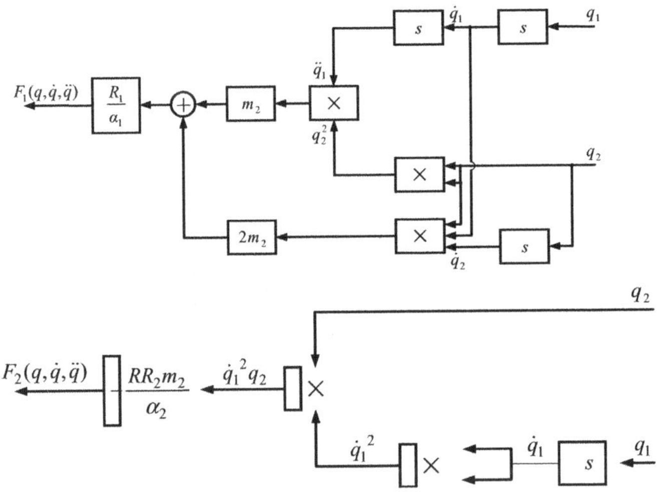

Figure 3.4. *Functional representation of functions* $F_1 = F_1(q,\dot{q},\ddot{q})$ *and* $F_2 = F_2(q,\dot{q},\ddot{q})$

3.3. Linearization by high gain

3.3.1. *Principle*

Linearization by high gain is based on the principle of high-gain control. It consists of integrating a cascaded high gain with the plant, then closing the entire loop by means of a linear transfer (which does not destabilize the loop).

Generally speaking, this linear transfer (of the feedback) is simply the inverse of the linear transfer resulting from passing all non-linear terms to the right-side of each of the plant's differential equations.

3.3.2. *Application*

As a nonlinear scalar plant, let us again consider the plant that is discussed in section 3.2.2.1.

The rewriting of its differential equation in the form,

$$(J+I)\ddot{q}(t)+f\dot{q}(t)+kq(t)=\Gamma^{'}(t), \qquad [3.26]$$

with

$$\Gamma^{'}(t)=\Gamma(t)+mgr\sin q(t), \qquad [3.27]$$

can be illustrated by the functional diagram in Figure 3.5, in which the linear transfer is defined by:

$$G^{'}(s)=\frac{Q(s)}{\Gamma^{'}(s)}=\frac{1}{k+fs+(J+I)s^2}. \qquad [3.28]$$

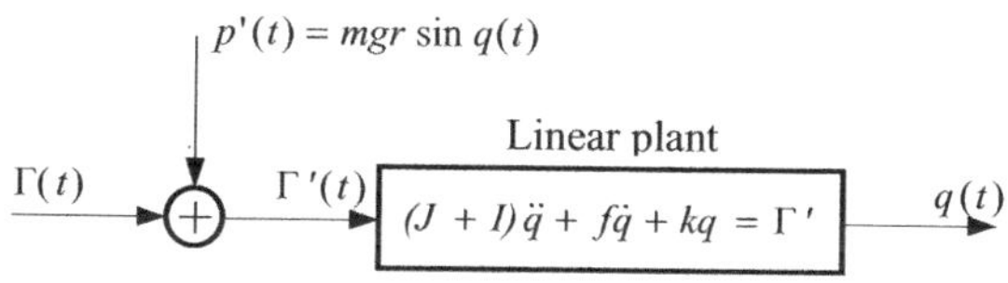

Figure 3.5. *The nonlinear plant is transformed in compliance with a disturbed linear plant: the linear plant, $G^{'}(s)=Q(s)/\Gamma^{'}(s)$, is indeed disturbed by the disturbance, p'(t) = mgr* sin *q(t)*

The linearization by high gain is then illustrated by the functional diagram in Figure 3.6.

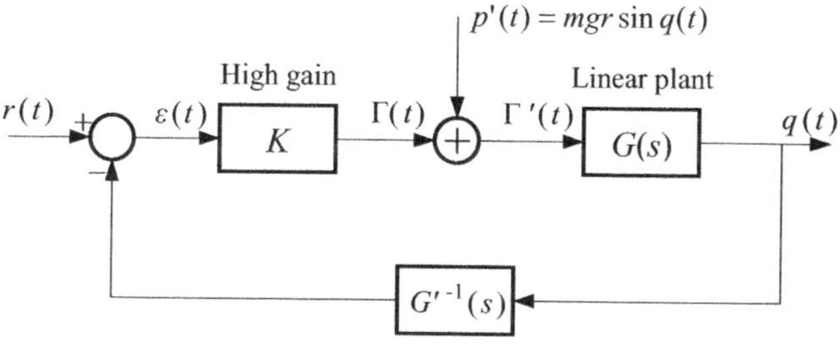

Figure 3.6. *Block diagram illustrating linearization by high gain*

The error signal, $\varepsilon(s)$, is written as:

$$\varepsilon(s) = R(s) - G^{'-1}(s)Q(s) \cdot \qquad [3.29]$$

While the output, $Q(s)$, is written as:

$$Q(s) = G^{'}(s)P^{'}(s) + KG^{'}(s)\varepsilon(s), \qquad [3.30]$$

from which we obtain:

$$\varepsilon(s) = \frac{Q(s) - G^{'}(s)P^{'}(s)}{KG^{'}(s)} \cdot \qquad [3.31]$$

By putting the expression of $\varepsilon(s)$ thus obtained into the first formulation of $\varepsilon(s)$, we obtain an equation of the form:

$$R(s) - G^{'-1}(s)Q(s) = \frac{Q(s) - G^{'}(s)P^{'}(s)}{KG^{'}(s)} \cdot \qquad [3.32]$$

Since $Q(s)$, $P^{'}(s)$ and $G^{'}(s)$ are finite and the gain, K, becomes infinitely large, the second member of the latter equation tends toward zero, thus we can write:

$$R(s) - G^{'-1}(s)Q(s) = 0, \qquad [3.33]$$

from which we obtain:

$$Q(s) = G^{'}(s)R(s) \; \forall P^{'}(s) = \mathcal{L}\left[mgr \sin q(t)\right] \cdot \qquad [3.34]$$

3.3.3. *Conclusion and comments*

Independent of the disturbance, $p^{'}(t)$, representing in this case the nonlinearity of the plant, the result thus obtained expresses a *total disturbance rejection*. That is to say, the result achieved is noteworthy, but unrealistic in terms of the input immunity as the input, $\Gamma^{'}(t)$, of the linear plant and *a fortiori* the input, $\Gamma^{'}(t)$, of the nonlinear plant are then subject to significant stresses.

Nevertheless, the principle of such a linearization is to be retained if it is used judiciously, notably by accepting a degradation of the linearity in favor of better input immunity (K then taking on reasonable values).

3.4. Linearization by disturbance rejection

3.4.1. *Principle*

Although the basic principle of linearization by high gain is expressed by a total disturbance rejection, linearization by disturbance rejection is reflected by a partial rejection, the rate of which is specified and directly guaranteed by the plant control.

Thus, linearization by disturbance rejection consists of:

– interpreting nonlinearities in *terms of disturbance* and notably *output disturbance*;

– integrating the plant into a control loop whose sensitivity ensures suitable attenuation of the modulus spectral density of the output disturbance.

3.4.2. *Application*

Represented in the form illustrated in Figure 3.5, the nonlinear scalar plant discussed in section 3.2.2.1 can be described by a symbolic equation of the form:

$$Q(s) = G'(s)\Gamma(s) + G'(s)P'(s), \qquad [3.35]$$

that is, by writing

$$P(s) = G'(s)P'(s): \qquad [3.36]$$

$$Q(s) = G'(s)\Gamma(s) + P(s), \qquad [3.37]$$

with this equation reflected by the functional diagram in Figure 3.7.

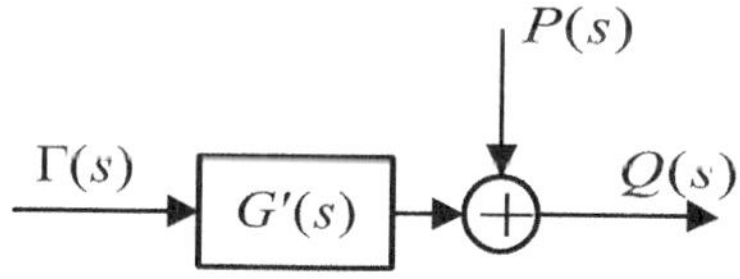

Figure 3.7. *Interpreting plant nonlinearities in terms of output disturbance*

With the plant represented like this, its integration into an elementary control loop featuring a cascaded regulator, $C(s)$, is reflected by the functional diagram in Figure 3.8.

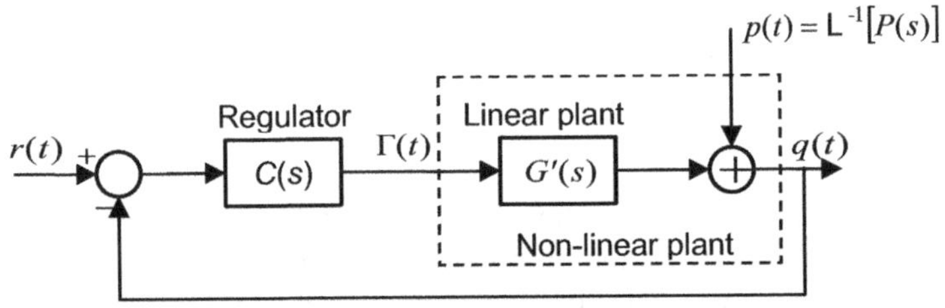

Figure 3.8. *Plant linearizing control*

The regulator, $C(s)$, is synthesized such that the sensitivity, $S(s)$, that is,

$$S(s) = \frac{1}{1 + C(s)G^{'}(s)}, \qquad [3.38]$$

satisfactorily filters the modulus spectral density of the output disturbance, that is,

$$|P(\mathrm{j}\omega)| = \left|G^{'}(\mathrm{j}\omega)\right|\left|P^{'}(\mathrm{j}\omega)\right|. \qquad [3.39]$$

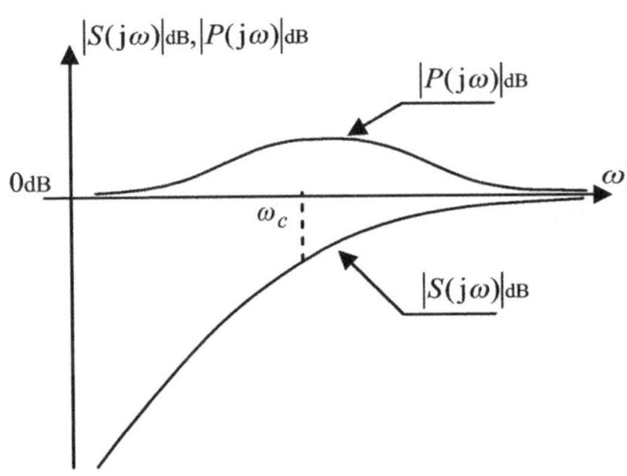

Figure 3.9. *The output disturbance rejection is all the better, the higher the regulation cut-off frequency,* ω_c

Figure 3.9 illustrates the $S(s)$ gain effect on the modulus spectral density of $p(t)$.

3.4.3. *Conclusion and comments*

For a given evolution of $q(t)$, $P^{'}(\mathrm{j}\omega) = \mathcal{F}\left[mgr\sin q(t)\right]$ is also known as $P(\mathrm{j}\omega)$ because $P(\mathrm{j}\omega) = G^{'}(\mathrm{j}\omega)P^{'}(\mathrm{j}\omega)$.

After determining $P(\mathrm{j}\omega)$, which then represents a piece of data, it is appropriate to choose the regulation cut-off frequency that offers the best compromise between the two contradictory requirements that are:

– the attenuation of $\left|P(\mathrm{j}\omega)\right|$ measuring the linearizing nature of the control;

– the plant input immunity defined by $1/\$(\omega)$ with

$$\$(\omega) = \left|\frac{C(\mathrm{j}\omega)}{1 + C(\mathrm{j}\omega)G^{'}(\mathrm{j}\omega)}\right|, \qquad [3.40]$$

$\$(\omega)$ designating the input sensitivity.

In fact, the linearizing nature of a well-synthesized control is widely used in the industry to overcome nonlinearities.

3.5. Linearization of the plant around a nominal trajectory: tangent linearized

3.5.1. *Principle*

The nonlinear state representation (or state model) of the plant is defined by the system of nonlinear vectorial equations:

$$\underset{n\times 1}{\dot{x}(t)} = \underset{n\times 1}{f}\,(\underset{n\times 1}{x(t)},\underset{m\times 1}{u(t)},t) \qquad [3.41]$$

and

$$\underset{p\times 1}{y(t)} = \underset{p\times 1}{g}\,(\underset{n\times 1}{x(t)},\underset{m\times 1}{u(t)},t), \qquad [3.42]$$

f and g being multivariable, nonlinear vectorial functions, the components of f and g constituting multivariable nonlinear scalar functions; the first equation is the *state equation* and the second is the *output or observation equation.*

Let us define a *nominal trajectory* characterized by $x_0(t)$, called the *nominal solution* and corresponding to a control, $u_0(t)$, also said to be *nominal*.

We write $u(t)$, $x(t)$ and $y(t)$, respectively, in the form:

$$u(t) = u_0(t) + \tilde{u}(t), \qquad [3.43]$$

$$x(t) = x_0(t) + \tilde{x}(t) \qquad [3.44]$$

and

$$y(t) = y_0(t) + \tilde{y}(t), \qquad [3.45]$$

we will then seek to determine linear relations between the different infinitesimal vectorial variations that get defined.

Putting the expressions [3.43], [3.44] and [3.45] into the equations [3.41] and [3.42] enables us to write:

$$\dot{x}_0(t) + \dot{\tilde{x}}(t) = f\left(x_0(t) + \tilde{x}(t), u_0(t) + \tilde{u}(t), t\right) \qquad [3.46]$$

and

$$y_0(t) + \tilde{y}(t) = g\left(x_0(t) + \tilde{x}(t), u_0(t) + \tilde{u}(t), t\right), \qquad [3.47]$$

that is, by applying the exact total differential:

$$\dot{x}_0(t) + \dot{\tilde{x}}(t) = f\left(x_0(t), u_0(t), t\right) + \left[\frac{\partial f}{\partial x}\right]_{\substack{x(t)=x_0(t)\\u(t)=u_0(t)}} \tilde{x}(t) + \left[\frac{\partial f}{\partial u}\right]_{\substack{x(t)=x_0(t)\\u(t)=u_0(t)}} \tilde{u}(t) \qquad [3.48]$$

and

$$y(t) + \tilde{y}(t) = g\left(x_0(t), u_0(t), t\right) + \left[\frac{\partial g}{\partial x}\right]_{\substack{x(t)=x_0(t)\\u(t)=u_0(t)}} \tilde{x}(t) + \left[\frac{\partial g}{\partial u}\right]_{\substack{x(t)=x_0(t)\\u(t)=u_0(t)}} \tilde{u}(t), \qquad [3.49]$$

from which we obtain, by identifying the two members of each of these equations:

$$\dot{x}_0(t) = f\left(x_0(t), u_0(t), t\right) \qquad [3.50]$$

and

$$y_0(t) = g\left(x_0(t), u_0(t), t\right), \qquad [3.51]$$

then

$$\underset{n\times 1}{\dot{\tilde{x}}(t)} = \underset{n\times n}{A(t)}\,\underset{n\times 1}{\tilde{x}(t)} + \underset{n\times m}{B(t)}\,\underset{m\times 1}{\tilde{u}(t)} \qquad [3.52]$$

and

$$\underset{p\times 1}{\tilde{y}(t)} = \underset{p\times n}{C(t)}\,\underset{n\times 1}{\tilde{x}(t)} + \underset{p\times m}{D(t)}\,\underset{m\times 1}{\tilde{u}(t)}, \qquad [3.53]$$

the state model of the plant linearized around the nominal trajectory, $x_0(t)$, called *tangent linearized*, in which the different matrices, said to be Jacobian, are defined as follows:

$$A(t) = \left[\frac{\partial f}{\partial x}\right]_{\substack{x_0(t)\\ u_0(t)}} = \left[\frac{\partial f_i}{\partial x_j}\right]_{\substack{x_0(t)\\ u_0(t)}} = \mathrm{J}_{f_x}, \qquad [3.54]$$

Jacobian matrix of f with respect to x;

$$B(t) = \left[\frac{\partial f}{\partial u}\right]_{\substack{x_0(t)\\ u_0(t)}} = \left[\frac{\partial f_i}{\partial u_j}\right]_{\substack{x_0(t)\\ u_0(t)}} = \mathrm{J}_{f_u}, \qquad [3.55]$$

Jacobian matrix of f with respect to u;

$$C(t) = \left[\frac{\partial g}{\partial x}\right]_{\substack{x_0(t)\\ u_0(t)}} = \left[\frac{\partial g_i}{\partial x_j}\right]_{\substack{x_0(t)\\ u_0(t)}} = \mathrm{J}_{g_x}, \qquad [3.56]$$

Jacobian matrix of g with respect to x;

$$D(t) = \left[\frac{\partial g}{\partial u}\right]_{\substack{x_0(t)\\ u_0(t)}} = \left[\frac{\partial g_i}{\partial u_j}\right]_{\substack{x_0(t)\\ u_0(t)}} = \mathrm{J}_{g_u}, \qquad [3.57]$$

Jacobian matrix of g with respect to u.

3.5.2. *Application*

3.5.2.1. *Tangent linearized at a current (operating) point*

As an example of a nonlinear scalar plant, let us return to the plant discussed in section 3.2.2.1, described by the second-order, nonlinear, differential equation:

$$(J+I)\ddot{q}(t) = \Gamma(t) + mgr\sin q(t) - kq(t) - f\dot{q}(t) \cdot \quad [3.58]$$

In order to establish its nonlinear state model, let us rewrite the differential equation in the form:

$$\ddot{q}(t) = \frac{\Gamma(t)}{J+I} + \frac{mgr\sin q(t)}{J+I} - \frac{kq(t)}{J+I} - \frac{f\dot{q}(t)}{J+I}, \quad [3.59]$$

then let us write:

$$q(t) = x_1(t) \quad [3.60]$$

and

$$\dot{q}(t) = \dot{x}_1(t) = x_2(t), \quad [3.61]$$

meaning the equation of $\ddot{q}(t)$ could then be written as:

$$\dot{x}_2(t) = \frac{\Gamma(t)}{J+I} + \frac{mgr}{J+I}\sin x_1(t) - \frac{k}{J+I}x_1(t) - \frac{f}{J+I}x_2(t) \cdot \quad [3.62]$$

Thus, we obtain the state model:

$$\begin{cases} \dot{x}_1(t) = x_2(t) \\ \dot{x}_2(t) = -\dfrac{k}{J+I}x_1(t) - \dfrac{f}{J+I}x_2(t) + \dfrac{mgr}{J+I}\sin x_1(t) + \dfrac{\Gamma(t)}{J+I} \end{cases} \quad [3.63]$$

and

$$y(t) = x_1(t), \quad [3.64]$$

given that we are considering, as the output quantity, $y(t)$, the angular position, $q(t)$, of the motor shaft (i.e. $q(t)$).

The condensed form of the state model defined by the relationships [3.41] and [3.42] is obtained by writing:

$$x(t) = \left[x_1(t) \quad x_2(t)\right]^T \qquad [3.65]$$

$$u(t) = \Gamma(t) \qquad [3.66]$$

$$f_1\left(x(t), u(t), t\right) = x_2(t) \qquad [3.67]$$

$$f_2\left(x(t), u(t), t\right) = -\frac{k}{J+I}x_1(t) - \frac{f}{J+I}x_2(t) + \frac{mgr}{J+I}\sin x_1(t) + \frac{\Gamma(t)}{J+I} \qquad [3.68]$$

$$g_1(x(t), u(t), t) = x_1(t). \qquad [3.69]$$

Having obtained the nonlinear state model, it thus becomes possible to determine the Jacobian matrices of the tangent linearized around the nominal trajectory defined by:

$$x_0(t) = \left[q_0(t) \quad \dot{q}_0(t)\right]^T \qquad [3.70]$$

and

$$u_0(t) = \Gamma_0(t) \cdot \qquad [3.71]$$

Indeed, these matrices are expressed by:

$$\mathrm{J}_{f_x} = \begin{bmatrix} \dfrac{\partial f_1}{\partial x_1} = 0 & \dfrac{\partial f_1}{\partial x_2} = 1 \\ \dfrac{\partial f_2}{\partial x_1} = -\dfrac{k}{J+I} + \dfrac{mgr}{J+I}\cos x_1(t) & \dfrac{\partial f_2}{\partial x_2} = -\dfrac{f}{J+I} \end{bmatrix}_{\substack{x_0(t) \\ u_0(t)}}$$

$$= \begin{bmatrix} 0 & 1 \\ -\dfrac{k}{J+I} + \dfrac{mgr}{J+I}\cos q_0(t) & -\dfrac{f}{J+I} \end{bmatrix} \qquad [3.72]$$

$$\mathrm{J}_{f_u} = \begin{bmatrix} \dfrac{\partial f_1}{\partial u_1} = 0 \\ \dfrac{\partial f_2}{\partial u_1} = \dfrac{1}{J+I} \end{bmatrix}_{\substack{x_0(t) \\ u_0(t)}} = \begin{bmatrix} 0 \\ \dfrac{1}{J+I} \end{bmatrix} \qquad [3.73]$$

$$\mathrm{J}_{g_x} = \begin{bmatrix} \dfrac{\partial g_1}{\partial x_1} = 1 & \dfrac{\partial g_1}{\partial x_2} = 0 \end{bmatrix}_{\substack{x_0(t) \\ u_0(t)}} = \begin{bmatrix} 1 & 0 \end{bmatrix} \qquad [3.74]$$

$$\mathrm{J}_{g_u} = \begin{bmatrix} \dfrac{\partial g_1}{\partial u_1} = 0 \end{bmatrix}_{\substack{x_0(t) \\ u_0(t)}} = 0\,. \qquad [3.75]$$

3.5.2.2. *Tangent linearized at an operating point*

The symbolic representation of the tangent linearized requires it to be stationary through the temporal independence of the Jacobian matrices. Such a condition implies (even if there is a loss in the modeling) the determination of the tangent linearized, not at a *current (operating) point* on the trajectory, but at a *resting* (or *operating*) *point*, that is, in this case, $\dot{q}_0(t) = 0$ (null speed).

Indeed, at the resting point defined by:

$$q_0(t) = q_0 = \text{ct}\,, \qquad [3.76]$$

the state model representing the tangent linearized, that is,

$$\dot{\tilde{x}}(t) = \mathrm{J}_{f_x}(t)\tilde{x}(t) + \mathrm{J}_{f_u}\tilde{u}(t) \qquad [3.77]$$

and

$$\tilde{y}(t) = \mathrm{J}_{g_x}\tilde{x}(t)\,, \qquad [3.78]$$

can be particularized according to:

$$\dot{\tilde{x}}(t) = \mathrm{J}_{f_x}\tilde{x}(t) + \mathrm{J}_{f_u}\tilde{u}(t) \qquad [3.79]$$

and

$$\tilde{y}(t) = \mathrm{J}_{g_x}\tilde{x}(t) \tag{3.80}$$

3.5.2.3. *Operational, then frequency, representation of the tangent linearized at an operating point (arbitrary then nominal)*

The stationarity of the state model obtained at an (arbitrary) operating point enables us to determine a transmittance between the output gap and the control gap, that is, in this case:

$$G^{'}(s) = \frac{\tilde{Y}(s)}{\tilde{U}(s)} = \mathrm{J}_{g_x}\left(s\mathbb{I} - \mathrm{J}_{f_x}\right)^{-1}\mathrm{J}_{f_u}, \tag{3.81}$$

an expression in which:

$$\mathbb{I} = \begin{bmatrix} 1 & 0 \\ 0 & 1 \end{bmatrix} \text{ (identity matrix)}, \tag{3.82}$$

$$\mathrm{J}_{f_x} = \begin{bmatrix} 0 & 1 \\ a & b \end{bmatrix}, \tag{3.83}$$

$$\mathrm{J}_{f_u} = \begin{bmatrix} 0 \\ c \end{bmatrix} \tag{3.84}$$

and

$$\mathrm{J}_{g_x} = \begin{bmatrix} 1 & 0 \end{bmatrix}, \tag{3.85}$$

with:

$$a = \frac{mgr}{J+I}\cos q_0 - \frac{k}{J+I}, \tag{3.86}$$

$$b = -\frac{f}{J+I} \tag{3.87}$$

and

$$c = \frac{1}{J+I}. \qquad [3.88]$$

In order to determine $G^{'}(s)$, the following matrix needs to be computed, that is:

$$s\mathbb{I} - \mathbf{J}_{f_x} = \begin{bmatrix} s & 0 \\ 0 & s \end{bmatrix} - \begin{bmatrix} 0 & 1 \\ a & b \end{bmatrix} = \begin{bmatrix} s & -1 \\ -a & s-b \end{bmatrix}, \qquad [3.89]$$

with the inversion of the matrix obtained, this leads to:

$$\left(s\mathbb{I} - \mathbf{J}_{f_x}\right)^{-1} = \frac{1}{s(s-b)-a}\begin{bmatrix} s-b & 1 \\ a & s \end{bmatrix}. \qquad [3.90]$$

$G^{'}(s)$ may then be written in the form:

$$G^{'}(s) = \frac{1}{s(s-b)-a}\begin{bmatrix} 1 & 0 \end{bmatrix}\begin{bmatrix} s-b & 1 \\ a & s \end{bmatrix}\begin{bmatrix} 0 \\ c \end{bmatrix}, \qquad [3.91]$$

that is:

$$G^{'}(s) = \frac{1}{s(s-b)-a}\begin{bmatrix} 1 & 0 \end{bmatrix}\begin{bmatrix} c \\ sc \end{bmatrix}, \qquad [3.92]$$

or:

$$G^{'}(s) = \frac{c}{s(s-b)-a}. \qquad [3.93]$$

Given the respective expressions of a, b and c, $G^{'}(s)$ becomes:

$$G^{'}(s) = \frac{\dfrac{1}{J+I}}{s^2 + \dfrac{f}{J+I}s + \dfrac{k - mgr\cos q_0}{J+I}}. \qquad [3.94]$$

The stability condition is expressed by:

$$k - mgr\cos q_0 > 0, \qquad [3.95]$$

or

$$\frac{k}{mgr} > \cos q_0, \qquad [3.96]$$

that is, given that

$$q_0 \in [0, 2\pi] \qquad [3.97]$$

and that therefore,

$$\cos q_0 \in [-1, 1]: \qquad [3.98]$$

$$\frac{k}{mgr} > 1. \qquad [3.99]$$

The frequency response, $G'(\mathrm{j}\omega)$, gives an expression of the form:

$$G'(\mathrm{j}\omega) = \frac{\frac{1}{J+I}}{(\mathrm{j}\omega)^2 + \frac{f}{J+I}(\mathrm{j}\omega) + \frac{k - mgr\cos q_0}{J+I}}. \qquad [3.100]$$

Numerical application:

$$\begin{aligned} m &= 0.1 \text{ kg} \\ g &= 10 \text{ m.s}^{-2} \\ r &= 0.5 \text{ m} \\ k &= 0.6 \text{ N.m} \\ f &= 0.5 \text{ N.m.s} \\ J + I &= 10^{-3} \text{ kg.m}^2 \\ \Gamma_0 &= 0 \text{ N.m} \\ q_{0\,\text{nominal}} &= \frac{\pi}{2} \end{aligned} \qquad [3.101]$$

For a value of q_0 that varies between 0 and 2π, nine transfers have been considered to construct uncertainty domains (see Figure 3.10).

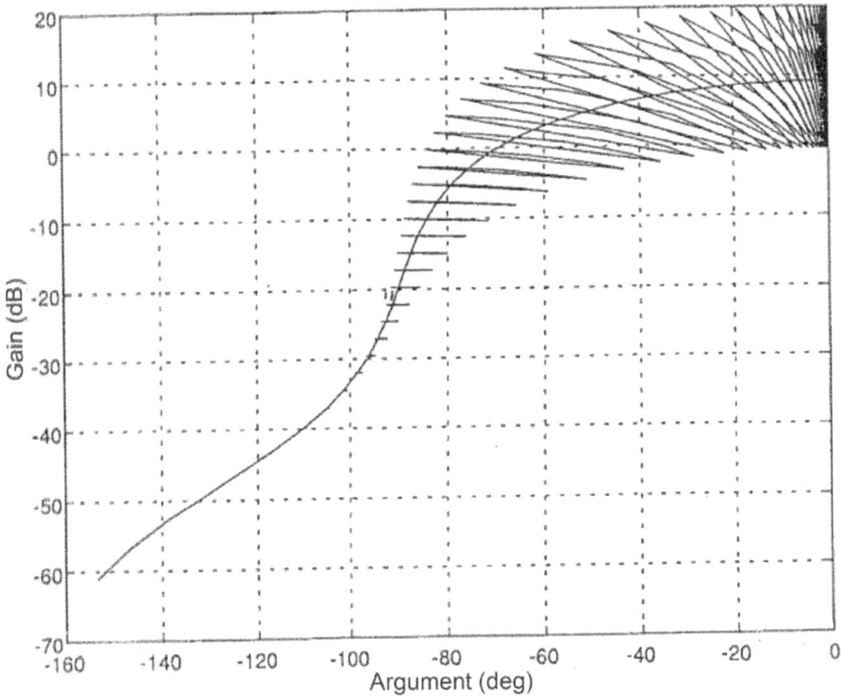

Figure 3.10. *Nominal plant Nichols locus and associated uncertainty domains*

3.6. Additional discussion provided by Brigitte d'Andréa-Novel

With simple examples of mechanical systems comprising nonlinearities, the author has presented different "linearization" methods, with a particularly pragmatic approach, from linearization by immersion to linearization around an operating point, via high-gain control and disturbance rejection, the nonlinear dynamics being interpreted as an output disturbance. With the exception of linearization by immersion, the control laws proposed are linear with respect to the state and are generally sufficient in order to solve the control problem considered.

With regard to linearization by immersion, it is not always possible, whether by conventional state feedback (see, for example, the work of Marino [MAR 86]), or by endogenous feedbacks with dynamic state extension (see, for example, [DES 85] and [CHA 91] for notions of dynamic extension and linearization by dynamic state feedback). In the latter scenario, when linearization is possible, we speak of a *flat system*, as described in Fliess *et al.* [FLI 95].

To date, no necessary, sufficient condition exists to determine the platitude property of a nonlinear dynamic system, and this remains an open problem. Charlet *et al.* [CHA 91] proposed a sufficient condition, and Rouchon [ROU 94] gives a necessary condition.

In the case of mechanical systems, when these are fully actuated (that is, when they have as many actuators as they have degrees of freedom), total linearization by immersion and static feedback is possible, giving rise to what is customarily known in robotics as "computed torque control", in the sense that in the case of known parameters, the nonlinear command ensuring exact linearization is directly obtained by computing the dynamic model linking the torques and accelerations (see [DAN 88]). Certain classes of underactuated mechanical systems are also fully linearizable by nonlinear dynamic state feedback (see [AOU 93, DAN 91]). This is also the case for all wheeled mobile robots under rolling constraints without slipping (see [DAN 92]).

This linearization approach makes it possible to ensure an exponentially stable behavior of the trajectory error and thus guarantees a "global" stability domain, aside from the singularities of the decoupling matrix (see [DAN 88, ISI 95] for this concept). The increase in the stability domain represents a considerable advantage in comparison with the approach by tangent linearization around the operating point or the reference trajectory, since the initial conditions are no longer restricted to a small vicinity around the operating point or the reference trajectory. It is worth noting that in the case of flat systems, the robustness to parametric variations can be improved by synthesizing the feedback part of the control law, in compliance with robust control methods taken from linear automation, such as the H_∞ or CRONE controls presented in this book.

This last, "hybrid" point of view, consisting, for flat systems, of taking advantage of exact linearization to eliminate structural nonlinearities and a robust approach to adjust the gains of the feedback term, would therefore appear to offer a very good compromise, leading to highly efficient control laws. This was highlighted, for example, in the work conducted by the CRONE team researchers.

4

High-Gain, Feedforward, Internal-Model, Quadratic-Criterion and Predictive Controls: From Principle to Control Loop

4.1. Introduction

Having so far studied the control loop in the form of exhaustive developments, our objective in this chapter is, in summary, to present different control schemes, highlighting their *point in common*, notably by showing that they are implicitly based on an elementary control loop (the regulator of which is then analyzed).

Given that the study domain, frequency or time, differs depending on the nature of the control, it is difficult to conduct a comparative study of the main control principles. More generally, this difficulty is exacerbated by the difference in the *study framework*, which incorporates both the study domain (frequency or time) and the structure of the control scheme. As such, this chapter aims for a *common study framework*, to develop a unified approach to control.

As developed in the first two chapters, the in-depth study of the control loop, notably in the frequency domain, has seen the development of a tried-and-tested control methodology over time, capable of meeting several temporal dynamic performance specifications (via frequency dynamic performances significant of these performances). Moreover, such a methodology should be considered as a *reference* among the main control methods proposed in this chapter, and indeed this book.

It is true that the idea aims to reduce each control scheme to an *equivalent control loop*, in order to come under the same study framework (that of the control loop), and thus conduct an objective comparative study of the different control strategies: indeed, the nature of the regulator of each equivalent control loop allows an objective appreciation of the richness of each control strategy. One initial conclusion is that *the sophistication of the control scheme is not always synonymous with the enriching of the control strategy.*

As presented in this chapter, the principle of high-gain control (see section 4.2) is defined and highlighted, considering a proportional, high-gain regulator. While this case study presents the benefit of rapidly showing the interest of the high-gain control through its insensitivity (or robustness), ensured by a transfer imposed by that of the reaction chain, $H^{-1}(s)$, this case is nevertheless unrealistic, as the sensitivity of the input (of the plant) at high frequency clearly shows: indeed, this input sensitivity is given directly by K in tracking and by $K\left|H^{-1}(j\omega)\right|$ in regulation; that is to say that a large value of K comes at the cost of a large input sensitivity, therefore low input immunity. In order to make judicious use of a high-gain principle such as this, it is then appropriate to aim, not for an insensitization of the control, but simply for a desensitization, by choosing a gain high enough for a sufficient desensitization, but not too high for an acceptable input immunity. Finally, as for the change to the control loop, rewriting the error signal of the high-gain control in an appropriate form makes it possible to transform its functional diagram in accordance with a transfer, $H^{-1}(s)$, in cascade with an elementary control loop.

Feedforward control (see section 4.3) differs from the elementary control loop through the addition of the inverse model of the plant, $\tilde{G}^{-1}(s)$, between the reference, $r(t)$, and the input, $u(t)$, of the plant, via a summator located between the controller output and the plant input. By thus ensuring a unit transfer in tracking, this control enables differentiation between the dynamics in tracking and regulation, as indeed a reference prefilter does. But does feedforward not essentially come down to such a prefilter? The response to this question is positive, as appropriately rewriting the plant input enables the functional diagram of the control to be transformed in accordance with a reference prefilter in cascade with the elementary control loop. Finally, it is worth highlighting that the unitary nature of the transmittance in tracking comes at a predictable price to say the least: it is true that this character is accompanied by an input sensitivity in tracking, which tends toward infinity with the frequency.

The internal-model control (see section 4.4) stands out from the elementary control loop through the addition of the plant model (or internal model), $\tilde{G}(s)$, to generate, via a substractor, a feedback signal equal to the difference between the outputs of the plant, $G(s)$, and its model, $\tilde{G}(s)$, with their common input, $u(t)$, being the controller output. In the absence of a disturbance, and in the case where the plant is perfectly modeled ($\tilde{G}(s) = G(s)$), the plant control is conducted without feedback and the controller synthesis is then remarkably simple. Appropriately rewriting the plant input enables the internal model scheme to be restructured in the form of an elementary control loop, the regulator of which is explained by a transmittance dependent on the controller of the internal-model control. In the case of a "robustification", which consists of ensuring a reference-output transfer independent of the plant for the control, the (equivalent) elementary control loop thus obtained has the advantage that it recognizes the *infinite character* underlying the control, not only through the regulator gain, but also through the input sensitivity in tracking and regulation. It is true that directly equating (purely mathematically) the internal-modern scheme has the disadvantage of masking the existence of any infinite gain, thus rendering any non-automated approach risky.

As a preamble to quadratic and predictive controls, which are based on an optimization with respect to the input, $u(t)$, of the plant, a property that is essential for this optimization is established, a property according to which the output, $y(t)$, and its derivatives depend on the past of the input, $u(t)$, and not its instantaneous value. Such a property originates from convolution, in the sense that the output, $y(t)$, of a linear system is given by the convolution product of the impulse response and the input, $u(t)$.

The quadratic-criterion control (see section 4.5) is presented based on a second-order study plant, with input $u(t)$ or $u(k)$ depending on whether the model is continuous or discrete. Following the determination of the continuous and discrete forms of the model, then the presentation of the objective and control strategy, the synthesis of the regulator (which assures a solid tracking of the output of a reference model at the plant output) is performed through the minimization of a quadratic criterion, continuous then discrete, which introduces a control weighting factor, δ. It is in this manner that the continuous and discrete versions of the control and regulator are established and then represented functionally. A continuous, direct approach based on continuous control and a discrete then continuous approach based on discrete control make it possible to result in a same elementary control loop between the outputs of the plant and the reference model. One result that is remarkable to say the least, is that the action of the regulator of this loop is reduced to a *proportional action*, $1/\delta$, which is none other than the inverse of the control weighting factor. When δ tends toward zero, with $1/\delta$ then tending toward infinity, the loop becomes a high-gain control whose reaction chain transfer is unitary.

Predictive control (see section 4.6), the principle of which is presented here in a continuous-time approach, makes a significant contribution through the highlighting, in the operational domain, of an elementary control loop equipped with a phase-lead regulator. The study of the regulator frequency placement with respect to unit gain frequency in open loop shows that the regulator cannot respond to the optimal frequency placement ensured by the direct frequency synthesis of a phase-lead regulator (as developed in Chapter 2): this frequency non-optimality of the regulator comes from the fact that the regulator results, here, from a temporal optimization of the control, which does not imply any time–frequency relationship. The answers to the problem of regulator frequency placement are provided through two problems, the principles of resolution of which are given as ideas for problems or exercises to be solved by students.

4.2. High-gain control

4.2.1. *Principle*

The principle of high-gain control is based on a transfer between the output and the reference, which comes down to the inverse of the reaction chain, using a high gain in open loop further to the introduction of a sufficiently high gain, K, in the action chain (see Figure 4.1).

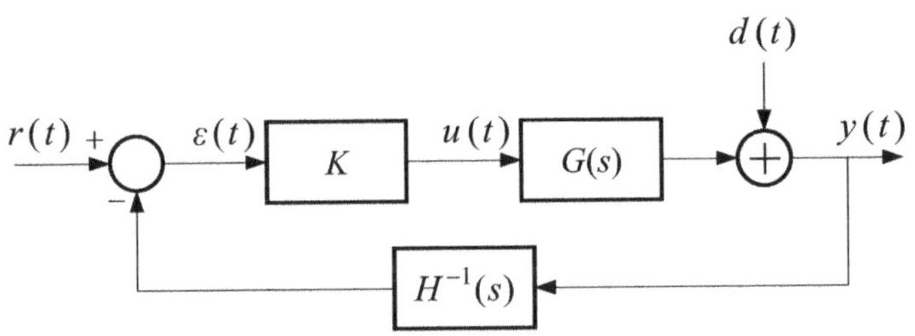

Figure 4.1. *General structure of the high-gain control*

The error signal, $\varepsilon(s)$, is written as:

$$\varepsilon(s) = R(s) - H^{-1}(s)Y(s) \cdot \quad [4.1]$$

The output, $Y(s)$, meanwhile, is written as:

$$Y(s) = D(s) + KG(s)\varepsilon(s), \quad [4.2]$$

from which we obtain:

$$\varepsilon(s) = \frac{Y(s) - D(s)}{KG(s)}. \qquad [4.3]$$

By applying the expression of $\varepsilon(s)$ thus obtained in relation [4.1], we obtain an equation of the form:

$$R(s) - H^{-1}(s)Y(s) = \frac{Y(s) - D(s)}{KG(s)}. \qquad [4.4]$$

Since $Y(s)$, $D(s)$ and $G(s)$ are finite and the gain, K, becomes infinitely large, the second member of equation [4.4] tends toward zero, thus enabling us to write:

$$R(s) - H^{-1}(s)Y(s) = 0, \qquad [4.5]$$

from which we obtain:

$$Y(s) = H(s)R(s) \ \forall \ G(s) \text{ and } D(s). \qquad [4.6]$$

4.2.2. *From high-gain control to control loop*

By expressing the error signal of the high-gain control illustrated in Figure 4.1, that is,

$$\varepsilon(s) = R(s) - H^{-1}(s)Y(s), \qquad [4.7]$$

then rewriting it in the form:

$$\varepsilon(s) = H^{-1}(s)\left[H(s)R(s) - Y(s)\right], \qquad [4.8]$$

it is possible to restructure the control scheme in accordance with that given by Figure 4.2, in which

$$C(s) = KH^{-1}(s). \qquad [4.9]$$

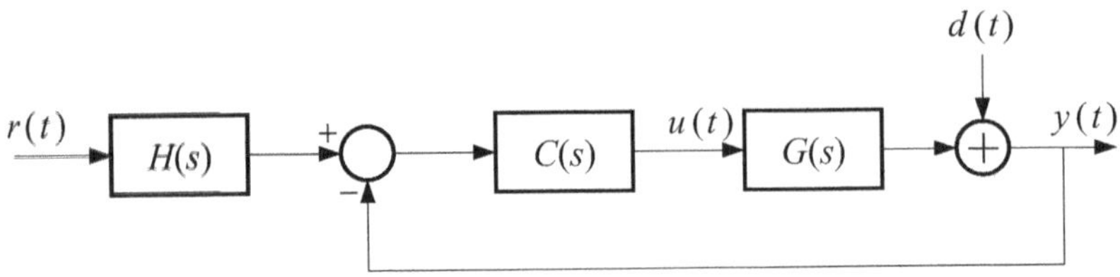

Figure 4.2. *Highlighting an elementary control loop by restructuring the high-gain scheme*

This leads us to a transfer, $H(s)$, in cascade with an elementary control loop. In tracking, it is characterized by the transmittance

$$\frac{\beta(s)}{1+\beta(s)}, \tag{4.10}$$

and in regulation by the transmittance:

$$\frac{1}{1+\beta(s)}, \tag{4.11}$$

expressions in which $\beta(s)$ denotes the open-loop transmittance defined by

$$\beta(s) = C(s)G(s)\,. \tag{4.12}$$

Such a vision of high-gain control enables us to determine the result [4.6] from the general form of transmittances in tracking and regulation of the elementary control loop. Indeed, Figure 4.2 enables the output Laplace transform to be expressed in the form:

$$Y(s) = H(s)\frac{\beta(s)}{1+\beta(s)}R(s) + \frac{D(s)}{1+\beta(s)}\,. \tag{4.13}$$

When the gain, K, becomes infinitely high, $\beta(s) = KG(s)H^{-1}(s)$ then tending toward infinity:

$$\frac{\beta(s)}{1+\beta(s)} = 1 \quad \text{and} \quad \frac{1}{1+\beta(s)} = 0\,, \tag{4.14}$$

which ensures an immediate reduction of equation [4.13], that is:

$$Y(s) = H(s)R(s)\ \forall\ G(s) \text{ and } D(s), \qquad [4.15]$$

a result that clearly confirms that expressed by relation [4.6].

Such a result expresses the *insensitivity* of the control (and therefore its robustness). Yet, this insensitivity comes at the price of significant (plant) input sensitivity, as demonstrated in section 4.2.3.

Furthermore, one way to make judicious use of a high-gain principle such as this consists of reasoning, in terms of *gain that is sufficiently large, to desensitize the control*. We then speak not of insensitivity, but of *desensitization*.

4.2.3. *Input sensitivity*

4.2.3.1. *Input sensitivity in tracking*

For $D(j\omega) = 0$, the functional diagram in Figure 4.1 enables us to write:

$$U(\mathrm{j}\omega) = K\left[R(\mathrm{j}\omega) - H^{-1}(\mathrm{j}\omega)G(\mathrm{j}\omega)U(\mathrm{j}\omega)\right], \qquad [4.16]$$

or indeed:

$$U(\mathrm{j}\omega)\left[1 + KH^{-1}(\mathrm{j}\omega)G(\mathrm{j}\omega)\right] = KR(\mathrm{j}\omega), \qquad [4.17]$$

from which we obtain:

$$\frac{U(\mathrm{j}\omega)}{R(\mathrm{j}\omega)} = \frac{K}{1 + KH^{-1}(\mathrm{j}\omega)G(\mathrm{j}\omega)}, \qquad [4.18]$$

that is:

$$\frac{U(\mathrm{j}\omega)}{R(\mathrm{j}\omega)} = \frac{K}{1 + \beta(\mathrm{j}\omega)} \quad \text{with} \quad \beta(\mathrm{j}\omega) = KH^{-1}(\mathrm{j}\omega)G(\mathrm{j}\omega), \qquad [4.19]$$

with the input sensitivity in tracking then expressed by:

$$\$_r(\omega) = \left|\frac{U(j\omega)}{R(j\omega)}\right| = \left|\frac{K}{1+\beta(j\omega)}\right|. \qquad [4.20]$$

At low frequencies ($\omega << \omega_u$ and $|\beta(j\omega)| >> 1$), $\$_r(\omega)$ is reduced to:

$$\$_r(\omega) = \left|\frac{K}{\beta(j\omega)}\right| = \left|\frac{K}{KH^{-1}(j\omega)G(j\omega)}\right| = \left|\frac{1}{H^{-1}(j\omega)G(j\omega)}\right|, \qquad [4.21]$$

and, at high frequencies ($\omega >> \omega_u$ and $|\beta(j\omega)| << 1$), to:

$$\$_r(\omega) = K, \qquad [4.22]$$

results expressing that the input sensitivity in tracking is independent of K at low frequency, and given directly by K at high frequency.

4.2.3.2. *Input sensitivity in regulation*

For $R(j\omega) = 0$, the functional diagram in Figure 4.1 enables us to write:

$$U(j\omega) = -KH^{-1}(j\omega)\left[D(j\omega) + G(j\omega)U(j\omega)\right], \qquad [4.23]$$

or indeed:

$$U(j\omega)\left[1 + KH^{-1}(j\omega)G(j\omega)\right] = -KH^{-1}(j\omega)D(j\omega), \qquad [4.24]$$

from which we obtain:

$$\frac{U(j\omega)}{D(j\omega)} = \frac{-KH^{-1}(j\omega)}{1+KH^{-1}(j\omega)G(j\omega)}, \qquad [4.25]$$

that is:

$$\frac{U(j\omega)}{D(j\omega)} = \frac{-KH^{-1}(j\omega)}{1+\beta(j\omega)} \quad \text{with} \quad \beta(j\omega) = KH^{-1}(j\omega)G(j\omega), \qquad [4.26]$$

with the input sensitivity in regulation then expressed by:

$$\$_d(\omega) = \left|\frac{U(\mathrm{j}\omega)}{D(\mathrm{j}\omega)}\right| = \left|\frac{KH^{-1}(\mathrm{j}\omega)}{1+\beta(\mathrm{j}\omega)}\right|. \qquad [4.27]$$

At low frequencies ($\omega << \omega_u$ and $|\beta(\mathrm{j}\omega)| >> 1$), $\$_d(\omega)$ admits, as a reduced expression:

$$\$_d(\omega) = \left|\frac{KH^{-1}(\mathrm{j}\omega)}{\beta(\mathrm{j}\omega)}\right| = \left|\frac{KH^{-1}(\mathrm{j}\omega)}{KH^{-1}(\mathrm{j}\omega)G(\mathrm{j}\omega)}\right| = \left|\frac{1}{G(\mathrm{j}\omega)}\right|, \qquad [4.28]$$

and, at high frequencies ($\omega >> \omega_u$ and $|\beta(\mathrm{j}\omega)| << 1$):

$$\$_d(\omega) = K\left|H^{-1}(\mathrm{j}\omega)\right|, \qquad [4.29]$$

results conveying that the input sensitivity in regulation is independent of K at low frequency, and given by $K\left|H^{-1}(j\omega)\right|$ at high frequency.

4.3. Feedforward control

4.3.1. *Principle*

Based on a *feedback* control, constituted by an elementary control loop, the *feedforward* control differs from this control with the introduction of the *inverse plant model*, $\tilde{G}^{-1}(s)$, between the reference, $r(t)$, and the input, $u(t)$, of the plant, via a *summator* of the internal variables, $u_1(t)$ and $u_2(t)$ (see Figure 4.3).

The feedforward control itself (as a complement to the elementary control loop), is therefore an *inverse model control of the plant*:

– based on the idea of a *unit transfer in tracking* (demonstrated below);

– that thus *differentiates* between dynamics in tracking and regulation;

– as a *reference prefilter* does;

– but is feedforward not the *equivalent of such a prefilter*? (response in section 4.3.2 below).

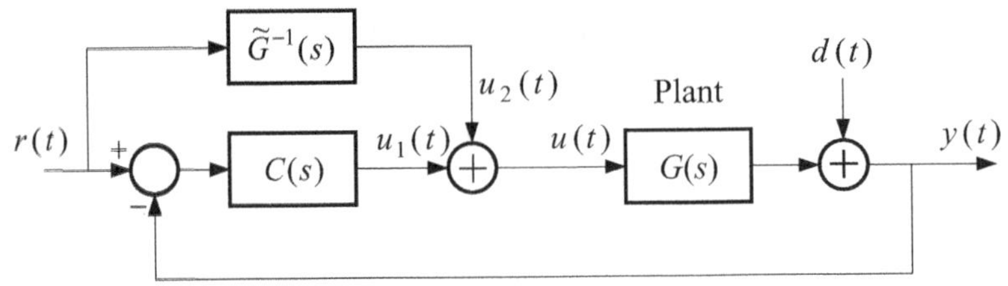

Figure 4.3. *Elementary control loop featuring a feedforward: $G(s)$ is the exact plant model (or plant); $\tilde{G}(s)$ is the approximate model of the plant (or plant model)*

Some previous studies associate feedforward control with the idea of control by anticipation. But it would seem that this idea is a result of confusing time and space. Anticipation would in fact imply a time-based lead, whereas the lead expressed by feedforward is not manifested in terms of time, but *space*, through a direct action on the plant input (further ahead than the reference). More precisely, in the frequency domain, the *feedforward* provides a phase lead that compensates for the phase lag of the plant (as a low-pass system).

In tracking ($D(s) = 0$), Figure 4.3 enables us to write:

$$U_1(s) = C(s)\left[R(s) - G(s)\left(\tilde{G}^{-1}(s)R(s) + U_1(s)\right)\right], \qquad [4.30]$$

or:

$$U_1(s) = C(s)\left[R(s)\left(1 - G(s)\tilde{G}^{-1}(s)\right) - G(s)U_1(s)\right], \qquad [4.31]$$

or even:

$$U_1(s)\left(1 + C(s)G(s)\right) = C(s)\left(1 - G(s)\tilde{G}^{-1}(s)\right)R(s), \qquad [4.32]$$

or indeed:

$$U_1(s) = C(s)\frac{1 - G(s)\tilde{G}^{-1}(s)}{1 + C(s)G(s)}R(s), \qquad [4.33]$$

that is:

$$U_1(s) = 0 \text{ for } \tilde{G}(s) = G(s). \qquad [4.34]$$

Note that $u(t) = u_1(t) + u_2(t)$ is reduced to $u(t) = u_2(t)$, thus expressing that the inverse model of the plant, $\tilde{G}^{-1}(s)$, is directly in cascade with the plant, $G(s)$, with the control then reduced to a *direct chain*, such that:

$$Y(s) = G(s)\tilde{G}^{-1}(s)R(s), \tag{4.35}$$

that is:

$$Y(s) = R(s) \text{ for } \tilde{G}(s) = G(s), \tag{4.36}$$

a result that clearly expresses a *unit transfer* between the output and the reference.

4.3.2. *From feedforward to reference filtering of the elementary control loop*

Figure 4.3 enables us to write:

$$U(s) = \frac{1}{\tilde{G}(s)}R(s) + C(s)\left(R(s) - Y(s)\right), \tag{4.37}$$

that is, by multiplying and dividing by $C(s)$:

$$U(s) = C(s)\left[\frac{1}{C(s)\tilde{G}(s)}R(s) + R(s) - Y(s)\right], \tag{4.38}$$

or:

$$U(s) = C(s)\left[\left(\frac{1}{C(s)\tilde{G}(s)} + 1\right)R(s) - Y(s)\right], \tag{4.39}$$

or indeed:

$$U(s) = C(s)\left[\left(\frac{1 + C(s)\tilde{G}(s)}{C(s)\tilde{G}(s)}\right)R(s) - Y(s)\right]. \tag{4.40}$$

The symbolic equation thus obtained is conveyed by the functional diagram in Figure 4.4. This diagram clearly expresses that the feedforward comes down to *filtering the reference* with a *prefilter* whose output is none other than the reference of the elementary control loop, which, initially, is subject to feedforward through $\tilde{G}^{-1}(s)$. Such a result is a positive answer to the question posed in the previous section.

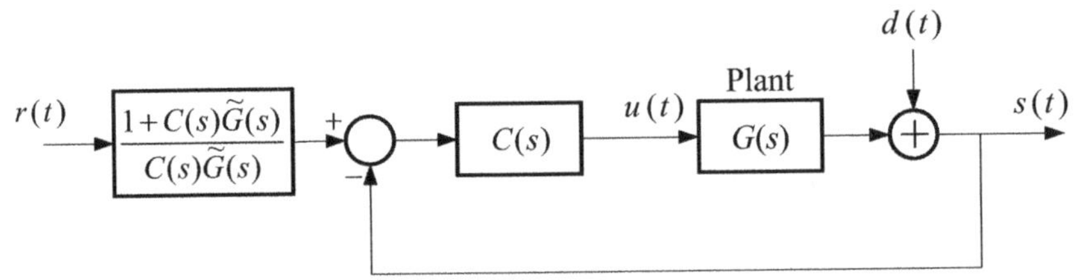

Figure 4.4. *Functional diagram illustrating the change to the elementary control loop and showing that the feedforward is equivalent to a reference prefilter*

4.3.3. *Tracking and regulation*

Using $\beta(s)$ to denote the transmittance in open loop of the elementary control loop, that is

$$\beta(s) = C(s)G(s), \qquad [4.41]$$

transmittance in tracking admits an expression of the form:

$$F_r(s) = \left[\frac{Y(s)}{R(s)}\right]_{D(s)=0} = \frac{1+C(s)\tilde{G}(s)}{C(s)\tilde{G}(s)}\frac{C(s)G(s)}{1+C(s)G(s)}, \qquad [4.42]$$

that is, for $\tilde{G}(s) = G(s)$:

$$F_r(s) = 1, \qquad [4.43]$$

a result that clearly confirms the *unit transfer*, as demonstrated in section 4.3.1, by $Y(s) = R(s)$ (relation [4.36]).

Regulation transmittance, meanwhile, is given by:

$$F_d(s) = \left[\frac{Y(s)}{D(s)}\right]_{R(s)=0} = \frac{1}{1+C(s)G(s)}. \qquad [4.44]$$

With regard to *input sensitivities in tracking and regulation*, these are expressed, respectively, by:

$$\$_r(\omega) = \left|\frac{U(j\omega)}{R(j\omega)}\right|_{D(j\omega)=0} = \left|\frac{1 + C(j\omega)\tilde{G}(j\omega)}{\tilde{G}(j\omega)(1 + C(j\omega)G(j\omega))}\right|$$

$$= \left|\frac{1}{G(j\omega)}\right| \quad \textit{for} \quad \tilde{G}(j\omega) = G(j\omega) \qquad [4.45]$$

and

$$\$_d(\omega) = \left|\frac{U(j\omega)}{D(j\omega)}\right|_{R(j\omega)=0} = \left|\frac{C(j\omega)}{1 + C(j\omega)G(j\omega)}\right|, \qquad [4.46]$$

that is, at low frequency:

$$\$_r(\omega) = \$_d(\omega) = \left|\frac{1}{G(j\omega)}\right|, \qquad [4.47]$$

and, at high frequency:

$$\$_r(\omega) = \left|\frac{1}{G(j\omega)}\right| \text{ and } \$_d(\omega) = \left|C(j\omega)\right|, \qquad [4.48]$$

the sensitivity in tracking, $\$_r(\omega)$, *thus tending toward infinity with the frequency,* since $G(j\omega)$ then tends toward zero (owing to the low-pass nature of the plant). Moreover, such a result is consistent with the *unitary nature* of transmittance in tracking ($F_r(s) = 1$).

4.4. Internal-model control

4.4.1. *Principle*

If $G(s)$ represents the exact model of the plant (or, simply, the plant), we use $\tilde{G}(s)$ to designate an approximated model, called the *plant model* (in the general scenario) or *internal model* (in the particular case of the internal-model control).

The principle of internal-model control is based on the development of a feedback signal, $v(t)$, from the gap between the outputs of the plant, $G(s)$, and of its model, $\tilde{G}(s)$, (or internal model), with their input being attacked by the same signal, $u(t)$ (see Figure 4.5).

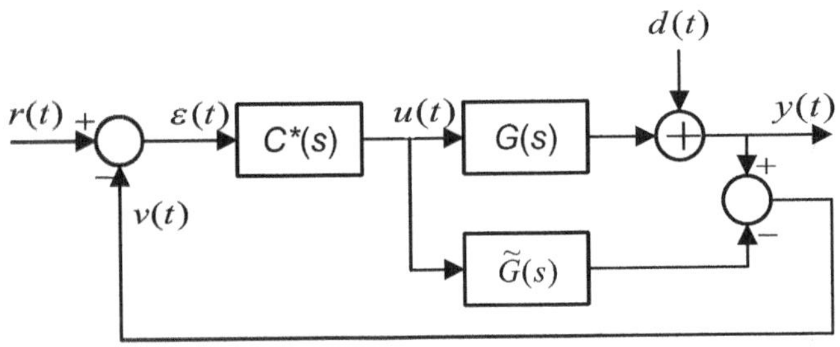

Figure 4.5. *Principle of internal-model control*

In the case where the plant is perfectly modeled, that is, $\tilde{G}(s) = G(s)$, and in the absence of a disturbance, that is, $d(t) = 0$, the feedback signal, $v(t)$, is nil. The error signal, $\varepsilon(t) = r(t) - v(t)$, is thus reduced to the reference, $r(t)$, conveying that the plant control is then produced in open loop and that the transmittance in tracking is none other than the product,

$$C^*(s)G(s). \quad [4.49]$$

If we wish for the output, $y(t)$, of the plant controlled by $u(t)$ to respond to the reference, $r(t)$, with a given dynamic set by a sought transfer, $H(s)$, the synthesis of the regulator, $C^*(s)$, proves noteworthy in its simplicity. Indeed, it is sufficient for $C^*(s)$ to satisfy the equation:

$$H(s) = C^*(s)G(s), \quad [4.50]$$

from which we immediately obtain:

$$C^*(s) = G^{-1}(s)H(s). \quad [4.51]$$

We can thus understand the interest that this type of control holds for manufacturers. But what about in the case of $\tilde{G}(s) \neq G(s)$ and the behavior in regulation ($d(t) \neq 0$)?

4.4.2. *From internal-model control to control loop*

The observation in Figure 4.5 enables the plant control, $U(s)$, to be expressed in the form:

$$U(s) = C^*(s)\left[R(s) - \left(Y(s) - \tilde{G}(s)U(s)\right)\right], \qquad [4.52]$$

that is:

$$U(s) = C^*(s)\left(R(s) - Y(s)\right) + C^*(s)\tilde{G}(s)U(s), \qquad [4.53]$$

from which we obtain:

$$U(s) = \frac{C^*(s)}{1 - C^*(s)\tilde{G}(s)}\left(R(s) - Y(s)\right), \qquad [4.54]$$

or:

$$U(s) = C(s)\left(R(s) - Y(s)\right), \qquad [4.55]$$

by writing:

$$C(s) = \frac{C^*(s)}{1 - C^*(s)\tilde{G}(s)}. \qquad [4.56]$$

The functional translation of equation [4.55] enables the transformation of the functional diagram in Figure 4.5, in compliance with that of Figure 4.6(a), which illustrates the most elementary version of the control loop. With regard to the functional translation of equation [4.56], it leads to the block diagram in Figure 4.6(b), which explains the regulator, $C(s)$.

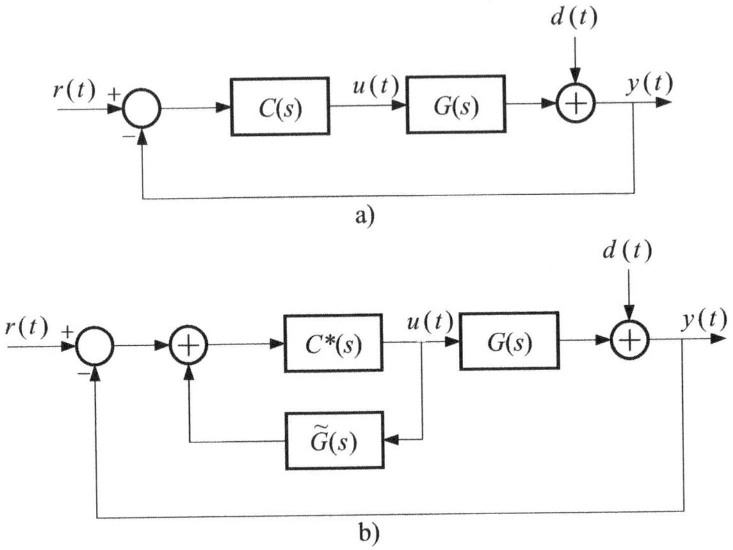

Figure 4.6. *Highlighting an elementary control loop by restructuring the internal-model scheme*

The synthesis of $C(s)$ (see Figure 4.6(a)) occurrs through the well-known arsenal of the synthesis methods of a regulator in cascade. The determination of $C^*(s)$ is performed based on $C(s)$ according to the change relationship that can be directly deduced from [4.56], that is:

$$C^*(s) = \frac{C(s)}{1 + C(s)\tilde{G}(s)}. \quad [4.57]$$

However, in principle, there is no reason for the regulator, $C^*(s)$, thus obtained to satisfy equation [4.50] and thus enable the desired transfer, $H(s)$, in tracking to be obtained. One solution then consists of filtering the reference, $r(t)$, using a prefilter, $C^{**}(s)$, which satisfies the equation:

$$H(s) = C^{**}(s)C^*(s)G(s), \quad [4.58]$$

from which we obtain:

$$C^{**}(s) = C^{*-1}(s)G^{-1}(s)H(s). \quad [4.59]$$

The control can then simultaneously meet the tracking and regulation objectives.

4.4.3. *Robustifying strategies*

The "robustification" that is the focus of this section consists of offering the control a reference-output transfer that is independent of the plant (transmittance and disturbance). That is to say, the robustness objective is severe, but unrealistic with respect to the input immunity in tracking and regulation, with the plant input then being subjected to significant stresses.

Nevertheless, such an objective has the merit of stimulating the search for robust strategies, the principle of which is to be retained if it is judiciously exploited, in particular by accepting degradation in robustness performance in favor of better input immunity.

4.4.3.1. *First strategy: unit reference-output transfer*

As conveyed by Figure 4.7, a first strategy consists of performing:

$$C^*(s) = \tilde{G}^{-1}(s). \qquad [4.60]$$

Given relation [4.56], the regulator, $C(s)$, of the control loop (see Figure 4.6(a)) then presents an infinite gain that ensures an infinite gain for open-loop transmittance, $\beta(s) = C(s)G(s)$. The output, $Y(s)$, expressed according to the functional diagram in Figure 4.6(a), that is,

$$Y(s) = \frac{\beta(s)}{1+\beta(s)}R(s) + \frac{D(s)}{1+\beta(s)}, \qquad [4.61]$$

is thus reduced to:

$$Y(s) = R(s) \ \forall \ G(s) \text{ and } D(s), \qquad [4.62]$$

a result that conveys a unit transfer of the control.

Equating the block diagram of Figure 4.7 enables us to find equation [4.62]. Indeed, the control, $U(s)$, of the plant is written as:

$$U(s) = \tilde{G}^{-1}(s)\left[R(s) - \left(Y(s) - \tilde{G}(s)U(s)\right)\right], \qquad [4.63]$$

that is, by developing:

$$U(s) = \tilde{G}^{-1}(s)\big(R(s) - Y(s)\big) + U(s), \qquad [4.64]$$

from which we obtain:

$$\tilde{G}^{-1}(s)\big(R(s) - Y(s)\big) = 0, \qquad [4.65]$$

or, for $\tilde{G}^{-1}(s) \neq 0$:

$$Y(s) = R(s) \ \forall \ G(s) \text{ and } D(s). \qquad [4.66]$$

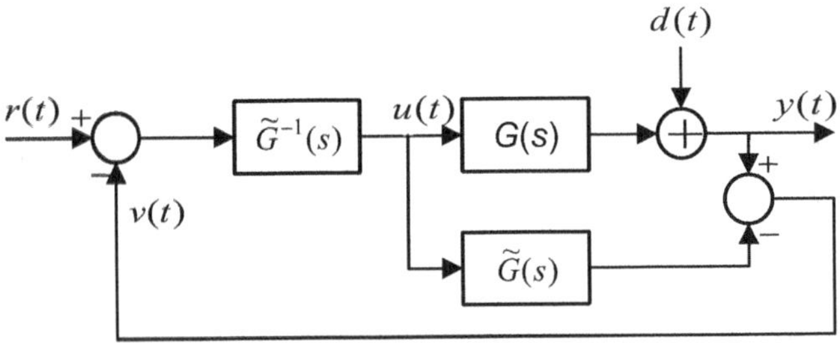

Figure 4.7. *Unit transfer control between output and reference*

We thus obtain a unit transfer of the control, in accordance with the result expressed by relation [4.62], except that *this transfer is established here without introducing or revealing any infinite gain.* In fact, while this equation has the *advantage* of directly finding equation [4.62], it nevertheless has the *disadvantage* of obscuring the existence of an infinite gain, hence the incongruity (by loss of essential information) of not returning to the control loop.

It is true that the control loop clearly detects the *infinite character* underlying the control, not only through the *regulator gain,* $C(s)$ (see the beginning of section 4.4.3.1), but also through the *sensitivity of the input in tracking and regulation,* $\$(\omega)$, which is expressed simply based on the functional diagram in Figure 4.6(a), that is:

$$\$(\omega) = \left| \frac{C(j\omega)}{1 + C(j\omega)G(j\omega)} \right|, \qquad [4.67]$$

or:

$$\$(\omega) = \left| \frac{1}{\frac{1}{C(j\omega)} + G(j\omega)} \right|, \qquad [4.68]$$

or, given the expression of $C(s)$ already established (relation [4.56]):

$$\$(\omega) = \left| \frac{1}{\frac{1 - C^{*}(j\omega)\tilde{G}(j\omega)}{C^{*}(j\omega)} + G(j\omega)} \right|, \qquad [4.69]$$

or indeed, for $C^{*}(j\omega) = \tilde{G}^{-1}(j\omega)$:

$$\$(\omega) = \left| \frac{1}{G(j\omega)} \right|. \qquad [4.70]$$

Knowing that the plant gain tends toward zero when the frequency tends towards infinity, *the input sensitivity in tracking and regulation tends toward infinity with the frequency*. Thus, any infinitely rapid dynamics (generating spectral components of infinite frequency) are entirely transmitted in tracking and rejected in regulation: this property clearly confirms the *unit transfer of the control* ($Y(s) = R(s)\ \forall\ G(s)$ and $D(s)$), which simultaneously expresses a *unit transfer in tracking* and a *nil transfer in regulation*.

4.4.3.2. *Second strategy: reference-output transfer set by the internal model*

Knowing that the transfer of the first version of the control is reduced to one (see Figure 4.7), a transfer set by the plant in its nominal parametric state, that is, $\tilde{G}(s)$, can be obtained by cascading the transfer, $\tilde{G}(s)$, in accordance with the functional diagram in Figure 4.8.

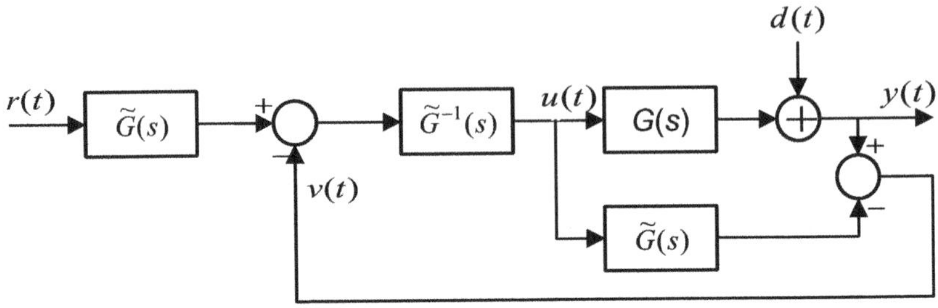

Figure 4.8. *Control with internal model, $\tilde{G}(s)$, determining the reference-output transfer*

The control, $U(s)$, of the plant then satisfies an equation of the form:

$$U(s) = \tilde{G}^{-1}(s)\left[\tilde{G}(s)R(s) - V(s)\right], \tag{4.71}$$

that is:

$$U(s) = R(s) - \tilde{G}^{-1}(s)V(s), \tag{4.72}$$

this equation enabling a reduction of the block diagram in Figure 4.8, in accordance with that recommended in Figure 4.9.

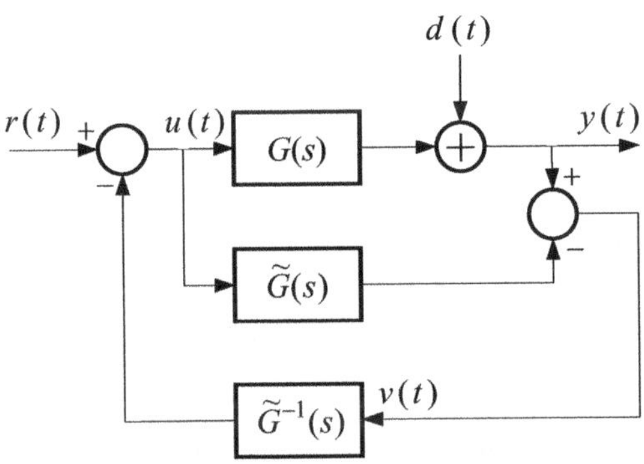

Figure 4.9. *Control scheme equivalent to that in Figure 4.8*

4.4.3.3. *Third strategy: reference-output transfer set by an arbitrary model*

Given that the transfer of the second version of the control is reduced to the internal model, $\tilde{G}(s)$ (see Figure 4.9), one possible extension consists of replacing

$\tilde{G}(s)$ with an arbitrary transfer, $H(s)$, which nevertheless ensures the control stability (see Figure 4.10).

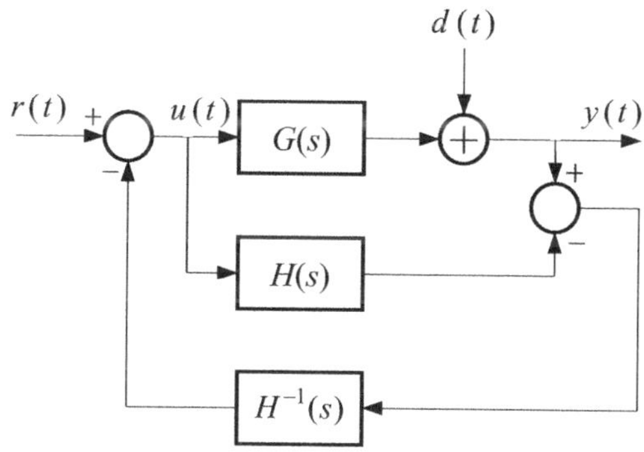

Figure 4.10. *Control with arbitrary model, $H(s)$, determining the reference-output transfer*

The control, $U(s)$, of the plant satisfies an equation of the form:

$$U(s) = R(s) - H^{-1}(s)\big(Y(s) - H(s)U(s)\big), \tag{4.73}$$

that is:

$$U(s) = R(s) - H^{-1}(s)Y(s) + U(s), \tag{4.74}$$

from which we obtain:

$$R(s) - H^{-1}(s)Y(s) = 0, \tag{4.75}$$

or:

$$Y(s) = H(s)R(s)\ \forall\ G(s) \text{ and } D(s). \tag{4.76}$$

4.4.4. *From internal model to high-gain control*

By introducing a constant, k, infinitely close to one (by lower values), that is,

$$k = 1^{-}, \tag{4.77}$$

the functional diagram in Figure 4.10 can be modified in accordance with the diagram presented in Figure 4.11.

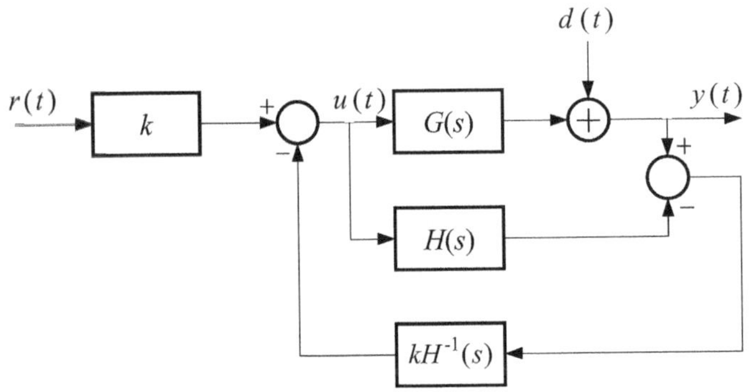

Figure 4.11. *Equivalent version (for $k = 1^-$) of the control scheme in Figure 4.10*

The control, $U(s)$, of the plant complies with the equation:

$$U(s) = kR(s) - kH^{-1}(s)\big(Y(s) - H(s)U(s)\big), \qquad [4.78]$$

or, by developing:

$$U(s) = kR(s) - kH^{-1}(s)Y(s) + kU(s), \qquad [4.79]$$

or:

$$(1-k)U(s) = k\left[R(s) - H^{-1}(s)Y(s)\right], \qquad [4.80]$$

or, by writing:

$$k = \frac{K}{1+K} \text{ with an infinitely large value of } K: \qquad [4.81]$$

$$\frac{1}{1+K}U(s) = \frac{K}{1+K}\left[R(s) - H^{-1}(s)Y(s)\right], \qquad [4.82]$$

that is, ultimately:

$$U(s) = K\left[R(s) - H^{-1}(s)Y(s)\right]. \qquad [4.83]$$

This last equation is conveyed by the functional diagram represented in Figure 4.12.

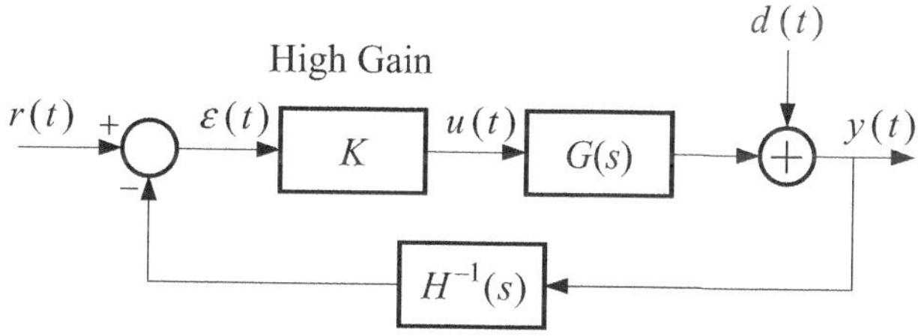

Figure 4.12. *High-gain strategy deduced from the control scheme in Figure 4.11*

Given Figure 4.1, it appears that this diagram is none other than that of the high-gain control, thus leading, owing to equation [4.6], to:

$$Y(s) = H(s)R(s)\ \forall\ G(s) \text{ and } D(s). \qquad [4.84]$$

4.5. Quadratic-criterion control

4.5.1. *On control law synthesis*

For quadratic-criterion control, as for predictive control, *temporal synthesis of the control law* is performed without directly taking into account the plant output disturbance, $d(t)$, which is also the case for the *frequential synthesis of a regulator*, for which the disturbance, like the reference, is not directly taken into account (but only through the input sensitivities in tracking and regulation, which are relative to these stresses).

4.5.2. *A property of linear systems that is decisive in this area*

The output of a linear system, $y(t)$, is given by the convolution product of the impulse response, $y_{imp}(t)$, and the input, $u(t)$, that is:

$$y(t) = y_{imp}(t) * u(t), \qquad [4.85]$$

or, given the definition of the convolution product:

$$y(t) = \int_{-\infty}^{+\infty} y_{imp}(t-\theta)u(\theta)d\theta \,, \qquad [4.86]$$

or, due to the causality of the impulse response ($y_{imp}(t) = 0$ for $t < 0$):

$$y(t) = \int_{-\infty}^{t} y_{imp}(t-\theta)u(\theta)d\theta \,, \qquad [4.87]$$

or indeed, in the case of a nil input for $t < 0$ (causal input):

$$y(t) = \int_{0}^{t} y_{imp}(t-\theta)u(\theta)d\theta \,. \qquad [4.88]$$

Knowing that θ varies from *0* to t in the integration that determines $y(t)$, the output depends on $u(t)$

– through the values of $u(t)$ at all instants in the past (*strictly* comprised between 0 and t);

– not through the value of $u(t)$ at the instant, t.

In other words, the output, $y(t)$, depends on the past of the input, $u(t)$, and not its instantaneous value.

Thus $y(t)$, and therefore its derivatives, do not depend on $u(t)$, a remarkable property that proves decisive in terms of optimization with respect to the input, $u(t)$. *The independence of these functions with respect to* $u(t)$ is indeed widely exploited in the partial derivation with respect to $u(t)$ of criteria to be minimized in quadratic-criterion control and predictive control (or any other control using a criterion of this type).

4.5.3. *Study plant*

The plant considered as an illustrative example is a second-order stationary, linear scalar system, the input of which is $u(t)$ or $u(k)$, and the output $y(t)$ or $y(k)$, depending on whether the model is continuous or discrete.

4.5.3.1. *Continuous model*

The continuous model is represented by the transmittance in s:

$$G(s)=\frac{Y(s)}{U(s)}=\frac{G_0}{1+2\zeta_0\dfrac{s}{\omega_0}+\dfrac{s^2}{\omega_0^2}}, \qquad [4.89]$$

where G_0 is the static gain, ζ_0 is the damping ratio and ω_0 is the undamped natural frequency.

The following differential equation corresponds to this:

$$\frac{1}{\omega_0^2}\ddot{y}(t)+\frac{2\zeta_0}{\omega_0}\dot{y}(t)+y(t)=G_0u(t), \qquad [4.90]$$

which is rewritten in the form:

$$y(t)=G_0u(t)-2\frac{\zeta_0}{\omega_0}\dot{y}(t)-\frac{1}{\omega_0^2}\ddot{y}(t), \qquad [4.91]$$

that is, by writing

$$a_1=\frac{2\zeta_0}{\omega_0} \text{ and } a_2=\frac{1}{\omega_0^2}: \qquad [4.92]$$

$$y(t)=G_0u(t)-a_1\dot{y}(t)-a_2\ddot{y}(t). \qquad [4.93]$$

4.5.3.2. *Discrete model*

The discrete model resulting from a discretization of the blocked, sampled type is represented by the transmittance in q^{-1}:

$$G\left(q^{-1}\right)=\frac{y(k)}{u(k)}=\frac{b_0q^{-1}+b_1q^{-2}}{1+a_1q^{-1}+a_2q^{-2}}, \qquad [4.94]$$

with $y(k)$ and $u(k)$ being the output and input of the plant at the kth sampling instant, that is, at the instant, $t=kT$, that is $y(kT)$ and $u(kT)$, with T denoting the sampling period. With regard to the four transmittance parameters, these are defined by the following relations:

$$b_0 = G_0\left[1-\frac{e^{-\zeta_0\omega_0 T}}{\sqrt{1-\zeta_0^2}}\cos\left(\omega_0\sqrt{1-\zeta_0^2}\,T-\arcsin\zeta_0\right)\right]; \qquad [4.95]$$

$$b_1 = G_0 e^{-\zeta_0\omega_0 T}\left[e^{-\zeta_0\omega_0 T}-\frac{1}{\sqrt{1-\zeta_0^2}}\cos\left(\omega_0\sqrt{1-\zeta_0^2}\,T+\arcsin\zeta_0\right)\right]; \quad [4.96]$$

$$a_1 = -2e^{-\zeta_0\omega_0 T}\cos\left(\omega_0\sqrt{1-\zeta_0^2}\,T\right) \text{ and } a_2 = e^{-2\zeta_0\omega_0 T}. \qquad [4.97]$$

From the expression of the transmittance in q^{-1}, we obtain the recurrent equation:

$$y(k)+a_1 y(k-1)+a_2 y(k-2) = b_0 u(k-1)+b_1 u(k-2), \qquad [4.98]$$

that is:

$$y(k) = b_0 u(k-1)+b_1 u(k-2)-a_1 y(k-1)-a_2 y(k-2), \qquad [4.99]$$

or, replacing k with $k+1$:

$$y(k+1) = b_0 u(k)+b_1 u(k-1)-a_1 y(k)-a_2 y(k-1). \qquad [4.100]$$

4.5.4. *Control objective and strategy*

We want the output, $y(t)$ or $y(k)$, of the plant controlled by $u(t)$ or $u(k)$ to respond to the reference, $r(t)$ or $r(k)$, with the given dynamics (see Figure 4.13).

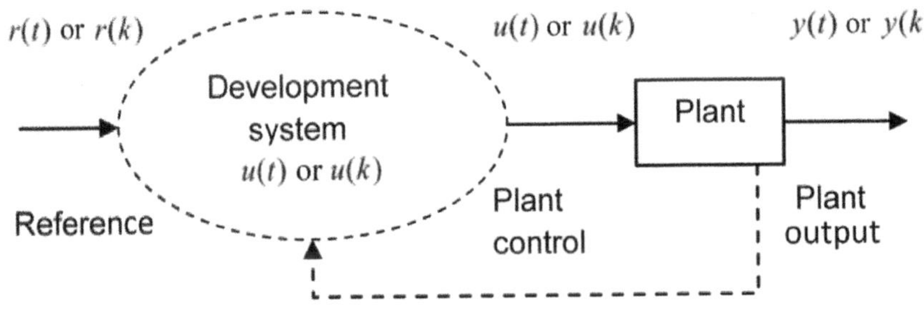

Figure 4.13. *Illustration of the control objective*

A means of achieving such an objective consists (see Figure 4.14):

– first, of imposing the dynamics sought by a reference model, $H(s)$, setting the dynamics of its output quantity, $y_M(t)$ or $y_M(k)$;

– second, of designing the regulator making the output, $y(t)$ or $y(k)$, of the plant as close as possible to the output, $y_M(t)$ or $y_M(k)$, of the reference model, which comes down to developing the control, $u(t)$ or $u(k)$, of the plant ensuring

$$y(t) \approx y_M(t) \text{ or } y(k) \approx y_M(k). \qquad [4.101]$$

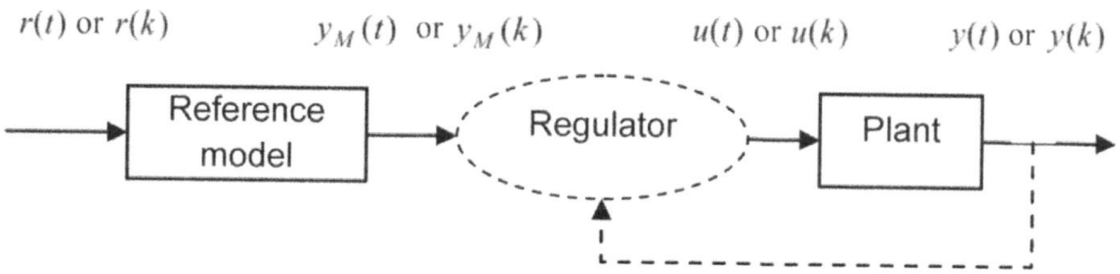

Figure 4.14. *Illustration of the control strategy*

4.5.5. *Regulator synthesis through minimization of a quadratic criterion*

In order to achieve the objective, $y(t) \approx y_M(t)$ or $y(k) \approx y_M(k)$, we need to develop the control, $u(t)$ or $u(k)$, which minimizes the quadratic criterion:

$$J(t) = \left[y(t) - y_M(t)\right]^2 + \delta G_0 u^2(t) \qquad [4.102]$$

or

$$J(k+1) = \left[y(k+1) - y_M(K+1)\right]^2 + \delta b_0 u^2(k); \qquad [4.103]$$

with δ as a weighting factor of the control.

The minimization of the criterion with respect to the control ensures an overall minimization of the gap, $y(t) - y_M(t)$ or $y(k+1) - y_M(k+1)$, and of the control, $u(t)$ or $u(k)$, such that the latter does not present excessively high values (realism). The presence of the term, $\delta G_0 u^2(t)$ or $\delta b_0 u^2(k)$, leads to a less satisfactory minimization of the gap, $y(t) - y_M(t)$ or $y(k+1) - y_M(k+1)$. Indeed, the control weighting comes at the cost of a greater output error (which is not at all surprising in itself).

4.5.5.1. *Continuous control*

Given the expression of $y(t)$ deduced from the continuous model of the plant, the criterion, $J(t)$, becomes:

$$J(t) = \left[G_0 u(t) - a_1 \dot{y}(t) - a_2 \ddot{y}(t) - y_M(t)\right]^2 + \delta G_0 u^2(t). \qquad [4.104]$$

Its minimization with respect to $u(t)$ is conveyed by:

$$\frac{\partial J(t)}{\partial u(t)} = 0, \qquad [4.105]$$

that is:

$$2G_0\left[G_0 u(t) - a_1 \dot{y}(t) - a_2 \ddot{y}(t) - y_M(t)\right] + 2\delta G_0 u(t) = 0, \qquad [4.106]$$

or:

$$2G_0\left[\left(G_0 + \delta\right)u(t) - a_1 \dot{y}(t) - a_2 \ddot{y}(t) - y_M(t)\right] = 0, \qquad [4.107]$$

from which we obtain the continuous control equation, $u(t)$, of the plant:

$$u(t) = \frac{y_M(t) + a_1 \dot{y}(t) + a_2 \ddot{y}(t)}{G_0 + \delta} \qquad [4.108]$$

or

$$U(s) = \frac{Y_M(s) + \left(a_1 s + a_2 s^2\right) Y(s)}{G_0 + \delta}; \qquad [4.109]$$

such control equations are conveyed, respectively, by the functional diagrams in Figure 4.15.

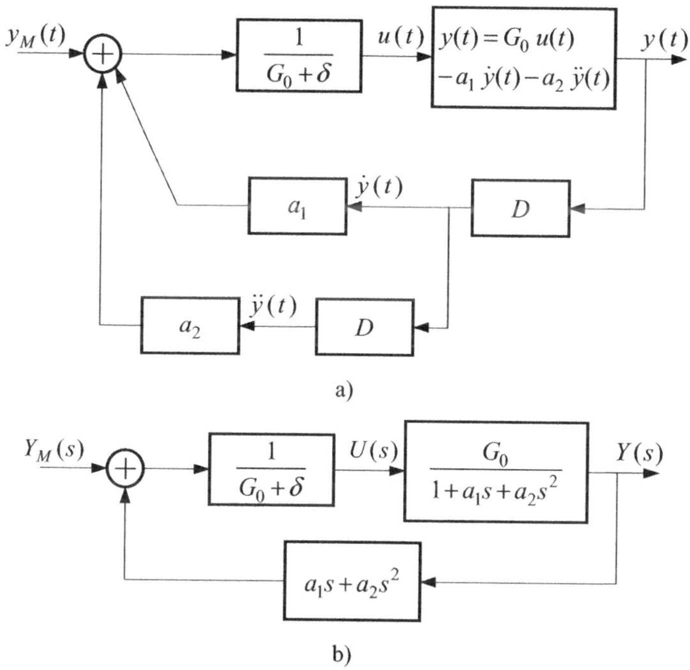

Figure 4.15. *Functional diagrams conveying the continuous-regulator equation: a) in time domain; b) in the operational domain*

4.5.5.2. *Discrete control*

Given the expression of $y(k+1)$ deduced from the discrete model of the plant, the criterion, $J(k+1)$, becomes:

$$J(k+1) = \begin{bmatrix} b_0 u(k) + b_1 u(k-1) \\ -a_1 y(k) - a_2 y(k-1) \\ -y_M(k+1) \end{bmatrix}^2 + \delta b_0 u^2(k) . \qquad [4.110]$$

Its minimization with respect to $u(k)$ is conveyed by:

$$\frac{\partial J(k+1)}{\partial u(k)} = 0 , \qquad [4.111]$$

that is:

$$2b_0\left[b_0u(k)+b_1u(k-1)-a_1y(k)-a_2y(k-1)-y_M(k+1)\right], \\ +2\delta b_0u(k)=0 \quad [4.112]$$

or:

$$2b_0\begin{bmatrix}(b_0+\delta)u(k)+b_1u(k-1)-a_1y(k)-a_2y(k-1)\\ -y_M(k+1)\end{bmatrix}=0, \quad [4.113]$$

from which we obtain the discrete control equation, $u(k)$, of the plant:

$$u(k)=\frac{y_M(k+1)-b_1u(k-1)+a_1y(k)+a_2y(k-1)}{b_0+\delta} \quad [4.114]$$

or

$$u(k)=\frac{y_M(k+1)-b_1q^{-1}u(k)+a_1y(k)+a_2q^{-1}y(k)}{b_0+\delta}; \quad [4.115]$$

such a control equation is conveyed by the functional diagram in Figure 4.16.

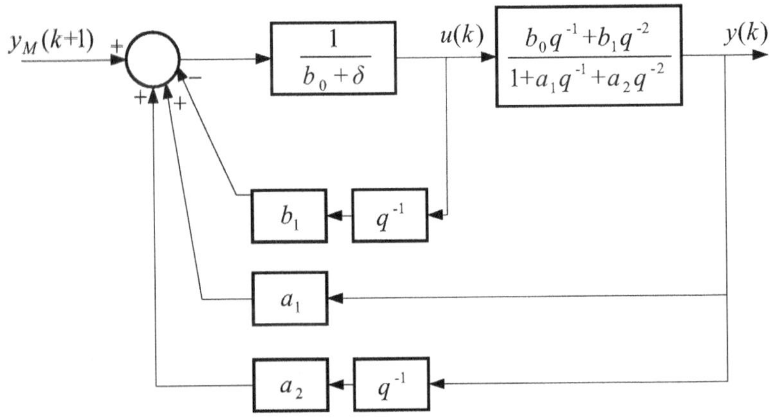

Figure 4.16. *Functional diagram conveying the discrete-regulator equation*

4.5.6. *From quadratic-criterion control to control loop*

In order for the change from quadratic-criterion control to control loop to occur continuously to remain consistent with all of our developments, two distinct approaches msut be used:

– a *direct continuous approach*, which consists of directly conducting the study continuously, based on one of the control schemes in Figure 4.15, and in particular, the second one, which relates to the operational domain;

– a *discrete then continuous approach*, which consists of moving from discrete to continuous, based on the discrete control scheme in Figure 4.16, and returning to continuous by allowing the sampling period to tend toward zero.

4.5.6.1. *Direct continuous approach from the continuous control*

Let us return to the second control scheme of Figure 4.15 and present it in a more conventional form, as required by the comparator (see Figure 4.17), a form enabling the well-known, general expressions of the transmittances in open loop and closed loop to be exploited more easily.

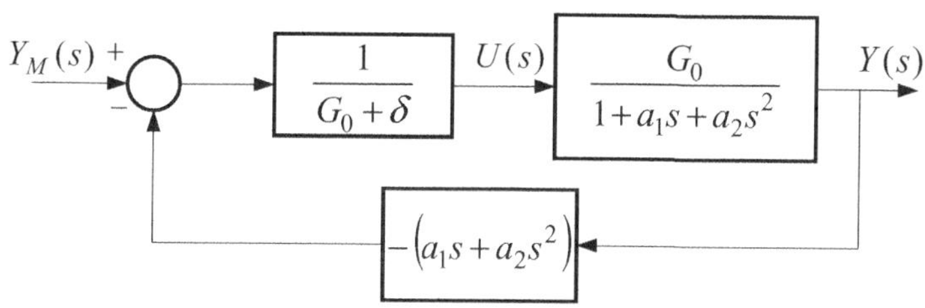

Figure 4.17. *Functional diagram involving a comparator*

Given Figure 4.17, the general expression of transmittance in tracking enables us to determine a transmittance in closed loop of the form:

$$\frac{Y(s)}{Y_M(s)}=\frac{1}{-\left(a_1 s+a_2 s^2\right)}\frac{-\dfrac{G_0}{G_0+\delta}\dfrac{a_1 s+a_2 s^2}{1+a_1 s+a_2 s^2}}{1-\dfrac{G_0}{G_0+\delta}\dfrac{a_1 s+a_2 s^2}{1+a_1 s+a_2 s^2}},\qquad [4.116]$$

or:

$$\frac{Y(s)}{Y_M(s)}=\frac{\dfrac{G_0}{G_0+\delta}\,\dfrac{1}{1+a_1s+a_2s^2}}{1-\dfrac{G_0}{G_0+\delta}\,\dfrac{a_1s+a_2s^2}{1+a_1s+a_2s^2}}, \qquad [4.117]$$

or:

$$\frac{Y(s)}{Y_M(s)}=\frac{G_0}{(G_0+\delta)\left(1+a_1s+a_2s^2\right)-G_0\left(a_1s+a_2s^2\right)}, \qquad [4.118]$$

or indeed:

$$\frac{Y(s)}{Y_M(s)}=\frac{G_0}{G_0+\delta+(G_0+\delta)\left(a_1s+a_2s^2\right)-G_0\left(a_1s+a_2s^2\right)}, \qquad [4.119]$$

that is:

$$\begin{aligned}\frac{Y(s)}{Y_M(s)}&=\frac{G_0}{G_0+\delta+\delta\left(a_1s+a_2s^2\right)}\\&=\frac{G_0}{G_0+\delta\left(1+a_1s+a_2s^2\right)}\end{aligned}, \qquad [4.120]$$

or, by revealing the plant transmittance:

$$\frac{Y(s)}{Y_M(s)}=\frac{\dfrac{G_0}{\delta\left(1+a_1s+a_2s^2\right)}}{1+\dfrac{G_0}{\delta\left(1+a_1s+a_2s^2\right)}}, \qquad [4.121]$$

therefore, ultimately:

$$\frac{Y(s)}{Y_M(s)}=\frac{\dfrac{1}{\delta}G(s)}{1+\dfrac{1}{\delta}G(s)}, \qquad [4.122]$$

a transmittance that is none other than that of the elementary control loop represented in Figure 4.18.

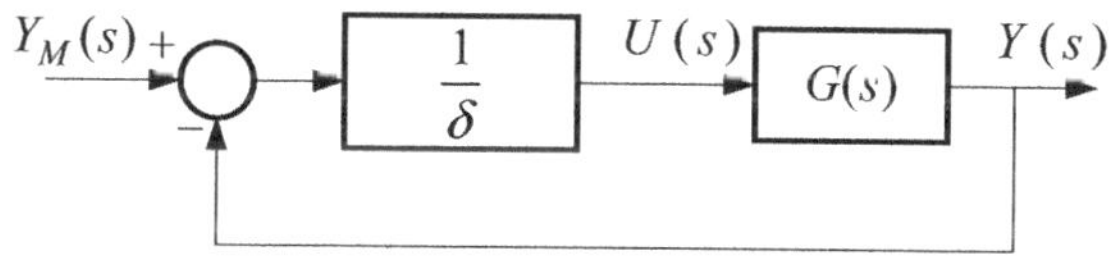

Figure 4.18. *Elementary control loop deduced from the quadratic-criterion control*

4.5.6.2. *Discrete then continuous approach based on discrete control*

Let us return to the discrete control scheme of Figure 4.16 and transform it, to reduce it to a continuous control scheme resulting in the sampling period, T, tending toward zero.

Coefficients b_0, b_1, a_1 and a_2 are then reduced to:

$$b_0 = G_0\left[1 - \frac{1}{\sqrt{1-\zeta_0^2}}\cos\left(\arcsin\zeta_0\right)\right],$$

$$b_1 = G_0\left[1 - \frac{1}{\sqrt{1-\zeta_0^2}}\cos\left(\arcsin\zeta_0\right)\right],$$

$$a_1 = -2 \text{ and } a_2 = 1, \qquad [4.123]$$

that is, knowing that

$$\cos(\arcsin\zeta_0) = \sqrt{1-\zeta_0^2}: \qquad [4.124]$$

$$b_0 = 0;\ b_1 = 0;\ a_1 = -2;\ a_2 = 1. \qquad [4.125]$$

Moreover, with the sampling period tending toward zero, we can interchange $y_M(k+1)$ and $y_M(k)$, then $y(k)$ and $y(k-1)$. The feedback signal, that is,

$$\begin{aligned} v(k) &= a_1 y(k) + a_2 q^{-1} y(k) \\ &= a_1 y(k) + a_2 y(k-1) \end{aligned}, \qquad [4.126]$$

can then be reduced to:

$$v(k) = (a_1 + a_2)\,y(k), \qquad [4.127]$$

or, given the reduced values of a_1 and a_2:

$$v(k) = -y(k), \qquad [4.128]$$

a result that conveys the presence of a unit feedback.

We thus have all of the elements at our disposal to transform the functional diagram in Figure 4.18 in accordance with that presented in Figure 4.19, which quite simply illustrates the most elementary form of the control loop.

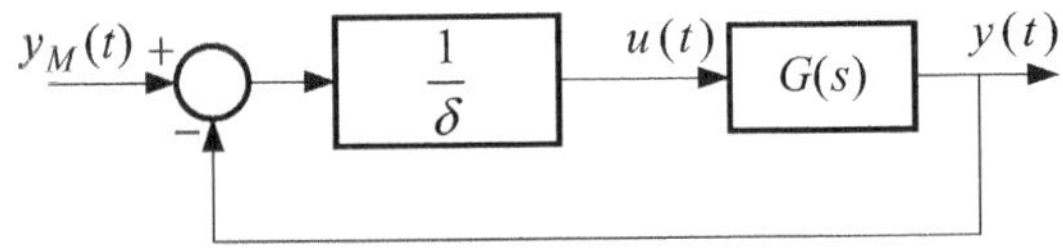

Figure 4.19. *Elementary control loop deduced from the quadratic-criterion control*

4.5.6.3. *Conclusion*

As previously mentioned, the conclusion focuses on the *proportional action* that the regulator, $1/\delta$, is reduced, *an action that is reduced to say the least.* It is highlighted by the two elementary control loops in Figures 4.18 and 4.19, respectively.

Indeed, these two loops clearly demonstrate that the presence of the weighting factor, δ, of the control, $u(t)$, is reduced to a gain, $1/\delta$, in the action chain of the control loop.

Modifying δ will merely give a variation in the gain in open loop, and will in no case result in any deformation of the locus of $\beta(j\omega)$ that might improve the stability degree–precision compromise. Therefore, setting the control through the adjustment of δ does not offer the best stability degree–precision compromise. This offers an indication of how a sophisticated control scheme (such as that in Figure 4.16) does not necessarily imply an advanced control strategy and that therefore "scheme" and "strategy" must always be distinguished between, with *scheme sophistication not synonymous with strategy enrichment.*

In the specific scenario where $\delta = 0$, the gain, $1/\delta$, tends toward infinity. The functional diagrams in Figures 4.18 and 4.19 then identify with that of a high-gain control, for which the transfer of the reaction chain, $H^{-1}(s)$, in this case is unitary, that is:

$$H^{-1}(s) = 1, \qquad [4.129]$$

thus leading, in view of equation [4.6], to:

$$Y(s) = Y_M(s) \; \forall \; G(s). \qquad [4.130]$$

4.6. Predictive control

The predictive control strategy underlying this section was proposed by Demircioglu and Gawthrop [DEM 91].

Beyond the principle of control, which takes the form of a continuous-time approach here, a significant contribution brought about because of the valuable help of Nolwenn Monot and François Levron highlights an elementary control loop featuring a phase-lead regulator.

4.6.1. *Principle*

As its name implies, predictive control is based on the prediction of the future output.

In the *generalized predictive control* (GPC), or *continuous-time generalized predictive control* (CGPC) approach, the development of the control law, $u(t)$, of the plant is conducted by minimizing, with respect to $u(t)$, a quadratic criterion of the form:

$$J(t) = \int_{T_1}^{T_2} \left[\hat{y}(t+T) - w(t+T)\right]^2 dT + \lambda \int_0^{T_2 - T_1} \left[\hat{u}(t+T)\right]^2 dT, \qquad [4.131]$$

a quadratic criterion in which:

– T is a dummy variable;

– T_1 is a minimum prediction horizon;

– T_2 is a maximum prediction horizon;

– $\hat{y}(t+T)$ is the predicted (or estimated) future output or the output estimate at instant, $t+T$, or the estimate of $y(t+T)$, with $y(t+T)$ as the real (or exact) future output;

– $w(t+T)$ is the future reference or the reference at instant, $t+T$, known or precalculated (at least on one horizon), with the notation, $w(t)$, assigned here to the reference, $r(t)$, in accordance with that frequently used within the framework of this control;

– $\hat{u}(t+T)$ is the predicted (or estimated) future control, or the control estimate at instant, $t+T$, or the estimate of $u(t+T)$, with $u(t+T)$ as the real (or exact) future control;

– λ is a weighting factor of the control.

As defined, this criterion comprises two distinct terms:

– one represents the energy associated with the evolution of the error between the (estimated) output and the reference;

– the other represents, to the nearest weighting factor, λ, the energy associated with the evolution of the control (estimated); the lower the weighting factor, λ, the less the energy associated with the control is penalized in the cost function, and the higher the control rapidity; *indeed, the control is all the more rapid, the lower the value of* λ.

Concerning the upper integration limit, T_2, it transpires *that the control is all the more rapid, the lower the value of* T_2. A more rapid control can therefore be obtained by *jointly reducing* λ *and* T_2.

With regard to the lower integration limit, T_1, it is chosen to be nil in the absence of pure plant delay. This is why the minimum prediction horizon, T_1, is canceled in our study framework, *exempt of any pure delay*, that is, $T_1 = 0$.

The estimates of the output and of the plant control at instant, $t+T$, that is, $\hat{y}(t+T)$ and $\hat{u}(t+T)$, result from the estimation of $y(t+T)$ and $u(t+T)$ by limited Taylor expansions:

$$\hat{y}(t+T) = \sum_{k=0}^{N_y} y^{(k)}(t)\frac{T^k}{k!} \qquad [4.132]$$

and

$$\hat{u}(t+T) = \sum_{k=0}^{N_u} u^{(k)}(t)\frac{T^k}{k!}, \qquad [4.133]$$

with the orders, N_y and N_u, of these expansions enabling the orders of the output and the control to be defined.

In the absence of a horizon of knowledge of the reference, $w(t)$, beyond instant t, $w(t+T)$ is no longer known. $w(t+T)$ can then only be predicted with an estimation given by a limited Taylor expansion, that is:

$$\hat{w}(t+T) = \sum_{k=0}^{N_w} w^{(k)}(t)\frac{T^k}{k!}, \qquad [4.134]$$

with the order, N_w, of this expansion defining the order of the reference.

Expressed in this way, the limited Taylor expansions of orders N_y, N_u and N_w define (in terms of prediction), *the predictors of horizon*, *T*, the orders of which are N_y, N_u and N_w, respectively.

4.6.2. *From CGPC to LQ then LQG control*

If the estimates $\hat{y}(t+T)$ and $\hat{u}(t+T)$ of $y(t+T)$ and $u(t+T)$ are now determined by unlimited Taylor expansions, the orders N_y and N_u then tend toward infinity, and $\hat{y}(t+T)$ and $\hat{u}(t+T)$ tend toward $y(t+T)$ and $u(t+T)$, respectively, that is, to the limit: $\hat{y}(t+T) = y(t+T)$ and $\hat{u}(t+T) = u(t+T)$, the $J(t)$ criterion then becomes:

$$J(t) = \int_0^{T_2}\left[y(t+T) - w(t+T)\right]^2 dT + \lambda\int_0^{T_2}\left[u(t+T)\right]^2 dT. \qquad [4.135]$$

By writing $t+T=\tau$, therefore $dT = d\tau$, τ varying from t to $t+T_2$ (since T varies from 0 to T_2), $J(t)$ becomes:

$$J(t) = \int_t^{t+T_2} [y(\tau) - w(\tau)]^2 \, d\tau + \lambda \int_t^{t+T_2} u^2(\tau) d\tau \, . \quad [4.136]$$

In the case of a *natural (or relaxation) response*, characterized by a relaxation between a *nil initial instant*, that is, $t = 0$, and an *infinite end instant*, that is, $T_2 = \infty$, with *the reference then being nil between these instants*, that is $w(\tau) = 0$, the criterion, $J(t)$, is reduced to an expression of the form:

$$J = \int_0^\infty \left[y^2(\tau) + \lambda u^2(\tau) \right] d\tau \, , \quad [4.137]$$

or, by applying $\tau = t$:

$$J = \int_0^\infty \left[y^2(t) + \lambda u^2(t) \right] \, dt \, . \quad [4.138]$$

This result is noteworthy in that *this quadratic criterion on the infinite horizon* is none other than that relating to the *LQ control*, thus expressing that the CGPC and LQ control are then equivalent.

Indeed, the linear quadratic (LQ) control (or optimization) is a predictive control, whereby the quadratic criterion is particularized in accordance with the establishment of the reduced criterion, J.

J is an energy criterion (to be minimized with respect to $u(t)$) which, through the weighting factor, λ, penalizes the energy associated with the joint evolution of the output and of the control beyond the *initial instant*, $t = 0$.

This is the instant at which the output, $y(t)$, has a value of y_0, that is, $y(0) = y_0$, with y_0 as the initial value of the output, or the *initial output*.

The issue thus lies in the return to zero (or relaxation) of the output, $y(t)$, of a linear plant from a non-nil initial output, y_0, penalizing the energy associated with the joint evolution of the output and the control after the initial instant, $t = 0$. The smaller the value of λ, the less the control is penalized and the return to zero is rapid.

With regard to the *LQG control*, it stands out from the LQ control by introducing *Gaussian white noise* at the plant and, more specifically and in the general scenario, in its state equation and observation equation. The *Gaussian* nature of the white noise introduced by the "G" in "LQG". According to the *separation theory*, the optimal control law then results from the separate resolution of the two problems constituting, first, the *deterministic LQ* by minimizing criterion J and, second, the *optimal statistical estimation* by minimizing a statistical variance criterion [DEL 09].

4.6.3. *LQ and quadratic-criterion control*

Our aim is to compare the LQ control, whose criterion comprises an integral, and the quadratic-criterion control, whose criterion is without an integral. To this end, it is appropriate to apply the LQ control to the plant used in the quadratic-criterion control, with the continuous model of this plant being represented by the temporal equation expressing the output, $y(t)$, as a function of its derivatives and of the control, $u(t)$, that is:

$$y(t) = G_0 u(t) - a_1 \dot{y}(t) - a_2 \ddot{y}(t), \qquad [4.139]$$

with the notations $\dot{y}(t)$ and $\ddot{y}(t)$, taken from the section on quadratic-criterion control, being equivalent to $y^{(1)}(t)$ and $y^{(2)}(t)$.

The J criterion of the LQ control can then be particularized according to:

$$J = \int_0^{\infty} \left[\left(G_0 u(t) - a_1 \dot{y}(t) - a_2 \ddot{y}(t) \right)^2 + \lambda u^2(t) \right] dt. \qquad [4.140]$$

Its minimization with respect to $u(t)$ requires $\partial J / \partial u(t)$ to be calculated, that is:

$$\begin{aligned} \frac{\partial J}{\partial u(t)} &= \int_0^{\infty} \left[2\left(G_0 u(t) - a_1 \dot{y}(t) - a_2 \ddot{y}(t) \right) G_0 + \lambda 2 u(t) \right] dt \\ &= 2\int_0^{\infty} \left[G_0^2 u(t) - a_1 G_0 \dot{y}(t) - a_2 G_0 \ddot{y}(t) + \lambda u(t) \right] dt. \end{aligned} \qquad [4.141]$$

If we choose $u(t)$ to cancel the integrand for any positive value of t, that is

$$(G_0^2 + \lambda) u(t) - a_1 G_0 \dot{y}(t) - a_2 G_0 \ddot{y}(t) = 0 \qquad \forall t > 0, \qquad [4.142]$$

then

$$\frac{\partial J}{\partial u(t)} = 0, \qquad [4.143]$$

with the continuous control, $u(t)$, of the plant then satisfying the equation:

$$u(t) = \frac{a_1 G_0 \dot{y}(t) + a_2 G_0 \ddot{y}(t)}{G_0^2 + \lambda}, \qquad [4.144]$$

that is:

$$u(t) = \frac{a_1 \dot{y}(t) + a_2 \ddot{y}(t)}{G_0 + \lambda / G_0}, \qquad [4.145]$$

with such a control equation conveyed by the diagram in Figure 4.20.

It transpires that this diagram is none other than that relating to the quadratic-criterion control, except that δ is replaced by λ / G_0, and $y_M(t)$ is nil (this being required by the relaxation state for the J criterion of the LQ control, since $w(t) = 0$).

Such an equivalence of the LQ and quadratic-criterion controls admits an immediate explanation: the integrand of $\partial J / \partial u(t)$ being nil for $t > 0$, the integration from zero to infinity of this nil integrand of course has no effect, which explains this equivalence.

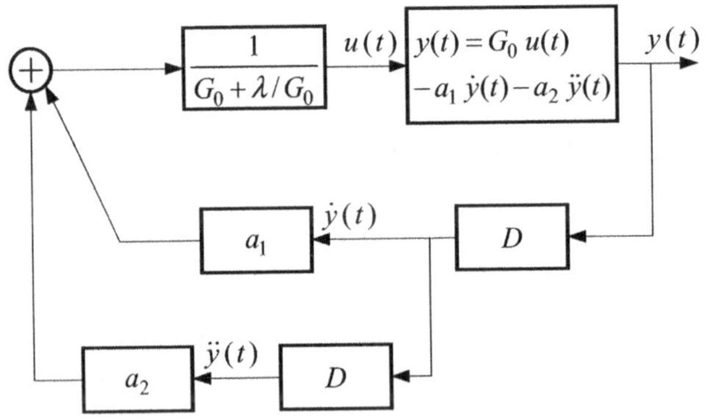

Figure 4.20. *Functional diagram conveying the continuous-control equation, $u(t)$: the absence of reference applied to the summator results from the natural response ($w(t) = 0$), to which the J criterion of the LQ control corresponds*

4.6.4. *Study plant*

Let us consider, as a study plant for predictive control, a second-order system resulting from the cascading of two first-order low-pass systems with the same transitional frequency, ω_0, that is:

$$G(s) = \frac{G_0}{\left(1 + \frac{s}{\omega_0}\right)^2}, \text{ with } G_0 = 10G_0 = 10 \text{ and } \omega_0 = 1rd / s, \qquad [4.146]$$

or, by developing the denominator:

$$G(s) = \frac{G_0}{1 + a_1 s + a_2 s^2}, \text{ with } a_1 = 2 \text{ and } a_2 = 1. \qquad [4.147]$$

Knowing that $G(s) = Y(s) / U(s)$, we can write:

$$Y(s) + a_1 sY(s) + a_2 s^2 Y(s) = G_0 U(s), \qquad [4.148]$$

a *symbolic equation* that admits, as a *concrete (or original) equation*, the differential equation:

$$y(t) + a_1 y^{(1)}(t) + a_2 y^{(2)}(t) = G_0 u(t), \qquad [4.149]$$

from which we obtain:

$$y^{(2)}(t) = \frac{G_0}{a_2} u(t) - \frac{1}{a_2} y(t) - \frac{a_1}{a_2} y^{(1)}(t). \qquad [4.150]$$

4.6.5. *Control law synthesis*

In order for the study plant defined above to particularize the limited development of $\hat{y}(t+T)$, such that $\hat{y}(t+T)$ is expressed (*in fine*) solely as a function of $y(t)$, $y^{(1)}(t)$ and $u(t)$, a second-order limited development must be chosen, that is, $N_y = 2$, with the second derivative of $y(t)$ then replaced by the expression of $y^{(2)}(t)$, as drawn from the plant model. This comes down to writing, for the first stage:

$$\hat{y}(t+T) = y(t) + Ty^{(1)}(t) + \frac{T^2}{2}y^{(2)}(t), \tag{4.151}$$

and for the second stage:

$$\hat{y}(t+T) = y(t) + Ty^{(1)}(t) + \frac{T^2}{2}\left(\frac{G_0}{a_2}u(t) - \frac{1}{a_2}y(t) - \frac{a_1}{a_2}y^{(1)}(t)\right), \tag{4.152}$$

therefore, ultimately:

$$\hat{y}(t+T) = y(t)\left(1 - \frac{T^2}{2a_2}\right) + y^{(1)}(t)\left(T - \frac{a_1 T^2}{2a_2}\right) + u(t)\frac{G_0 T^2}{2a_2}, \tag{4.153}$$

Similarly, for $\hat{u}(t+T)$ to be expressed solely as a function of $u(t)$ and $u^{(1)}(t)$, it is sufficient to choose a first-order limited development, that is, $N_u = 1$, therefore:

$$\hat{u}(t+T) = u(t) + Tu^{(1)}(t), \tag{4.154}$$

all the more so since $N_u = 1$ ensures a first-degree denominator for the regulator, arising from the change to the control loop (as confirmed below).

Expressed in this way, $\hat{y}(t+T)$ and $\hat{u}(t+T)$ particularize the $J(t)$ criterion in accordance with:

$$J(t) = \int_0^{T_2} \left[\begin{array}{l} y(t)\left(1-\dfrac{T^2}{2a_2}\right)+y^{(1)}(t)\left(T-\dfrac{a_1 T^2}{2a_2}\right) \\ +u(t)\dfrac{G_0 T^2}{2a_2}-w(t+T) \end{array} \right]^2 dT$$
$$+\lambda \int_0^{T_2} \left[u(t)+Tu^{(1)}(t)\right]^2 dT. \qquad [4.155]$$

Its minimization with respect to $u(t)$ is conveyed by:

$$\frac{\partial J(t)}{\partial u(t)} = 0, \qquad [4.156]$$

that is, having demonstrated (see section 4.5.2) that $y(t)$ and its derivatives, as well as $u^{(1)}(t)$ (output of a differentiator of input, $u(t)$) are not dependent on $u(t)$:

$$\frac{\partial J(t)}{\partial u(t)} = \int_0^{T_2} 2 \left[\begin{array}{l} y(t)\left(1-\dfrac{T^2}{2a_2}\right)+y^{(1)}(t)\left(T-\dfrac{a_1 T^2}{2a_2}\right) \\ +u(t)\dfrac{G_0 T^2}{2a_2}-w(t+T) \end{array} \right] \frac{G_0 T^2}{2a_2} dT$$
$$+\lambda \int_0^{T_2} 2\left[u(t)+Tu^{(1)}(t)\right] dT = 0, \qquad [4.157]$$

or:

$$\int_0^{T_2} \left[\begin{array}{l} y(t)\left(1-\dfrac{T^2}{2a_2}\right)+y^{(1)}(t)\left(T-\dfrac{a_1 T^2}{2a_2}\right) \\ +u(t)\dfrac{G_0 T^2}{2a_2}-w(t+T) \end{array} \right] \frac{G_0 T^2}{2a_2} dT$$
$$+\lambda \int_0^{T_2} \left[u(t)+Tu^{(1)}(t)\right] dT = 0, \qquad [4.158]$$

or:

$$\begin{aligned}
&y(t)\int_0^{T_2}\left(\frac{G_0T^2}{2a_2}-\frac{G_0T^4}{4a_2^2}\right)dT \\
&+y^{(1)}(t)\int_0^{T_2}\left(\frac{G_0T^3}{2a_2}-\frac{a_1G_0T^4}{4a_2^2}\right)dT \\
&+u(t)\int_0^{T_2}\frac{G_0^2T^4}{4a^2}dT-\int_0^{T_2}\frac{G_0T^2}{2a_2}w(t+T)dT \\
&+u(t)\lambda\int_0^{T^2}dT+u^{(1)}(t)\lambda\int_0^{T^2}TdT=0,
\end{aligned} \quad [4.159]$$

or indeed:

$$\begin{aligned}
&y(t)\left[\frac{1}{3}\frac{G_0T_2^3}{2a_2}-\frac{1}{5}\frac{G_0T_2^5}{4a_2^2}\right] \\
&+y^{(1)}(t)\left[\frac{1}{4}\frac{G_0T_2^4}{2a_2}-\frac{1}{5}\frac{a_1G_0T_2^5}{4a_2^2}\right] \\
&+u(t)\left[\frac{1}{5}\frac{G_0^2T_2^5}{4a_2^2}+\lambda T_2\right] \\
&+u^{(1)}(t)\lambda\frac{T_2^2}{2}-\frac{G_0}{2a_2}\int_0^{T_2}T^2w(t+T)dT=0.
\end{aligned} \quad [4.160]$$

By writing

$$\begin{aligned}
a&=\left[\frac{1}{3}\frac{G_0T_2^3}{2a_2}-\frac{1}{5}\frac{G_0T_2^5}{4a_2^2}\right], \\
b&=\left[\frac{1}{4}\frac{G_0T_2^4}{2a_2}-\frac{1}{5}\frac{a_1G_0T_2^5}{4a_2^2}\right], \\
c&=\left[\frac{1}{5}\frac{G_0^2T_2^5}{4a_2^2}+\lambda T_2\right]
\end{aligned} \quad [4.161]$$

and

$$d = \lambda \frac{T_2^2}{2},$$

as well as

$$r(t) = \frac{G_0}{2a_2} \int_0^{T_2} T^2 w(t+T) dT, \qquad [4.162]$$

we ultimately obtain an equation of the form:

$$ay(t) + by^{(1)}(t) + cu(t) + du^{(1)}(t) - r(t) = 0, \qquad [4.163]$$

an equation from which it is easy to extract the control law, $u(t)$, of the plant, that is:

$$u(t) = \frac{1}{c}\left[r(t) - du^{(1)}(t) - ay(t) - by^{(1)}(t)\right]; \qquad [4.164]$$

such a control equation is conveyed by the diagram in Figure 4.21.

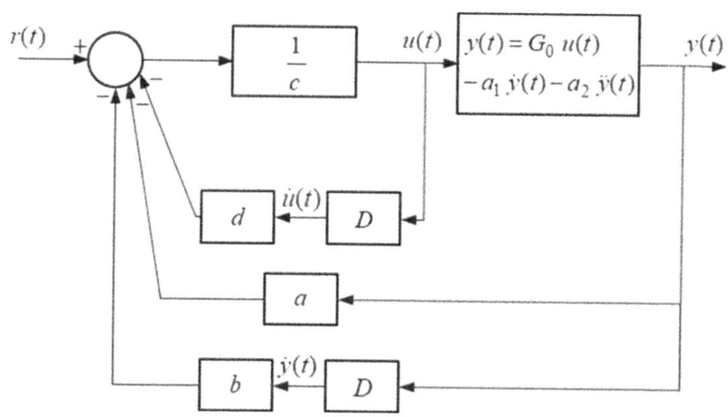

Figure 4.21. *Functional diagram conveying the control law,* $u(t)$

4.6.6. *From predictive control to control loop*

The Laplace transformation of the concrete equation,

$$ay(t) + by^{(1)}(t) + cu(t) + du^{(1)}(t) - r(t) = 0, \qquad [4.165]$$

leads to a symbolic equation of the form:

$$aY(s) + bsY(s) + cU(s) + dsU(s) - R(s) = 0, \qquad [4.166]$$

or:

$$(a + bs)Y(s) + (c + ds)U(s) - R(s) = 0, \qquad [4.167]$$

or, by writing

$$A(s) = a + bs \text{ and } B(s) = c + ds: \qquad [4.168]$$

$$A(s)Y(s) + B(s)U(s) - R(s) = 0, \qquad [4.169]$$

an equation from which we can draw the Laplace transform, $U(s)$, of the control, $u(t)$, that is:

$$U(s) = \frac{A(s)}{B(s)}\left[\frac{1}{A(s)}R(s) - Y(s)\right], \qquad [4.170]$$

a symbolic equation conveyed by the diagram in Figure 4.22.

This block diagram highlights an elementary control loop

– whose non-inverting comparator input results from the filtering of $r(t)$ by a (first-order) low-pass filter of transmittance

$$\frac{1}{A(s)} = \frac{1}{a + bs} = \frac{1/a}{1 + \frac{b}{a}s} \qquad [4.171]$$

– and whose regulator is defined by the transmittance

$$C(s) = \frac{A(s)}{B(s)} = \frac{a + bs}{c + ds} = \frac{a}{c}\frac{1 + \frac{b}{a}s}{1 + \frac{d}{c}s}, \qquad [4.172]$$

that is, by writing

$$C_0 = \frac{a}{c},\ \omega_1 = \frac{a}{b} \text{ and } \omega_2 = \frac{c}{d}: \qquad [4.173]$$

$$C(s) = C_0 \frac{1+\dfrac{s}{\omega_1}}{1+\dfrac{s}{\omega_2}}, \qquad [4.174]$$

this transmittance being none other than that of a *simple phase-lead controller* if C_0, ω_1 and ω_2 are positive and if ω_1 is less than ω_2; moreover, as the following section shows, these conditions are satisfied for the numerical values of the parameters of the study plant and the positive values of parameters λ and T_2 of the control.

Although the change to the control loop leads to a *proportional regulator* for the quadratic-criterion control, as covered in section 4.5, here, the change to the control loop leads to a *phase-lead regulator.* In other words, the prediction horizon, T_2, introduced by the predictive control enables a difference to be made using a richer regulator that can locally deform the frequency response locus in open loop to improve the stability degree–precision compromise.

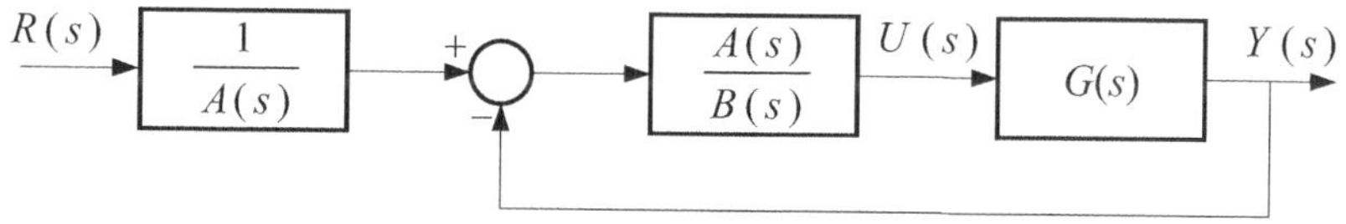

Figure 4.22. *Functional diagram highlighting an elementary control loop*

It should be noted that the transitional frequency, a / b, of the first-order low-pass filter (in cascade with the elementary control loop) is none other than the low transitional frequency, ω_1, of the regulator. Given that, as shown below, the frequency, ω_u, of this loop (which determines its transitional frequency) has proven to be barely higher than ω_1, the transfer, $Y(s) / R(s)$, of the control can be interpreted as a first approximation, resulting from the cascading of two low-pass filters of the same transitional frequency, ω_1. In other words, through $\omega_u \approx \omega_1$, *the frequency, ω_u, of the elementary control loop determines the order of magnitude of the transitional frequency of the control.*

4.6.7. *On the phase-lead regulator stemming from the change to the control loop*

4.6.7.1. *Frequency optimality and time optimality*

In the frequency synthesis of a phase-lead controller (simple, double or triple, etc.) of transitional frequencies, ω_1 and ω_2 with $\omega_1 < \omega_2$, the controller frequency placement (which is the first step in the synthesis process) is essential for its action (in this case, phase lead) to have the greatest effect on the phase margin.

The optimal frequency placement is that for which the transitional frequencies, ω_1 and ω_2, of the controller are geometrically distributed with respect to the unit gain frequency in open loop, ω_u, that is, $(\omega_1\omega_2)^{1/2} = \omega_u$. For a given ratio, α, between the transitional frequencies, ω_1 and ω_2, that is, $\alpha = \omega_2 / \omega_1 > 1$, such a placement enables the controller to bring its maximum phase lead (imposed by α) to the unit gain frequency in open loop, thus ensuring the best phase margin for the control loop.

This being so, it transpires that the simple phase-lead controller arising from the change to the control loop based on the predictive control cannot satisfy this optimal frequency placement. For the study plant as defined, and for a given value of $\omega_u > 0$, the resolution of a problem based on satisfying the set of relations [4.175], [4.181], [4.182], [4.183] and [4.187], does indeed show that there is no solution to this problem, given that it is impossible to find a (λ , T_2) couple ensuring

$$\omega_u = \sqrt{\omega_1\omega_2}. \qquad [4.175]$$

The failure to satisfy this relationship is all the more problematic since, for the same phase lead of the controller at frequency ω_u, the ratio, α, between frequencies ω_2 and ω_1 is then greater, thus increasing the dispersion of these frequencies (detrimental to the numerical stability) and therefore the input sensitivity (detrimental to its immunity).

It is unsurprising that the *frequency optimality* of the controller, $C(s)$, is not ensured, as $C(s)$ indirectly results from the *time optimality* of the control, $u(t)$, in the sense of the *minimization of the time criterion*, $J(t)$. Indeed, frequency optimization and time optimization are not to be confused when the optimization operations do not take account of the *time-frequency relations* or, more specifically, the frequency dynamic performances signifying time dynamic performances.

Where controller frequency optimality is not obtained, we lower our requirements by considering that the transitional frequencies, ω_1 and ω_2, of the controller only frame frequency ω_u, that is,

$$\omega_1 < \omega_u < \omega_2. \qquad [4.176]$$

Within this study context, we have demonstrated the feasibility of this simple framing through the resolution of a problem that consists, for a given value of $\omega_u > 0$, of finding a prediction horizon, $T_2 > 0$, and a weighting factor, $\lambda > 0$, satisfying the set of relations [4.176], [4.181], [4.182], [4.183] and [4.187]. There are then several solutions, among which it is possible to find an optimal solution in the sense of the minimization of the relative error between ω_u and $\sqrt{\omega_1\omega_2}$. For different ω_u frequencies, chosen above the transitional frequency, $\omega_0 = 1rd/s$, of the plant, we propose, in the following section, the numerical solutions to this problem with respect to the couple (λ, T_2), the static gain, C_0, the transitional frequencies, ω_1 and ω_2, the ratio $\alpha = \omega_2/\omega_1$, the maximum phase lead, ϕ_m, of the controller at the frequency, $\omega_m = \sqrt{\omega_1\omega_2}$, the phase lead, ϕ_u, of the controller at the frequency, ω_u, the phase margin, Φ_m, and the sensitivity of the input at high frequency, $C_0\alpha$, of the control loop.

4.6.7.2. *Numerical results*

Let us again use the expressions of C_0, ω_1 and ω_2, given by the ratios

$$C_0 = \frac{a}{c},\ \omega_1 = \frac{a}{b},\ \omega_2 = \frac{c}{d}, \qquad [4.177]$$

replacing *a*, *b* and *c* with their respective expressions, as determined in the synthesis of the control law, *u*(*t*). We thus obtain expressions that can be put in the form:

$$C_0 = \frac{G_0T_2^2}{6a_2}\frac{1-\dfrac{3T_2^2}{10a_2}}{\dfrac{G_0^2T_2^4}{20a_2^2}+\lambda}, \qquad [4.178]$$

$$\omega_1 = \frac{4}{3T_2} \frac{1 - \frac{3T_2^2}{10a_2}}{1 - \frac{2}{5}\frac{a_1}{a_2}T_2} \qquad [4.179]$$

and

$$\omega_2 = \frac{2}{\lambda T_2}\left(\frac{G_0^2 T_2^4}{20a_2^2} + \lambda\right). \qquad [4.180]$$

These literal expressions can be particularized by the numerical values of the parameters of the study plant, namely $G_0 = 10$, $a_1 = 2$ and $a_2 = 1$:

$$C_0 = \frac{5T_2^2\left(1 - \frac{3T_2^2}{10}\right)}{3(5T_2^4 + \lambda)}, \qquad [4.181]$$

$$\omega_1 = \frac{4\left(1 - \frac{3T_2^2}{10}\right)}{3T_2\left(1 - \frac{4T_2}{5}\right)} \qquad [4.182]$$

and

$$\omega_2 = \frac{10T_2^3}{\lambda} + \frac{2}{T_2}. \qquad [4.183]$$

These formulas of the controller can be completed by introducing the unit gain frequency in open loop, ω_u, through the relation that defines it, that is:

$$\left|\beta(j\omega_u)\right| = 1, \qquad [4.184]$$

or, in this case:

$$\left|C(j\omega_u)\right|\left|G(j\omega_u)\right| = 1, \qquad [4.185]$$

or, given the expressions of $C(j\omega)$ and $G(j\omega)$:

$$C_0\left(\frac{1+\left(\frac{\omega_u}{\omega_1}\right)^2}{1+\left(\frac{\omega_u}{\omega_2}\right)^2}\right)^{\frac{1}{2}}\frac{G_0}{1+\left(\frac{\omega_u}{\omega_0}\right)^2} = 1, \qquad [4.186]$$

or indeed, knowing that $G_0 = 10$ and $\omega_0 = 1rd/s$:

$$\left(\frac{1+\left(\frac{\omega_u}{\omega_2}\right)^2}{1+\left(\frac{\omega_u}{\omega_1}\right)^2}\right)^{\frac{1}{2}}\frac{1+\omega_u^2}{10} = C_0. \qquad [4.187]$$

Once introduced, the frequency, ω_u, can be compared to the frequency, ω_m, at which the controller brings the maximum phase lead, ϕ_m, with ω_m and ϕ_m, being defined, respectively, by:

$$\omega_m = \sqrt{\omega_1\,\omega_2} \qquad [4.188]$$

and

$$\phi_m = \phi(\omega_m) = \arg\left(C(j\omega_m)\right) = \arcsin\frac{\alpha-1}{\alpha+1} \text{ with } \alpha = \frac{\omega_2}{\omega_1} > 1, \qquad [4.189]$$

with the α ratio being representative of the dispersion of the controller's transitional frequencies.

The phase lead, ϕ_u, brought by the controller at the frequency, ω_u, can also be compared to the maximum phase lead, ϕ_m, with ϕ_u defined by:

$$\phi_u = \phi(\omega_u) = \arg\left(C(j\omega_u)\right) = \arctan\frac{\omega_u}{\omega_1} - \arctan\frac{\omega_u}{\omega_2}. \qquad [4.190]$$

With regard to the phase margin, Φ_m, defined by

$$\Phi_m = \arg\left(\beta(j\omega_u)\right) + \pi, \qquad [4.191]$$

that is,

$$\Phi_m = \arg\left(C(j\omega_u)\right) + \arg\left(G(j\omega_u)\right) + \pi, \qquad [4.192]$$

it is written, taking into account the expressions of $C(j\omega)$ and $G(j\omega)$:

$$\Phi_m = \arctan\frac{\omega_u}{\omega_1} - \arctan\frac{\omega_u}{\omega_2} - 2\arctan\frac{\omega_u}{\omega_0} + \pi, \qquad [4.193]$$

or, since $\omega_0 = 1$ rd/s:

$$\Phi_m = \arctan\frac{\omega_u}{\omega_1} - \arctan\frac{\omega_u}{\omega_2} - 2\arctan\omega_u + \pi. \qquad [4.194]$$

Lastly, the high-frequency input sensitivity is given by:

$$S(\omega)_{HF} = \left|C(j\omega)\right|_{HF} = C_0\frac{\omega_2}{\omega_1} = C_0\alpha, \qquad [4.195]$$

If we recall that the frequency, ω_u, of the control loop determines the order of magnitude of the transitional frequency of the predictive control (see the end of section 4.6.6), a frequency that certainly determines its rapidity, we can understand why the table shown in Figure 4.23 lists the numerical results of the problem for different values of the frequency, ω_u, above the transitional frequency, $\omega_0=1$rd/s, of the plant. It is true that the frequency, ω_u, is higher than this frequency, ω_0, for the control loop to be faster than the plant.

Frequency ω_u	**10**	**15**	**20**		**10**	**15**	**20**
λ	0.00246	0.00053	0.00017		0.00237	0.00047	0.00014
T_2	0.15835	0.1100	0.08437		0.19715	0.14369	0.11225
C_0	7.39362	15.834	27.418		6.45010	13.13637	22.19487
ω_1	9.56893	13.233	16.909		7.93554	10.41902	13.00052
ω_2	28.73140	43.026	57.375		42.45250	76.82367	112.90601
$\alpha = \dfrac{\omega_2}{\omega_1}$	3.00257	3.251	3.39309		5.34966	7.37340	8.68472
$\omega_m = \sqrt{\omega_1 \omega_2}$	16.58098	23.862	31.148		18.35439	28.29182	38.31236
$\varphi_m = \varphi(\omega_m)$	30.02124	31.976	33.007		43.23739	49.56555	52.51280
ω_u	10	15	20		10	15	20
$\varphi_u = \varphi(\omega_u)$	27.07139	29.361	30.569		38.31127	44.16796	46.92999
Φ_m^o	38.492	36.988	36.293		49.732	51.796	52.654
$S(\omega)_{HF}$	22.19987	51.481	93.031		34.50591	96.85981	192.75652
e_r	0.65809	0.591	0.55738		0.83543	0.88612	0.91561

Figure 4.23. *Table of the problem's numerical results*

The results collated in the table have been obtained by minimizing the *relative frequency deviation*, defined as the relative deviation between frequencies ω_m and ω_u, that is:

$$e_r = \frac{\omega_m - \omega_u}{\omega_u} = \frac{\omega_m}{\omega_u} - 1, \qquad [4.196]$$

this deviation being positive since ω_u is always below ω_m, that is, $\omega_m / \omega_u > 1$.

The results presented in the first three columns best minimize the relative frequency deviation.

Corresponding to a less-optimal minimization of this deviation, the results presented in the last three columns nevertheless favor the phase margin, but to the detriment of the dispersion of frequencies ω_1 and ω_2 (α larger) and of the input immunity (the sensitivity indeed being greater).

Finally, all of these results demonstrate that the increase in the rapidity of the predictive control, consecutive to the growth of frequency ω_u, is accompanied by a joint decrease in the parameters, λ and T_2, of the control.

4.6.8. *A specific scenario suitable as the basis of an exercise or a problem*

If (to generate a deterministic stress that is sufficiently rich in frequencies) the reference, $r(t)$, results from the superposing of a ramp of slope, a', and a step of amplitude, b', $r(t)$ is expressed by the sum:

$$r(t) = a' t\, u(t) + b' u(t), \qquad [4.197]$$

with $u(t)$ denoting the unit step function here.

Given the Laplace transform, $R(s)$, of $r(t)$, that is,

$$R(s) = \frac{a'}{s^2} + \frac{b'}{s} = \frac{1}{s^2}(a' + b's), \qquad [4.198]$$

the Laplace transform, $R(s) / A(s)$, of the non-inverting input of the comparator is written (see Figure 4.22):

$$\frac{R(s)}{A(s)} = \frac{1}{s^2}\frac{a' + b's}{a + bs} = \frac{1}{s^2}\frac{a'}{a}\frac{1 + \frac{b'}{a'}s}{1 + \frac{b}{a}s}, \qquad [4.199]$$

or, for

$$\frac{b'}{a'} = \frac{b}{a}: \qquad [4.200]$$

$$\frac{R(s)}{A(s)} = \left(\frac{a'}{a}\right)\frac{1}{s^2}, \qquad [4.201]$$

with the non-inverting input of the comparator then admitting an expression of the form:

$$\mathcal{L}^{-1}\left[\frac{R(s)}{A(s)}\right] = \left(\frac{a'}{a}\right) t\, u(t) \cdot \qquad [4.202]$$

This means that the response, $y(t)$, of the predictive control to the sum, $r(t)$, of a ramp of slope, a', and a step of amplitude, b', is reduced (for $b'/a' = b/a$) to the response, $y(t)$, of the elementary control loop to a simple ramp of slope, a'/a, thus simplifying thc determination of such a response.

5

On the Three Generations of CRONE Control

5.1. Introduction

Robustness is a very broad concept that always conveys the same idea, that is to say *insensitivity* (which may result from *insensitization*) or, if unattainable, *quasi-insensitivity* (which may result from *desensitization*). So, within the same scientific domain, there are as many robustness types as there are insensitive or quasi-insensitive quantities (generally performances) and quantities to which these robustness types are insensitive or quasi-insensitive (generally parameters). The control domain is no exception.

In this domain, it is common to consider *stability robustness* (this is the case of the H_∞ approach) whose aim is to maintain stability (*robustness in stability*) or, at best, the guarantee of a maximum value of the resonance ratio in tracking or in regulation.

In the non-integer approach used by CRONE control (*Commande Robuste d'Ordre Non Entier*, robust non-integer order command), the robustness is more severe in nature, since the considered robustness is *stability degree robustness* whose aim is to maintain or desensitize a time or frequency dynamic performance that measures stability degree (*robustness in performance*).

More precisely, the robustness in question is that of the stability degree vis-à-vis the plant uncertainties. The stability degree is measured by the resonance ratio in tracking or the damping ratio in tracking and regulation: the first factor or ratio is indeed significant for the reduced first overshoot of the step response in tracking or in regulation; the second is significant of the damping of its oscillatory mode. Uncertainties are taken into account via the *genuine uncertainty domains* that they define, unlike the case of the H_∞ approach, where the *overestimation* of uncertainty domains does not enable the best results to be obtained. With regard to the dual frequency dynamic performance of the resonance ratio in tracking, that is, the resonance ratio in regulation, the non-integer approach does not seek to ensure the robustness of this factor, with the latter not significant of any time dynamic performance.

With regard to the time dynamic performance, which is inseparable from the damping ratio, namely the natural frequency, the non-integer approach in no case seeks to ensure the robustness of this frequency. Indeed, allowing the control natural frequency to vary freely enables better "compliance" with the plant, notably by avoiding forcing its input when its dynamics becomes much slower than that imposed on the control.

In accordance with the strategies used by the three generations of CRONE control, for the first generation, a *non-integer differentiation order*, which is real, makes it possible to parameterize the *transmittance of the regulator* with few parameters. For the second and third generation, a *non-integer integration order*, which is real or complex depending on the generation (real for the second and complex for the third), makes it possible to parameterize the *open-loop transmittance* with few parameters and thus reduce its optimization: *the synthesis of the regulator in the form of a rational fraction is then conducted uniquely for optimal transmittance in open loop*, thus avoiding the iterative synthesis of a regulator with numerous parameters (required, notably, by the quantitative feedback theory (QFT) approach).

In the case of the relaxation of water on a porous dyke, where the fractal nature of the porosity enables the concept of *fractal robustness* to be introduced, the robustness in question is that of relaxation damping with respect to the mass of water in motion, thus bringing into question the *mass-damping dilemma* in mechanics. Such *robustness of the damping ratio*, which is paradoxical in the integer mechanics approach, where any relaxation presents a damping linked to the carried mass, can be illustrated

– by two *isodamping half-straight lines* in the operational plane;

– and by a *(vertical) frequency template* in the Nichols plane, whose *verticality* (for a nominal mass of water) and *sliding on itself* (when the water mass varies) ensure the *invariance of the phase margin*.

The transposition of this robustness in automatic control is at the origin of the initial CRONE control approach. Indeed, its first principles arise from the interpretation of the model of water relaxation on a porous dyke, in conformity with two legitimate interpretations which respectively define

– first-generation CRONE control through a *controller phase locking* (at a non-integer multiple of $\pi/2$) around the unit gain frequency in open loop, ω_u;

– second-generation CRONE control through an *open-loop phase locking* (at a non integer multiple of $-\pi/2$, between $-\pi/2$ and $-\pi$) around the frequency, ω_u, and for the nominal and non-nominal plant, this phase locking defining a *vertical template* in the Nichols plane.

The transition from second- to third-generation CRONE control results from the generalization of the *vertical template* in conformity with two generalization levels:

– in the first generalization level, for the nominal plant, the vertical template is replaced by a straight-line segment of any direction, called a *generalized template*, the direction of which is given by that of the main axis of the uncertainty domain (of the plant or the open loop) calculated at:

- the open-loop unit gain frequency, ω_u,

- or the resonance frequency in tracking, ω_r;

– in the second generalization level, for the nominal plant, the generalized template is replaced by a set of generalized templates, called a *multi-template*, which leads to a *curvilinear template* generalizing the *rectilinear template* formed by the generalized template; the tangent at each point of the curvilinear template is the main axis of the corresponding uncertainty domain.

5.2. From the porous dyke to first- and second-generation CRONE control

With the transition from porous dyke to first- and second-generation CRONE control being based on two legitimate interpretations of the model governing the relaxation of water on a porous dyke, it is convenient to recall here the two differential equations constituting this model [OUS 14].

The first differential equation represents the model of the water–dyke interface, namely:

$$Q(t) = \frac{1}{\omega_0^{\,n-1}} \left(\frac{d}{dt} \right)^{n-1} P(t) \text{ with } 1 < n < 2 \,, \qquad [5.1]$$

where $Q(t)$ denotes the water flow (rate) upstream of the dyke and $P(t)$ the pressure at the water–dyke interface.

The second differential equation stems from the application of the fundamental principle of dynamics to the motion water mass, M, namely:

$$\frac{M}{S^2}\frac{dQ(t)}{dt}+P(t)=0; \qquad [5.2]$$

the cross-section, S, of water flow upstream of the dyke is accompanied by the relations, $Q(t)=SV(t)$ and $F(t)=P(t)S$, with $V(t)$ and $F(t)$ denoting the water mass speed and the reaction force of the dyke, respectively.

5.2.1. *First interpretation of the relaxation model: first-generation CRONE control*

5.2.1.1. *Dynamic behavior of the water mass*

Introducing the water displacement $X(t)$ with respect to the resting position, thus enabling a new formulation of the flow according to

$$Q(t)=S\frac{dX(t)}{dt}, \qquad [5.3]$$

relation [5.2] becomes:

$$\frac{M}{S}\frac{d^2X(t)}{dt^2}+P(t)=0, \qquad [5.4]$$

a differential equation from which one can draw a symbolic equation of the form:

$$X(s)=-\frac{S}{Ms^2}P(s). \qquad [5.5]$$

5.2.1.2. *Dynamic behavior of the water–dyke interface*

The combination of relations [5.1] and [5.3] determines the differential equation:

$$\frac{1}{\omega_0{}^{n-1}}\left(\frac{d}{dt}\right)^{n-1}P(t)=S\frac{dX(t)}{dt}, \qquad [5.6]$$

from where one draws the symbolic equation:

$$\frac{1}{\omega_0^{n-1}} s^{n-1} P(s) = SsX(s), \qquad [5.7]$$

which can be written under the form:

$$P(s) = S\omega_0^{n-1} s^{2-n} X(s), \qquad [5.8]$$

namely, by putting $n = 2 - m$ with $0 < m < 1$:

$$P(s) = S\omega_0^{1-m} s^m X(s). \qquad [5.9]$$

This symbolic equation conveys that the pressure $P(t)$ at the water–dyke interface is proportional to the non-integer derivative (of order m) of the water displacement $X(t)$.

Knowing that the physics of the dyke limit the order m differentiation to a medium frequency range, it is convenient to replace the operator, s, with the *frequency bounded operator*,

$$\omega_b \frac{1 + s/\omega_b}{1 + s/\omega_h} \text{ with } \omega_b < \omega_h, \qquad [5.10]$$

ω_b and ω_h denoting the low and high transitional frequencies that define the frequency truncation of the operator, s; ω_b and ω_h are distributed geometrically with respect to the *unit frequency*, that is, $\omega_b\omega_h = 1$, in order to ensure a unit gain at this frequency.

Equation [5.9] can thus be restructured under a more realistic form:

$$P(s) = C_0 \left(\frac{1 + s/\omega_b}{1 + s/\omega_h} \right)^m X(s), \qquad [5.11]$$

by putting:

$$C_0 = S\omega_0^{1-m} \omega_b^m. \qquad [5.12]$$

5.2.1.3. *Functional representation highlighting a constant-phase CRONE controller*

Let us take again two symbolic equations [5.5] and [5.11], namely:

$$X(s) = -\frac{S}{Ms^2} P(s) \qquad [5.13]$$

and

$$P(s) = C_0 \left(\frac{1 + s/\omega_b}{1 + s/\omega_h} \right)^m X(s) \cdot \qquad [5.14]$$

This equation system can be translated by the functional diagram in Figure 5.1 that recalls that of a free control loop (nil reference input $R(s)$).

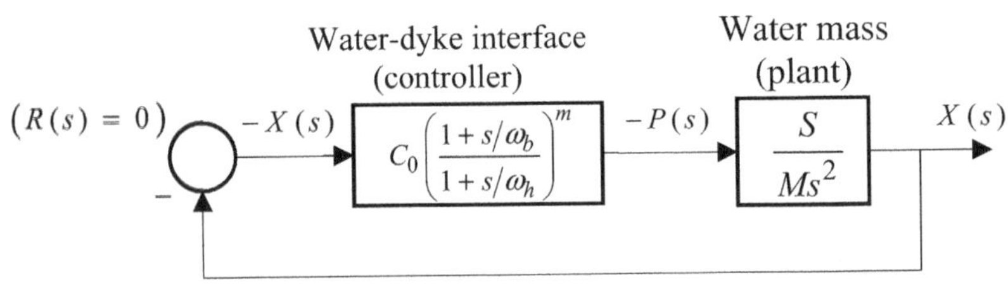

Figure 5.1. *Functional diagram governing relaxation*

The action chain structure successively reveals that:

– the transmittance that describes the motion of the water mass M, namely

$$\frac{X(s)}{-P(s)} = \frac{S}{Ms^2}, \qquad [5.15]$$

can be interpreted as that of a *plant* $G(s)$ of input $-P(s)$ and of output $X(s)$;

– the transmittance of the water–dyke interface, namely

$$\frac{-P(s)}{-X(s)} = C_0 \left(\frac{1 + s/\omega_b}{1 + s/\omega_h} \right)^m, \qquad [5.16]$$

can be interpreted as that of a *controller,* $C_m(s)$, of input $-X(s)$ and of output $-P(s)$.

The structure of the controller $C_m(s)$ so defined is none other than the *ideal structure* of a *constant-phase CRONE controller* whose *real structure* is obtained through the synthesis of a frequency-bounded real non-integer differentiator.

5.2.1.4. *Idea of the first-generation CRONE strategy*

Located in a medium frequency range, the open-loop unit gain frequency, ω_u, belongs to the *asymptotic behavior of order m* of the controller. The phase locking at $m\pi / 2$ that characterizes such behavior then ensures the invariance of the controller phase around the frequency, ω_u.

Thus, the controller presents at least the merit of not participating in the phase-margin variations; notably, it does not reinforce them in any case as a variable-phase controller would around the frequency, ω_u, among others the PID. Actually, on account of its transparency to the phase variations, a simple constant-phase CRONE controller generally ensures better robustness than a PID controller.

In automatic control, seeking the synthesis of such a controller defines the approach (said to be non-integer) that the *first-generation CRONE control* uses.

5.2.2. *Second interpretation of the relaxation model: second-generation CRONE control*

5.2.2.1. *Non-integer order differential equation as a relaxation model*

Putting the expression of $Q(t)$ given by [5.1] into relation [5.2] enables us to establish a linear differential equation of non-integer order n between 1 and 2, namely:

$$\frac{M}{S^2}\frac{1}{\omega_0{}^{n-1}}\left(\frac{d}{dt}\right)^n P(t)+P(t)=0\,, \qquad [5.17]$$

or, in a canonical form:

$$\tau^n\left(\frac{d}{dt}\right)^n P(t)+P(t)=0\,, \qquad [5.18]$$

by putting

$$\tau = \left(\frac{M}{s^2} \frac{1}{\omega_0{}^{n-1}} \right)^{1/n} . \qquad [5.19]$$

5.2.2.2. Functional representation leading to an open-loop frequency template

The Laplace transform of equation [5.18], namely

$$\tau^n s^n P(s) + P(s) = 0 , \qquad [5.20]$$

enables us to write the following relation:

$$P(s) = -\left(\frac{1}{\tau s} \right)^n P(s) . \qquad [5.21]$$

This symbolic equation is translated by the functional diagram in Figure 5.2, which recalls a free control loop (nil reference input $R(s)$).

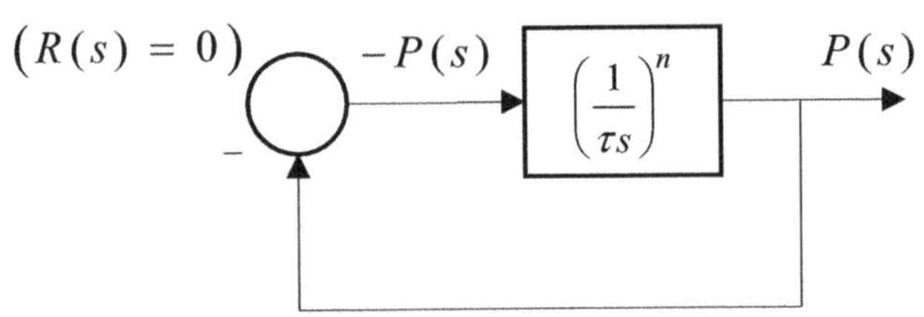

Figure 5.2. *Functional diagram enabling us to define an open-loop transfer*

Because of the unit feedback of the control loop, the action (or direct) chain determines an open-loop transmittance of the following form:

$$\beta(s) = \left(\frac{1}{\tau s} \right)^n = \left(\frac{\omega_u}{s} \right)^n , \qquad [5.22]$$

which is none other than the transmittance of a *non-integer integrator* of unit gain (or transition) frequency, $\omega_u = 1/\tau$.

Knowing that $\arg\beta(j\omega) = -n\pi/2$ and $1 < n < 2$, the Nichols locus of $\beta(j\omega)$ is a *vertical straight line* of abscissa between $-\pi/2$ and $-\pi$.

For the same physical reason as in the case of the first interpretation, the order n integration is limited to a medium frequency range. Thus, the vertical straight line of abscissa $-n\pi/2$ is reduced to a *vertical straight line segment* stretching around the unit gain frequency, ω_u (Figure 5.3). This segment is called an *open-loop frequency template* (or more simply, *template*).

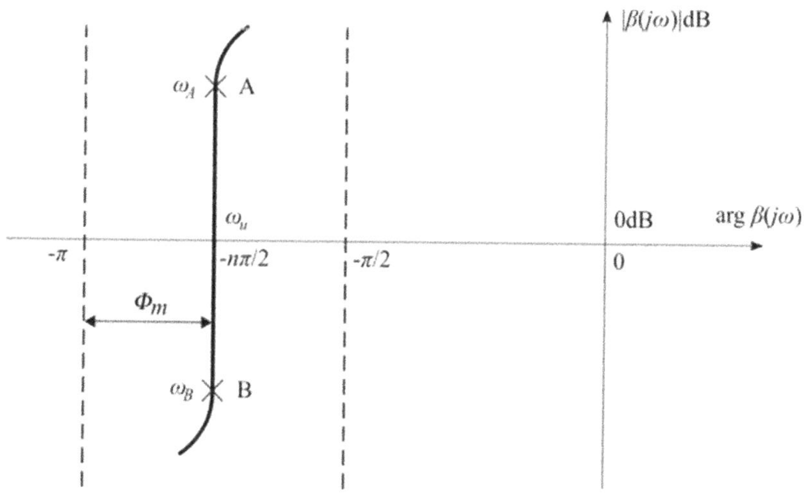

Figure 5.3. *A vertical straight line segment defines the template in the Nichols plane*

5.2.2.3. *Idea of the second-generation CRONE strategy*

When the water mass, *M*, changes, the frequency, ω_u, is modified in conformity with the relation

$$\omega_u = \frac{1}{\tau} = \left(\omega_0^{\,n-1}\frac{S^2}{M}\right)^{1/n}. \qquad [5.23]$$

This leads us to say that the template so defined slides on itself when there is variation in water mass. Given its form, this vertical displacement template ensures the constancy of the phase margin Φ_m (Figure 5.3) and, consequently, the constancy of the "distance to the critical point", which is then directly measured by Φ_m, thus conveying the stability degree robustness. It is easy to conceive that robustness improves as the length AB of the template increases, the template displacement being linked to the modification of the parameter, *M*.

In automatic control, the purpose is to obtain a similar behavior, that is to say (Figure 5.3):

– an open-loop Nichols locus which forms the vertical template so defined for the nominal parametric state of the plant;

– a *sliding of the template on itself at the time of a reparameterization of the plant* (this condition is verified when the reparameterization only leads to gain variations around ω_u).

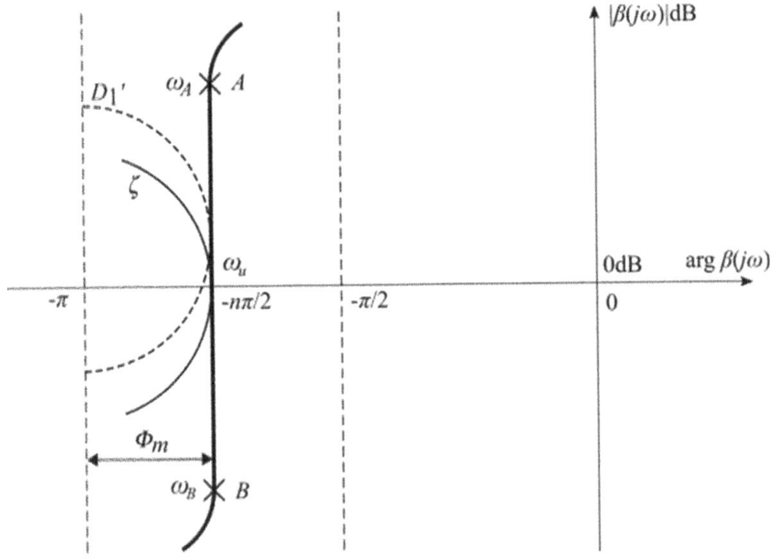

Figure 5.4. *The form and the vertical sliding of the template AB simultaneously ensure the robustness of the phase margin Φ_m, the reduced first overshoot D_1' and the damping ratio ζ*

The form and the vertical sliding of the template ensure not only the constancy of the phase margin, Φ_m, but also (Figure 5.4):

– the constancy of the reduced first overshoot, D_1', of the step response in tracking or in regulation, through the tangency of the template to the same iso-overshoot contour (of parameter D_1'), a performance contour which is a Nichols amplitude contour that we have validated as an iso-overshoot contour by using complex non-integer integration (Appendix 3);

– the *constancy of the damping ratio* ζ *in tracking and in regulation* through the tangency of the template to the same *isodamping contour* (of parameter ζ), a performance contour which is none other than an Oustaloup isodamping contour that we have *built* by the ingenious use of complex non-integer integration and the covering technique (Appendix 3).

Seeking the synthesis of such a template defines the approach (said to be non-integer) the *second-generation CRONE control* uses.

5.3. Second-generation CRONE control and uncertainty domains

5.3.1. *Uncertainty domains*

Case of the plant: an uncertainty domain of the plant is defined as the domain of the Nyquist or Nichols plane, which groups together the image points of the frequency responses of the plant obtained, at a given frequency, for its various parametric states (Figure 5.5).

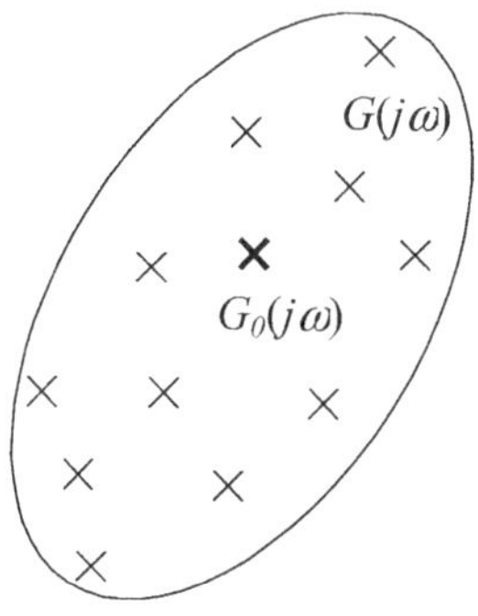

Figure 5.5. *Uncertainty domain of the plant:* $G_0(j\omega)$ *denotes the frequency response of the nominal plant;* $G(j\omega) = G_0(j\omega) + \Delta G(j\omega)$ *represents the frequency response of the plant for a parametric state different from the nominal one*

Case of the open loop: an uncertainty domain of the open loop is defined as the domain of the Nyquist or Nichols plane, which regroups the image points of the open-loop frequency responses obtained, at a given frequency, for the various parametric states of the plant (Figure 5.6).

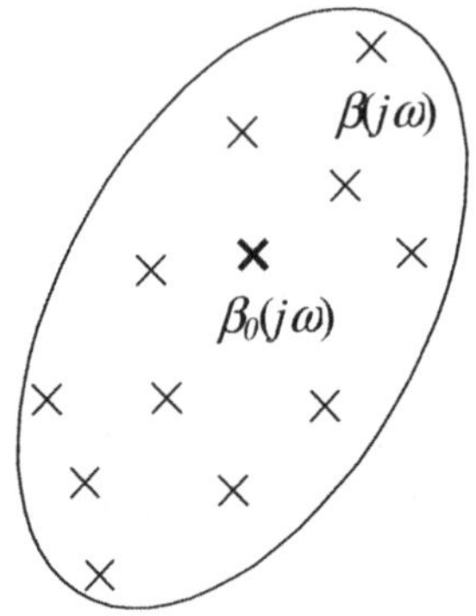

Figure 5.6. *Uncertainty domain of the open loop: $\beta_0(j\omega) = C(j\omega)G_0(j\omega)$ denotes the nominal open-loop frequency response; $\beta(j\omega) = \beta_0(j\omega) + \Delta\beta(j\omega)$ with $\Delta\beta(j\omega) = C(j\omega)\Delta G(j\omega)$, represents the open-loop frequency response for a plant parametric state different from the nominal one*

Given that (Figure 5.7)

$$\beta(j\omega) = C(j\omega)G(j\omega), \qquad [5.24]$$

it is possible to write:

$$|\beta(j\omega)|_{dB} = |C(j\omega)|_{dB} + |G(j\omega)|_{dB} \qquad [5.25]$$

and

$$\arg\beta(j\omega) = \arg C(j\omega) + \arg G(j\omega) \cdot \qquad [5.26]$$

Such relations express that the change from an image point of $G(j\omega)$ to an image point of $\beta(j\omega)$ is carried out, for each point, through the same translation operations $|C(j\omega)|_{dB}$ and $\arg C(j\omega)$, the controller $C(j\omega)$ providing the same gain and the same phase at a given frequency ω.

This leads us to say that the uncertainty domains of the open loop are identical to those of the plant except for their positions, which are different in conformity with the relation $\beta_0(j\omega) = C(j\omega)G_0(j\omega)$.

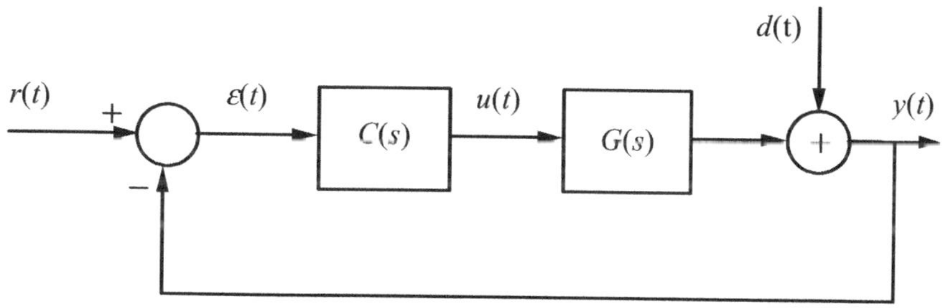

Figure 5.7. *Elementary control loop: r(t), reference input; d(t), disturbance; y(t), output; Σ(t), error signal; u(t), plant input; C(s), regulator; G(s), plant*

5.3.2. *Particular open-loop uncertainty domains*

When the relaxation water mass, *M*, changes, the frequency ω_u is modified according to relation [5.23], namely:

$$\omega_u = \frac{1}{\tau} = \left(\omega_0^{n-1} \frac{S^2}{M} \right)^{1/n} . \qquad [5.27]$$

If the phase in open loop remains constant, namely arg $\beta(j\omega) = -n\pi/2$, the open-loop gain varies in conformity with the extremal values:

$$\left| \beta(j\omega) \right|_{\min} = \left(\frac{\omega_{u_{\min}}}{\omega} \right)^n \qquad [5.28]$$

and

$$\left| \beta(j\omega) \right|_{\max} = \left(\frac{\omega_{u_{\max}}}{\omega} \right)^n , \qquad [5.29]$$

from which we draw:

$$\frac{\left| \beta(j\omega) \right|_{\max}}{\left| \beta(j\omega) \right|_{\min}} = \left(\frac{\omega_{u_{\max}}}{\omega_{u_{\min}}} \right)^n , \qquad [5.30]$$

namely in decibels:

$$\left|\beta(j\omega)\right|_{\max} dB - \left|\beta(j\omega)\right|_{\min} dB = 20n\log\left(\frac{\omega_{u_{\max}}}{\omega_{u_{\min}}}\right). \quad [5.31]$$

In the Nichols plane, the uncertainty domains are thus vertical straight-line segments of constant length, $20n\log\left(\omega_{u_{\max}}/\omega_{u_{\min}}\right)$ (Figure 5.8).

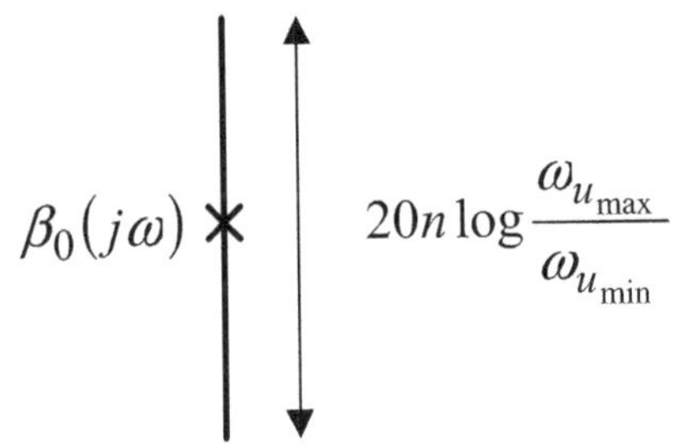

Figure 5.8. *Type of uncertainty domains corresponding to water mass variations*

5.3.3. *Adequacy of the second-generation CRONE control template to the particular uncertainty domains*

The frequency configuration defined by the vertical template and the uncertainty domains so particularized express the part played by the template in the positioning of the uncertainty domains.

Figure 5.9 shows that the template ensures optimal positioning of the uncertainty domains such that they do not overlap the low stability degree areas. The low stability degree area, illustrated as an example, is the area that is delimited by a Nichols amplitude contour.

It turns out that optimal positioning of the uncertainties ensured by the template results from the fact that the *direction of the template is the same as that of the uncertainty domains*. This property is at the origin of the generalization of the template, which less particular uncertainty domains require.

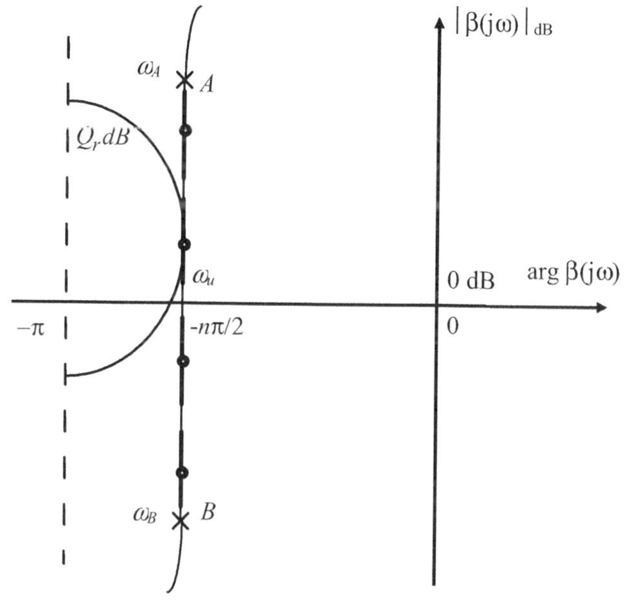

Figure 5.9. *The vertical template positions the uncertainty domains so that they do not overlap the amplitude contour $Q_r dB$*

5.4. Generalization of the vertical template through the third-generation CRONE control

5.4.1. *First level of generalization*

5.4.1.1. *Generalized template*

When the uncertainty domains are not reduced to vertical straight-line segments, it is convenient to generalize the vertical template, particularly by a template:

– *always defined as a straight-line segment for the nominal parametric state of the plant;*

– *but which no longer systematically determines phase locking around the frequency,* ω_u.

The *vertical straight-line segment* defining the template in the Nichols plane until now is thus replaced by a *straight-line segment of any direction*, called a *generalized template* (Figure 5.10).

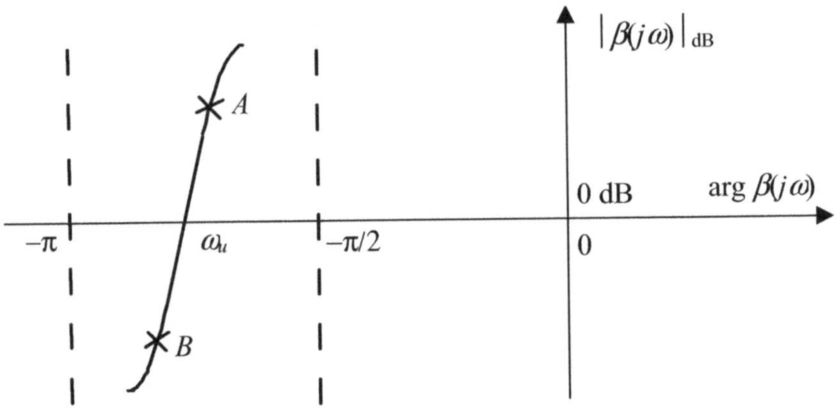

Figure 5.10. *Representation of the generalized template in the Nichols plane*

In order that the generalized template ensures optimal positioning of the uncertainty domains such that they overlap the low stability degree areas as little as possible, the direction of the generalized template can be given by the mean direction of the main axes of the uncertainty domains, calculated around the frequency ω_u (Figure 5.11).

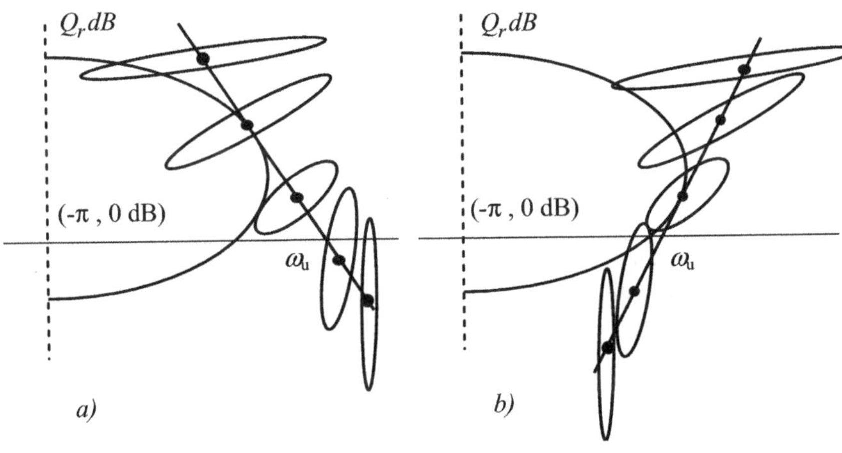

Figure 5.11. *The generalized template positions the uncertainty domains so that they overlap the amplitude contour, QrdB, as little as possible: a) any generalized template; b) optimal generalized template*

Given the continuity of the direction of the main axes of the uncertainty domains, a good approximation consists of giving to the generalized template the direction of the main axis of the uncertainty domain, calculated at the frequency ω_u (Figure 5.12(a)).

The direction of the generalized template can also be that of the main axis of the uncertainty domain, calculated at the resonance frequency in tracking ω_r; this choice appears preferential since the generalized template is tangent to the amplitude contour, $Q_r dB$, at this frequency (Figure 5.12(b)).

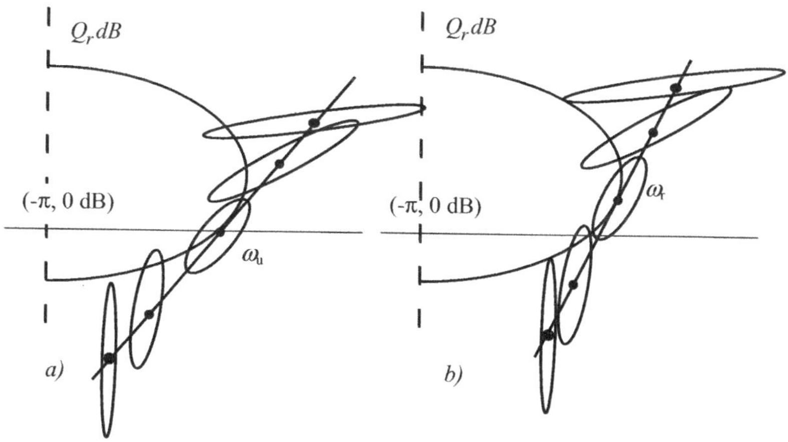

Figure 5.12. *Optimal generalized template whose direction is given by that of the main axis of the uncertainty domain corresponding to a) the frequency ω_u and b) the frequency ω_r.*

5.4.1.2. *Generalized template and complex non-integer integration*

The vertical template ($\omega \in [\omega_A, \omega_B]$) is characterized by a real non-integer integration order, n, which determines its phase placement at $-n\pi/2$ (Figure 5.4). It is indeed described by the transmittance of a real non-integer integrator, namely:

$$\tilde{\beta}(s) = \left(\frac{\omega_u}{s}\right)^n \text{ with } n \in \mathrm{R}^+ . \qquad [5.32]$$

In the study, which consists of extending the description of the vertical template [OUS 91], we have shown that the generalized template can be characterized by a complex non-integer integration order, n, whose real part, a, determines its phase placement at the frequency ω_u, namely $-a\pi/2$, and whose imaginary part, b, determines its incline to the vertical (Figure 5.13).

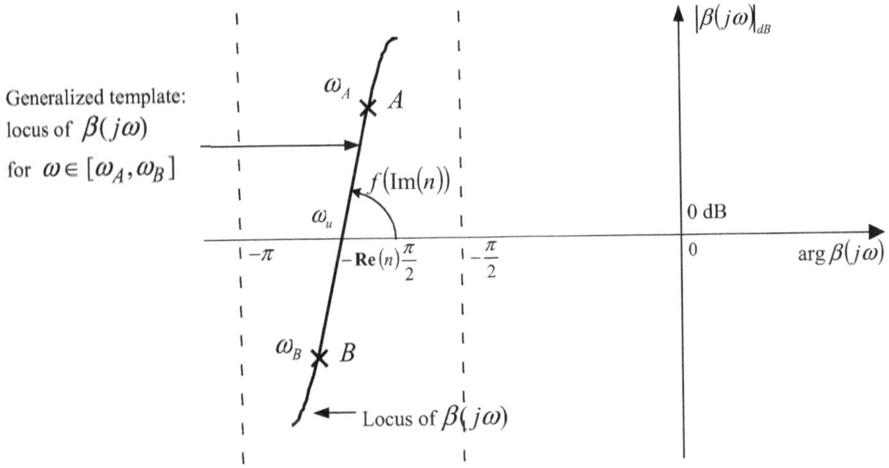

Figure 5.13. *The integration real order, $a = \mathrm{Re}(n)$, determines the phase placement at the frequency ω_u of the generalized template, namely $-a\pi/2$; the integration imaginary order, $b = \mathrm{Im}(n)$, determines its angle to the vertical*

The generalized template ($\omega \in [\omega_A, \omega_B]$) is indeed described by a transmittance based on the complex non-integer integration, namely (Appendix 3):

$$\tilde{\beta}(s) = \left(\cosh\left(b\frac{\pi}{2}\right)\right)^{sign(b)} \left(\frac{\omega_u}{s}\right)^a \left(\left[\left(\frac{\omega_u}{s}\right)^{ib}\right]_{C_j}\right)^{-sign(b)} \quad \text{with } a \in \mathrm{R}^+ \text{ and } b \in \mathrm{R}, \qquad [5.33]$$

or even, developing the part reduced to the operational plane C_j:

$$\tilde{\beta}(s) = \left(\cosh\left(b\frac{\pi}{2}\right)\right)^{sign(b)} \left(\frac{\omega_u}{s}\right)^a \left(\cos\left(b\ln\left(\frac{\omega_u}{s}\right)\right)\right)^{-sign(b)}. \qquad [5.34]$$

The parameter, n ($n = a + ib$), only represents (strictly speaking) the complex integration order of the transmittance, $\tilde{\beta}(s)$, [5.33] when the parameter, b, is negative. Nevertheless, in order to lessen subsequent developments, we will agree to conserve the integration order denomination when the b parameter is positive. Moreover, it is convenient to specify that the pure imaginary unit, i, featuring in the integration order, $n = a + ib$, is independent of the pure imaginary unit, j, featuring in the operational variable, $s = \sigma + j\omega$.

In the case when the description transmittance of the generalized template is the subject of a frequency truncation at low and high frequencies, it is convenient to replace relation [5.33] by a more general expression:

$$\tilde{\beta}(s) = C^{sign(b)}\left(C_0\frac{1+\frac{s}{\omega_h}}{1+\frac{s}{\omega_b}}\right)^a\left(\left[\left(C_0\frac{1+\frac{s}{\omega_h}}{1+\frac{s}{\omega_b}}\right)^{ib}\right]_{C_j}\right)^{-sign(b)}, \quad [5.35]$$

in which:

$$C = \cosh\left[b\left(\arctan\left(\frac{\omega_u}{\omega_b}\right) - \arctan\left(\frac{\omega_u}{\omega_h}\right)\right)\right] \quad [5.36]$$

and

$$C_0 = \left(\frac{1+\left(\frac{\omega_u}{\omega_b}\right)^2}{1+\left(\frac{\omega_u}{\omega_h}\right)^2}\right)^{\frac{1}{2}}, \quad [5.37]$$

the expression of C_0 being reduced to $C_0 = \omega_u / \omega_b$ in the particular case when the low and high transitional frequencies, ω_b and ω_h, are geometrically distributed in relation to ω_u, namely $(\omega_b\omega_h)^{1/2} = \omega_u$.

5.4.2. *Second level of generalization*

5.4.2.1. *Curvilinear template*

A rectilinear template whose direction is that of the main axis of the uncertainty domain, calculated at the frequency ω_r, only ensures optimal positioning of this uncertainty domain; the other domains can overlap the amplitude contour $Q_r dB$ too much.

In order to process each uncertainty domain as that calculated at the frequency ω_r (whose generalized template takes its direction), it is convenient to replace the generalized template by a *curvilinear template* whose tangent at each frequency point is the main axis of the corresponding uncertainty domain.

By best exploiting the property of fractal robustness, through which *the direction of the template is the same as that of the uncertainty domains*, the strategy consisting of putting into play a curvilinear template in fact makes it possible to obtain, around the frequency ω_r, an uncertain open-loop Nichols locus with as small a "thickness" as possible, a robustness criterion par excellence, the thickness in this case being measured by the gap between the extremal Nichols loci covering the set of uncertainty domains (Figure 5.14).

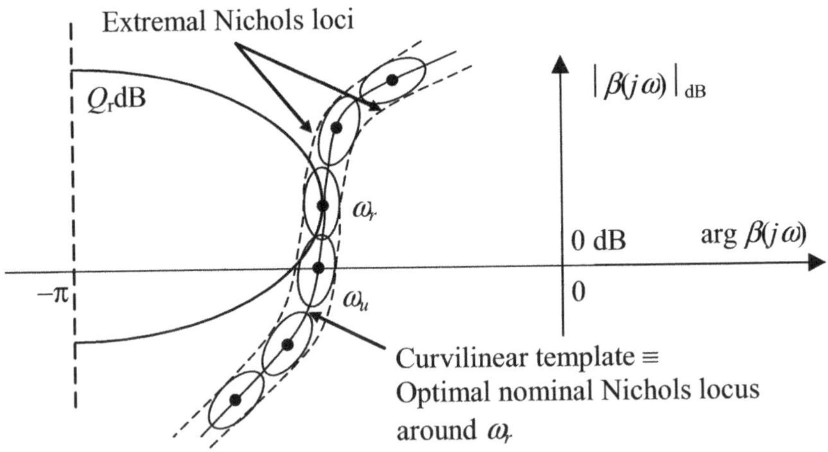

Figure 5.14. *The area delimited by the extremal Nichols loci covering the uncertainty domains positioned by the curvilinear template, defines, around ω_r, an uncertain Nichols locus whose "thickness" determines its uncertainty*

5.4.2.2. *Operational description of the curvilinear template*

To obtain a description of the curvilinear template by a non-integer transmittance, it is convenient to interpret the asymptotic locus of the curvilinear template as a *multi-template* defined by a set of generalized templates (Figure 5.15).

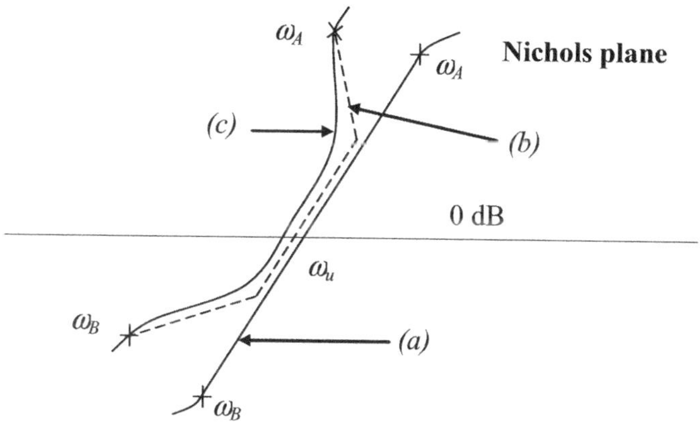

Figure 5.15. *From the generalized template (a) to the multi-template (b); from the rectilinear template (a) to the curvilinear template (c)*

The interpretation as defined through the multi-template makes it possible to describe the curvilinear template by a transmittance resulting from the product of the frequency-bounded transmittances of the same form as expression [5.35], namely:

$$\tilde{\beta}(s) = \prod_{k=-N^-}^{N^+} \tilde{\beta}_k(s) \,, \qquad [5.38]$$

with (for $k \neq 0$):

$$\tilde{\beta}_k(s) = C_k^{sign(b_k)} \left(\frac{\omega'_k}{\omega_k} \frac{1+\dfrac{s}{\omega_{k+1}}}{1+\dfrac{s}{\omega_k}} \right)^{a_k} \left(\left[\left(\frac{\omega'_k}{\omega_k} \frac{1+\dfrac{s}{\omega_{k+1}}}{1+\dfrac{s}{\omega_k}} \right)^{ib_k} \right]_{C_j} \right)^{-sign(b_k)} ; \qquad [5.39]$$

for $k = 0$, $\tilde{\beta}_k(s)$ is defined by relation [5.35] in which ω_b and ω_h are in this case replaced by ω_0 and ω_1.

ω_k and ω_{k+1} denote the low and high transitional frequencies of the generalized template of rank k. ω'_k represents the frequency with which this template is defined. More precisely, it represents its "central" frequency when k is different from zero, that is to say:

$$\omega'_k = \left(\omega_k \omega_{k+1}\right)^{1/2} \text{ for } k \neq 0 . \quad [5.40]$$

The rank 0 generalized template, said to be "median", is that which is defined in relation to $\omega'_0 = \omega_u$. ω'_0 does not necessarily represent its central frequency, considering that in order to ensure an additional degree of freedom, ω'_0 is not necessarily equal to $\left(\omega_0 \omega_1\right)^{1/2}$.

The gain, C_k, makes it possible to ensure a unit gain to $\tilde{\beta}_k(s)$ at the frequency ω_u, set by the designer.

N^- denotes the number of generalized templates defined in relation to the frequencies, ω'_k, lower than ω_u. N^+ denotes the number of generalized templates defined in relation to the frequencies, ω'_k, higher than ω_u. In the particular case when $N^- = N^+ = 0$, $\tilde{\beta}(s)$ is reduced to the frequency-bounded transmittance defined by [5.35].

5.4.3. *Open-loop transfer integrating the curvilinear template*

The aim is to describe analytically the open-loop behavior (for the nominal plant), which takes into account at once:

– the accuracy specifications at the low frequencies;

– the curvilinear template close to the frequency ω_u;

– the behavior of the plant at the high frequencies and the specification on the input sensitivity at these frequencies.

For phase minimum stable plants, it turns out the behavior so defined can be described by a transmittance based on frequency-bounded complex non-integer integration [OUS 99], namely:

$$\beta(s) = \beta_b(s)\tilde{\beta}(s)\beta_h(s) \cdot \quad [5.41]$$

The transmittance, $\tilde{\beta}(s)$, based on complex non-integer integration, is none other than the description transmittance of the curvilinear template defined by relations [5.38] and [5.39]. Its extremal transitional frequencies are ω_{-N^-} in low frequency and $\omega_{N^+ +1}$ in high frequency.

The transmittance $\beta_b(s)$ is that of an order n_b proportional-integrator, whose transitional frequency is equal to the low extremal transitional frequency of $\tilde{\beta}(s)$, in order that connecting $\beta_b(s)$ with $\tilde{\beta}(s)$ does not lead to introducing additional parameters. $\beta_b(s)$ is then defined by:

$$\beta_b(s) = C_b \left(\frac{\omega_{-N^-}}{s} + 1 \right)^{n_b} ; \qquad [5.42]$$

if n_{pb} represents the asymptotic behavior order of the plant in low frequency ($\omega < \omega_{-N^-}$), the order n_b is given by $n_b \geq 1$ if $n_{pb} = 0$ and $n_b \geq n_{pb}$ if $n_{pb} \geq 1$ knowing that $n_b = 1$ cancels the position error and $n_b = 2$ cancels the velocity error; finally C_b permits us to ensure a unit gain to $\beta_b(s)$ at the frequency ω_u.

The transmittance $\beta_h(s)$ is that of an order n_h low-pass filter, whose transitional frequency is equal to the high extremal transitional frequency of $\tilde{\beta}(s)$, in order that connecting $\beta_h(s)$ with $\tilde{\beta}(s)$ does not introduce additional parameters. $\beta_h(s)$ is then defined by:

$$\beta_h(s) = \frac{C_h}{\left(1 + \dfrac{s}{\omega_{N^+ +1}} \right)^{n_h}} ; \qquad [5.43]$$

if n_{ph} denotes the asymptotic behavior order of the plant in high frequency ($\omega > \omega_{N^+ +1}$), the order n_h is given by $n_h \geq n_{ph}$, knowing that (for $\omega > \omega_{N^+ +1}$) $n_h = n_{ph}$ ensures the constancy versus frequency of the input sensitivity and $n_h > n_{ph}$ ensures its decline versus frequency; finally C_h permits us to ensure a unit gain to $\beta_h(s)$ at the frequency ω_u.

5.4.4. *Optimization of the open-loop behavior*

In order to simplify the developments of this section, the curvilinear template is reduced to the generalized template by taking N^- and N^+ (relation [5.38]) according to $N^- = N^+ = 0$. $\tilde{\beta}(s)$ is then defined by expression [5.35], the extremal transitional frequencies ω_{-N^-} and ω_{N^++1} thus being reduced to ω_b and ω_h.

5.4.4.1. *Optimal template*

The optimization of the generalized template as for:

– its *position* (along the axis 0 dB);

– its *direction* (or angle to the vertical);

– and its *length*;

consists of determining the optimal values of the following four parameters:

– the *integration real order,* a , which determines the position of the template along the axis, 0 dB;

– the *integration imaginary order,* b, which determines its incline with respect to the vertical;

– the *low and high transitional frequencies,* ω_b and ω_h, which determine its length through their action on its extremal frequencies, ω_A and ω_B.

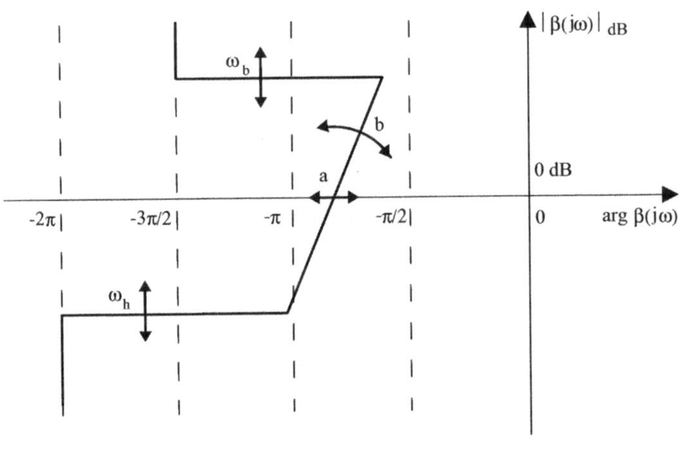

Figure 5.16. *Illustration of the effects of the parameters* a*,* b*,* ω_b *and* ω_h *on the Nichols asymptotic locus of* $\beta(j\omega)$

These parameters indeed "preparameterize" the open-loop transmittance $\beta(s)$. In the particular case when $n_b = 3$ and $n_h = 4$, Figure 5.16 shows how they act in the search for the optimal form of $\beta(s)$.

In fact, the tangency condition of the Nichols locus of $\beta(s)$ at an iso-overshoot or isodamping contour for a given frequency, ω_u, reduces the "preparameterization" with three independent parameters.

The "preparameterization" of the transmittance in open loop, enabled by the complex, non-integer integration order, considerably simplifies the synthesis. Indeed, optimal transmittance in open loop is sought not only based on a set of transmittances restricted to those satisfying the nominal performances, but also in a parametric space of description of reduced dimension, in this case to three. This means that the introduction of another generalized template leads to three additional independent parameters. The curvilinear template must therefore result only from a small number of rectilinear templates, in order not to lose the advantage presented by the non-integer integration in terms of preparameterization with few parameters.

Finally, given that the structure of the open-loop transmittance is known and the frequency, ω_u, is set by the designer, the four optimal parameters, a_{opt}, b_{opt}, ω_{bopt} and ω_{hopt}, completely determine the optimal form of $\beta(s)$. This one is indeed obtained by putting the optimal parameters so defined into expression [5.41], in which $\tilde{\beta}(s)$ is given by [5.35] and $\beta_b(s)$ and $\beta_h(s)$ by [5.42] and [5.43], the parameters C_b and C_h being deduced from the others.

5.4.4.2. *Criterion under constraints to be minimized by the optimal template*

The search for the optimal template is a matter of optimization under constraints. The optimal template must minimize a quadratic criterion of the form

$$J = (D'_{1\max} - D'_{1d})^2 + (D'_{1\min} - D'_{1d})^2 , \qquad [5.44]$$

namely, in the frequency domain [OUS 95a, OUS 99a]

$$J = (Q_{r_{\max}} - Q_{r_d})^2 + (Q_{r_{\min}} - Q_{r_d})^2 , \qquad [5.45]$$

while respecting the frequency constraints on the entirety of the sensitivity functions [OUS 99b] for the set of parametric states of the plant. Q_{r_d} designates the desired value of the resonance ratio in tracking for the nominal parametric state of the plant, and

$Q_{r_{max}}$ and $Q_{r_{min}}$ characterizing its extremal variations due to the reparameterization of the plant.

5.4.5. *Structure and parametric estimation of the controller*

The controller in cascade with the plant is defined by a *non-integer transmittance*, $C_{ne}(s)$, deduced from the ratio:

$$C_{ne}(s) = \frac{\beta(s)}{G_0(s)}, \qquad [5.46]$$

where $\beta(s)$ is given by [5.41] and $G_0(s)$ denotes the nominal transmittance of the plant.

The *integer transmittance* of the controller, $C_{en}(s)$, results from the identification of $C_{ne}(s)$ through a *reduced dimension rational integer model*, namely:

$$C_{en}(s) = \frac{N(s)}{D(s)}, \qquad [5.47]$$

where $N(s)$ and $D(s)$ are polynominals of specified degrees, n and d:

$$N(s) = \sum_{k=0}^{n} n_k s^k \text{ and } D(s) = \sum_{k=0}^{d} d_k s^k . \qquad [5.48]$$

Two parametric estimation techniques are used. The first technique is based on the elementary symmetrical functions of the roots of Viète [LAN 94, OUS 02]. The second technique is based on solving a linear programming problem [MAT 94].

5.4.6. *Application*

The study plant considered as an example is described by the transfer function:

$$G(s) = G_0 \frac{\left(1+\frac{s}{\omega_1'}\right)}{\left(1+\frac{s}{\omega_1}\right)\left(1+2\zeta\frac{s}{\omega_n}+\frac{s^2}{\omega_n^2}\right)} . \qquad [5.49]$$

Its nominal parametric state is defined by $G_{0_{nom}} = 50$, $\omega'_{1_{nom}} = 0.01$, $\omega_{1_{nom}} = 0.02$, $\omega_{n_{nom}} = 1$ and $\zeta_{nom} = 0.8$. The parameters G_0, ω'_1, ω_1 and ω_n are uncertain and can take any value within the following intervals:

$$\frac{G_{0_{nom}}}{2} \leq G_0 \leq 2G_{0_{nom}}, \qquad \frac{\omega'_{1_{nom}}}{2} \leq \omega'_1 \leq 2\omega'_{1_{nom}},$$

$$\frac{\omega_{1_{nom}}}{2} \leq \omega_1 \leq 2\omega_{1_{nom}} \qquad \text{and} \ \frac{\omega_{n_{nom}}}{1.5} \leq \omega_n \leq 1.5\omega_{n_{nom}}.$$

The aim of the control law to be synthesized is to ensure a tracking resonance ratio of 2.3 dB and an open-loop unit gain frequency of 20 rd/s.

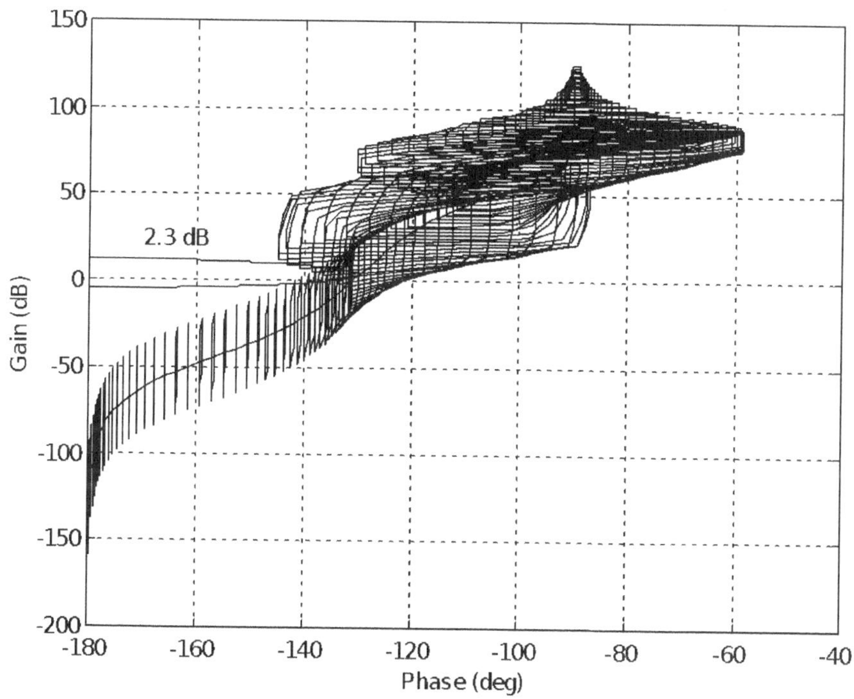

Figure 5.17. *Optimal open-loop Nichols locus and associated uncertainty domains*

The controller that responds to these objectives is given by relation [5.46] in which the optimal open-loop transmittance $\beta(s)$ is defined by relation [5.41] with:

$$N^- = N^+ = 0, \qquad n_b = 1, \qquad n_h = 2,$$

$$a = 1.44, \quad b = -0.209, \quad \omega_b = 0.12 \text{ rd/s and} \quad \omega_h = 955.44 \text{rd/s.} \qquad [5.50]$$

As illustrated in Figure 5.18, this optimal open-loop behavior positions the uncertainty domains so as to ensure the robustness of the closed-loop stability degree.

In the time domain, the stability degree robustness is illustrated by the desensitization to the parametric state of the plant, of the first overshoot of the response to a step output disturbance (Figure 5.18).

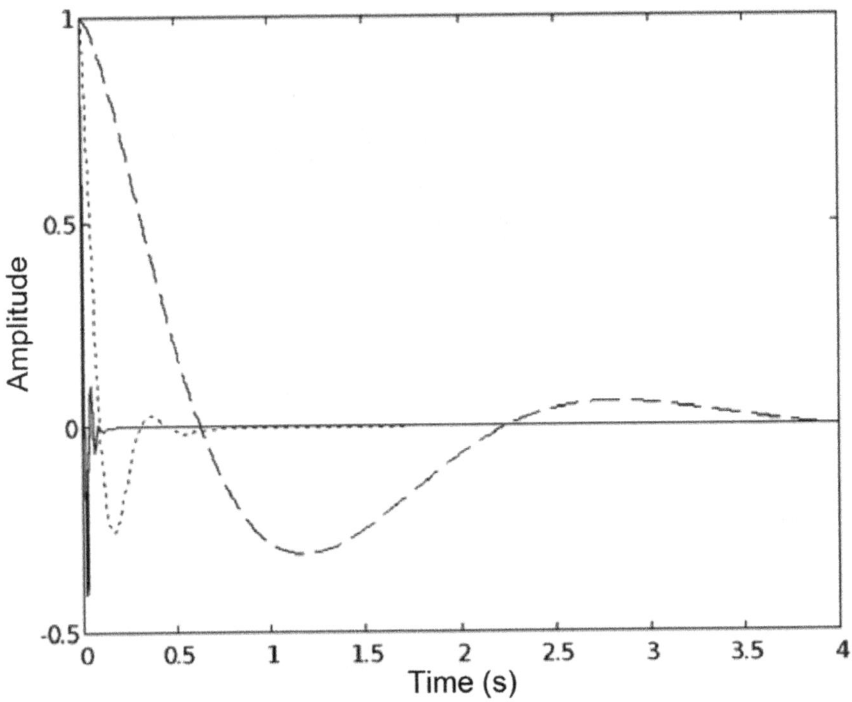

Figure 5.18. *Closed-loop responses to a step output disturbance for the nominal parametric state of the plant and two extremal parametric states*

5.5. An appendix on the frequency response describing the generalized template

In the frequency interval, $[\omega_A, \omega_B]$, corresponding to the rectilinear segment of any direction, defining the generalized template, the description transmittance of the template is given by relation [5.34], that is:

$$\tilde{\beta}(s) = \left(\cosh\left(b\frac{\pi}{2}\right)\right)^{sign(b)} \left(\frac{\omega_u}{s}\right)^a \left(\cos\left(b\ln\left(\frac{\omega_u}{s}\right)\right)\right)^{-sign(b)} . \qquad [5.51]$$

Considering $b < 0$ in accordance with Figure 5.16 (itself compliant with an elementary-form open-loop Nichols locus), this transmittance is reduced to:

$$\tilde{\beta}(s) = \left(\cosh\left(b\frac{\pi}{2}\right)\right)^{-1} \left(\frac{\omega_u}{s}\right)^a \cos\left(b\ln\left(\frac{\omega_u}{s}\right)\right), \qquad [5.52]$$

the corresponding frequency response being:

$$\tilde{\beta}(j\omega) = \left(\cosh\left(b\frac{\pi}{2}\right)\right)^{-1} \left(\frac{\omega_u}{j\omega}\right)^a \cos\left(b\ln\left(\frac{\omega_u}{j\omega}\right)\right), \qquad [5.53]$$

an expression that can be used to show that [OUS 91, OUS 95a, OUS 99b]:

$$\left|\tilde{\beta}(j\omega_u)\right| = 1 \text{ or} \left|\tilde{\beta}(j\omega_u)\right|\text{dB} = 0 ; \qquad [5.54]$$

$$\arg\tilde{\beta}(j\omega_u) = -a\frac{\pi}{2} ; \qquad [5.55]$$

$$\left.\frac{d}{d\log\omega}\left|\tilde{\beta}(j\omega)\right|\text{dB}\right|_{\omega=\omega_u} = -20a ; \qquad [5.56]$$

$$\left.\frac{d}{d\log\omega}\arg\tilde{\beta}(j\omega)\right|_{\omega=\omega_u} = -\ln(10)b\tanh\left(b\frac{\pi}{2}\right). \qquad [5.57]$$

In the case of a generalized, *narrow-band* template, which thus covers a frequency domain, $[\omega_A, \omega_B]$, that is not very extended, the modulus (in dB) and argument of $\tilde{\beta}(j\omega)$ admit, around the frequency ω_u, an excellent approximation by linear functions of $\Delta \log \omega = \log \omega - \log \omega_u = \log(\omega/\omega_u)$, that is:

$$\left|\tilde{\beta}(j\omega)\right| \mathrm{dB} = -20a \log \frac{\omega}{\omega_u} \quad [5.58]$$

and

$$\arg \tilde{\beta}(j\omega) = -a\frac{\pi}{2} - 2.3b \tanh\left(b\frac{\pi}{2}\right) \log \frac{\omega}{\omega_u}. \quad [5.59]$$

By drawing, for example, $\log(\omega/\omega_u)$ from equation [5.58], that is,

$$\log \frac{\omega}{\omega_u} = -\frac{1}{20a}\left|\beta(j\omega)\right| \mathrm{dB}, \quad [5.60]$$

then applying the expression thus obtained to equation [5.59], we obtain:

$$\arg \tilde{\beta}(j\omega) = -a\frac{\pi}{2} + 0.115\frac{b}{a} \tanh\left(b\frac{\pi}{2}\right)\left|\beta(j\omega)\right| \mathrm{dB}, \quad [5.61]$$

an equation (of an *affine function*), expressing the linearity between the values applied on the abscissa and ordinate in the Nichols plane, a linearity that clearly justifies the straight-line segment constituted by the generalized template.

In fact, as highlighted, this linearity of the Nichols locus of the frequency response that describes the generalized template results from a legitimate linearization of the gain (in dB) and of the phase, around ω_u, in accordance with a narrow-band template.

Yet, by an analysis of numerous Bode diagrams and Nichols loci of $\tilde{\beta}(j\omega)$, it transpires that this linearity is well conserved at frequencies sufficiently far from ω_u, thus expressing the *large linearity* of the Bode diagrams and the Nichols locus of this frequency response.

Moreover, the large linearity of the Nichols locus of $\tilde{\beta}(j\omega)$ is widely used, through the covering technique, in the *validation* of the Nichols amplitude contours as iso-overshoot contours and in the *building* of isodamping contours [OUS 95b].

SOLVED PROBLEMS

Presentation of Problem 1

Elementary Synthesis of a PID Regulator Based on the Single Phase-Lead Controller

The study plant that is the subject of this problem is achieved by a continuous-current motor, for which the output considered is the angular speed. Taking account of its electrical time constant, the study plant is described by a second-order transmittance of the form:

$$G(s)=\frac{G_0}{\left(1+\frac{s}{\omega_m}\right)\left(1+\frac{s}{\omega_e}\right)},$$

ω_m and ω_e denoting the mechanical transitional frequency and the electrical transitional frequency, respectively, with these frequencies admitting, in the absence of a mechanical charge, the following numerical values:

$$\omega_m = 50\ rd/s \text{ and } \omega_e = 250\ rd/s,$$

with these values to be completed with that for static gain, namely: $G_0 = 50$.

1) In order for the dynamics of the motor control to be sufficiently fast, the designer sets the unit gain frequency in open loop, ω_u, at a value 10 times that of the mechanical transitional frequency, that is, $\omega_u = 10\omega_m = 500\ rd/s$. Justify how this choice is made.

2) In the absence of phase correction, the regulator then being reduced to a simple proportional, determine the phase margin, ϕ_m, of the control. Would you say its value seems suitable?

3) At the same frequency, ω_u, what phase must be presented by the regulator that is to substitute the proportional in order to ensure a phase margin of 54° that may be considered as an optimal value ?

4) With the designer opting for a simple phase-lead controller, that is,

$$C(s) = C_0 \frac{1+\dfrac{s}{\omega_1}}{1+\dfrac{s}{\omega_2}} \text{ with } \omega_2 > \omega_1,$$

synthesize this controller by determining its three parameters.

5) Determine the high-frequency input sensitivity. How can the regulator be complexified in order to improve this sensitivity in accordance with a decrease of 6 dB/oct then 12 dB/oct? Let us use ω_h to denote the high transitional frequency introduced by this complexification.

6) How can the regulator again be complexified to transform it into a well-known regulator (a classic in its field), with the function of simultaneously improving the control's stability degree and precision at low frequency? Let ω_b denote the low transitional frequency introduced by this complexification.

7) After looking at the motor when idle, now let us turn to the motor under load. Through a factor of 10, increase of the motor-shaft inertia (consecutive to the introduction of a load), the mechanical transitional frequency of the plant passes from 50 to 5 rd/s, ω_m, so that its value now stands at 5 rd/s (that is, $\omega_m = 5\ rd/s$ for the motor under load). For a unit gain frequency in open loop, ω_u, again set at 10 times the mechanical transitional frequency, ω_m, that is, $\omega_u = 10\omega_m = 50\ rd/s$, a choice that you are asked to justify. Determine, successively, the control phase margin in the absence of phase correction, the phase that the regulator must present at the frequency, ω_u, to ensure a phase margin of 54°, as well as the synthesis of the regulator, then performed by a simple phase-lag controller, a synthesis from which we will draw the high-frequency input sensitivity.

8) How can the regulator be complexified to improve the high-frequency input sensitivity as well as the low-frequency precision? We will use ω_h and ω_b to denote the high and low transitional frequencies introduced by this complexification.

9) Does the regulator thus obtained (for the motor under load) produce a PID regulator per se?

Solution to Problem 1

1) Justification of the choice of frequency ω_u

The plant cut-off frequency at –3 dB, ω_c, would equal ω_m if ω_e did not exist (or would have a very high value), but in this case it proves to be slightly less than ω_m, given the presence and value of ω_e, therefore:

$$\omega_c = \omega_m^- = 50^- \ rd/s.$$

Knowing that $\omega_u = 500 \ rd/s$,

$$\frac{\omega_u}{\omega_c} = \frac{500}{50^-} = 10^+,$$

a ratio in accordance with the frequency range recommended in the field, that is,

$$5\omega_c \leq \omega_u \leq 20\omega_c.$$

2) Phase margin in the absence of phase correction

The phase margin is defined by:

$$\phi_m = \arg \beta(j\omega_u) - (-\pi) = \arg \beta(j\omega_u) + \pi,$$

that is:

$$\phi_m = \arg C(j\omega_u) + \arg G(j\omega_u) + \pi,$$

or, in this case (since $\arg C(j\omega_u) = 0$):

$$\phi_m = \arg G(j\omega_u) + \pi,$$

or, given the plant:

$$\phi_m = -arctg\frac{\omega_u}{\omega_m} - arctg\frac{\omega_u}{\omega_e} + \pi,$$

or indeed, numerically:

$$\phi_m = -arctg\frac{500}{50} - arctg\frac{500}{250} + \pi,$$

therefore:

$$\phi_m^{\circ} = -84.29 - 63.43 + 180 = 32.28.$$

The value thus obtained is not suitable as it does not satisfy the relevant recommendation, namely:

$$50 \leq \phi_m^{\circ} \leq 55\,.$$

3) Regulator phase at frequency, ω_u

As

$$\phi_m = \arg\beta(j\omega_u) + \pi\,,$$

$$\phi_m = \arg C(j\omega_u) + \arg G(j\omega_u) + \pi\,,$$

from which we obtain:

$$\arg C(j\omega_u) = \phi_m - \arg G(j\omega_u) - \pi\,,$$

that is, numerically:

$$\arg^{\circ} C(j\omega_u) = 54 + (84.29 + 63.43) - 180 = 21.72\,,$$

a result that can be confirmed by a more direct approach, which is written as:

$$\arg C(j\omega_u) = 54^{\circ} - 32.28^{\circ} = 21.72^{\circ}\,.$$

4) Synthesis of the simple phase-lead controller

4.1) *Transitional frequencies*

As

$$\omega_u = \sqrt{\omega_1 \omega_2} ,$$

$$\arg C(j\omega_u) = \varphi_m = \arcsin \frac{\alpha - 1}{\alpha + 1} \text{ with } \alpha = \frac{\omega_2}{\omega_1} > 1 ,$$

from which we obtain:

$$\sin \varphi_m = \frac{\alpha - 1}{\alpha + 1} ,$$

or:

$$\alpha - 1 = (\alpha + 1) \sin \varphi_m ,$$

or:

$$\alpha - 1 = \alpha \sin \varphi_m + \sin \varphi_m ,$$

or indeed:

$$\alpha (1 - \sin \varphi_m) = 1 + \sin \varphi_m ,$$

therefore:

$$\alpha = \frac{1 + \sin \varphi_m}{1 - \sin \varphi_m} = \frac{1 + \sin 21.27^\circ}{1 - \sin 21.27^\circ} = \frac{1 + 0.37}{1 - 0.37} = 2.17 .$$

Distributed geometrically with respect to the frequency, ω_u, the transitional frequencies, ω_1 and ω_2, admit the following expressions:

$$\omega_1 = \frac{\omega_u}{\sqrt{\alpha}} - \frac{500}{\sqrt{2.17}} = 339.44 \; rd / s$$

and

$$\omega_2 = \omega_u \sqrt{\alpha} = 500\sqrt{2.17} = 736.5 \; rd \, / \, s \, .$$

4.2) *Static gain*

As

$$|\beta(j\omega_u)| = 1 \, ,$$

we can write:

$$|C(j\omega_u)||G(j\omega_u)| = 1 \, ,$$

therefore:

$$C_0 \frac{\sqrt{1+\left(\dfrac{\omega_u}{\omega_1}\right)^2}}{\sqrt{1+\left(\dfrac{\omega_u}{\omega_2}\right)^2}} |G(j\omega_u)| = 1 \, ,$$

from which we obtain:

$$C_0 = \frac{1}{|G(j\omega_u)|} \frac{\sqrt{1+\left(\dfrac{\omega_u}{\omega_2}\right)^2}}{\sqrt{1+\left(\dfrac{\omega_u}{\omega_1}\right)^2}} \, ,$$

or, given the expression of $|G(j\omega_u)|$, that is,

$$|G(j\omega_u)| = \frac{G_0}{\sqrt{1+\left(\dfrac{\omega_u}{\omega_m}\right)^2}\sqrt{1+\left(\dfrac{\omega_u}{\omega_e}\right)^2}} :$$

$$C_0 = \frac{\sqrt{1+\left(\frac{\omega_u}{\omega_m}\right)^2}\sqrt{1+\left(\frac{\omega_u}{\omega_e}\right)^2}\sqrt{1+\left(\frac{\omega_u}{\omega_2}\right)^2}}{G_0\sqrt{1+\left(\frac{\omega_u}{\omega_1}\right)^2}},$$

or indeed, numerically:

$$C_0 = \frac{\sqrt{1+100}\sqrt{1+4}\sqrt{1+0.46}}{50\sqrt{1+2.17}} = 0.3 .$$

5) High-frequency input sensitivity

With the high-frequency input sensitivity being reduced to the regulator gain at these frequencies, that is,

$$S(\omega)_{HF} = \left|C(j\omega)\right|_{HF} = C_0\frac{\omega_2}{\omega_1} = C_0\alpha ,$$

we can write, numerically:

$$S(\omega)_{HF} = (0.3)(2.17) = 0.65 .$$

To improve the high-frequency input sensitivity through its decrease, it is appropriate to integrate a low-pass filter (in cascade) into the regulator, whose transparency at the frequency, ω_u, is ensured by a transitional frequency, ω_u, significantly higher than ω_u, that is:

$$PB(s) = \frac{1}{1+\frac{s}{\omega_h}} \text{,with } \omega_h >> \omega_u ,$$

for a decrease in sensitivity of 6 dB/oct, and

$$PB(s) = \left(\frac{1}{1+\frac{s}{\omega_h}}\right)^2 \text{,with } \omega_h >> \omega_u ,$$

for a decrease in sensitivity of 12 dB/oct.

6) Transformation of the regulator into a PID regulator

To transform the simple phase-lead controller (which may or not be filtered by a low-pass filter) into a non-academic PID regulator (whose differential action is truncated at high frequency), it is appropriate for it to be completed by a proportional-integral controller (in cascade), whose transparency at frequency, ω_u, is ensured by a transitional frequency, ω_b significantly lower than ω_u, that is:

$$PI(s) = 1 + \frac{\omega_b}{s} = \frac{\omega_b}{s}\left(1 + \frac{s}{\omega_b}\right) \text{with } \omega_b \ll \omega_u \,.$$

7) New answers for a new plant

7.1) *Justification of the choice of frequency,* ω_u

Given that now,

$$\omega_m = 5 \; rd \,/\, s \text{ and } \omega_e = 250 \; rd \,/\, s \,,$$

ω_e then being 50 times greater than ω_m, the plant cut-off frequency at –3 dB, ω_c, identifies, in a first approximation, with ω_m, that is

$$\omega_c = \omega_m = 5 \; rd \,/\, s \,.$$

Knowing that $\omega_u = 10\omega_m = 50 \; rd \,/\, s$, the $\omega_u / \omega_c = 50/5 = 10$ ratio satisfies the relevant recommendation, that is,

$$5\omega_c \le \omega_u \le 20\omega_c \,,$$

Since $\omega_u = 10\omega_c$ is situated geometrically in the middle of the interval, $\left[5\omega_c, 20\omega_c\right]$, with ω_u indeed the geometric mean of the interval terminals.

7.2) *Phase margin in the absence of phase correction*

The phase margin is written as:

$$\phi_m = \arg \beta\left(j\omega_u\right) + \pi \,,$$

or, since $\arg C(j\omega_u) = 0$:

$$\phi_m = \arg G(j\omega_u) + \pi ,$$

or, given the plant:

$$\phi_m = -arctg\frac{\omega_u}{\omega_m} - arctg\frac{\omega_u}{\omega_e} + \pi ,$$

or indeed, numerically:

$$\phi_m = -arctg\frac{50}{5} - arctg\frac{50}{250} + \pi = -arctg10 - arctg\frac{1}{5} + \pi ,$$

therefore:

$$\phi_m^{\circ} = -84.29 - 11.31 + 180 = 84.40 .$$

7.3) *Regulator phase at the frequency,* ω_u

Given that the regulator must make the phase margin change from 84.40° to 54°, it must present a phase at the frequency, ω_u, such that:

$$\arg C(j\omega_u) = 54° - 84.40° = -30.40° ,$$

in other words, the regulator must provide a maximum phase lag, φ_m, at the frequency, ω_u, that is,

$$\varphi_m = -\arg C(j\omega_u) = 30.40° .$$

7.4) *Synthesis of the simple phase-lag controller*

Knowing that the maximum phase lag, φ_m, of the simple phase-lag controller is given by:

$$\varphi_m = \arcsin\frac{\alpha - 1}{\alpha + 1} \text{ with } \alpha = \frac{\omega_1}{\omega_2} > 1 ,$$

we can write:

$$\sin\varphi_m = \frac{\alpha - 1}{\alpha + 1},$$

from which we obtain:

$$\alpha = \frac{1 + \sin\varphi_m}{1 - \sin\varphi_m},$$

that is, numerically:

$$\alpha = \frac{1 + \sin 30.40}{1 - \sin 30.40} = \frac{1 + 0.506}{1 - 0.506} = 3.048\,.$$

The transitional frequencies, ω_1 and ω_2, are then given by:

$$\omega_1 = \omega_u\sqrt{\alpha} = 50\sqrt{3.048} = 87.292$$

and

$$\omega_2 = \frac{\omega_u}{\sqrt{\alpha}} = \frac{50}{\sqrt{3.048}} = 28.639\,.$$

As for the static gain, C_0, as already established in literal terms for the simple phase-lead controller, which is of identical structure, is given by:

$$C_0 = \frac{\sqrt{1 + \left(\frac{\omega_u}{\omega_m}\right)^2}\sqrt{1 + \left(\frac{\omega_u}{\omega_e}\right)^2}\sqrt{1 + \left(\frac{\omega_u}{\omega_2}\right)^2}}{G_0\sqrt{1 + \left(\frac{\omega_u}{\omega_1}\right)^2}},$$

or, numerically:

$$C_0 = \frac{\sqrt{1 + \left(\frac{50}{5}\right)^2}\sqrt{1 + \left(\frac{50}{250}\right)^2}\sqrt{1 + \left(\frac{50}{28.639}\right)^2}}{50\sqrt{1 + \left(\frac{50}{87.292}\right)^2}},$$

therefore:

$$C_0 = 0.357 .$$

7.5) *High-frequency input sensitivity*

Reduced to the high-frequency regulator gain, the input sensitivity at these frequencies is given by:

$$S(\omega)_{HF} = \left|C(j\omega)\right|_{HF} = C_0 \frac{\omega_2}{\omega_1} = \frac{C_0}{\alpha},$$

or, numerically,

$$S(\omega)_{HF} = \frac{0.357}{3.048} = 0.117 .$$

8) Complexification of the regulator

To improve both the high-frequency input sensitivity and the low-frequency precision, it is appropriate to integrate into the regulator:

– a low-pass filter (in cascade), whose transparency at the frequency, ω_u, is ensured by a transitional frequency, ω_h, that is large in relation to ω_u, that is,

$$PB(s) = \frac{1}{1 + \dfrac{s}{\omega_h}} \text{ with } \omega_h >> \omega_u$$

– a proportional-integral controller (in cascade), whose transparency at the frequency, ω_u, is ensured by a transitional frequency, ω_b, that is small in relation to ω_u, that is,

$$PI(s) = 1 + \frac{\omega_b}{s} = \frac{\omega_b}{s}\left(1 + \frac{s}{\omega_b}\right) \text{ with } \omega_b << \omega_u .$$

9) On the nature of the regulator

The answer to the question "Does the regulator thus obtained produce a PID regulator per se?" is negative, since (asymptotically) the regulator does not present any differential action, but, strictly, two proportional actions and three integral actions.

Presentation of Problem 2

Improvement of the Elementary Synthesis of a PID Regulator by Reducing Transitional Frequency Dispersion

Let there be a second-order plant defined by a transmittance of the form:

$$G(s) = \frac{\omega_0'}{s\left(1 + \frac{s}{\omega_0}\right)},$$

with $\omega_0' = 5$ rd/s and $\omega_0 = 10$ rd/s.

This transmittance can be interpreted as the model of a continuous-current motor whose mechanical transitional frequency, ω_m, is ω_0 here, the electrical transitional frequency of which, ω_e, is not taken into account as it is assumed to be very large in relation to $\omega_m = \omega_0$, and for which the output sconsidered is the *angular position (integral of the angular speed)*, hence the presence of the integration term, ω_0'/s, with ω_0' denoting the transition frequency linked to this integration term.

1) In the absence of phase correction, with the action of the regulator therefore being uniquely proportional, determine the control phase margin at the unit gain frequency in open loop, ω_u, set, by the designer, at five times the transitional frequency, ω_0, of the plant, that is: $\omega_u = 5\omega_0$.

2) At the same frequency, ω_u, what phase must be presented by the regulator to substitute the proportional in order to ensure a phase margin of 51.31°?

3) For the same plant and the same frequency, ω_u, synthesize a simple phase-lead controller that provides at ω_u a maximum phase lead of 40°.

4) Given the C_0, ω_1 and ω_2 parameters of the simple phase-lead controller, as synthesized in response to Question 3, determine the literal expression and the numerical value of the high-frequency input sensitivity.

5) How can the regulator thus synthesized be completed to transform it into a well-known regulator (a classic in its field), with the function of simultaneously improving the control's stability degree and precision at low frequency? We will use ω_b ($\omega_b << \omega_u$) to denote the low transitional frequency introduced by this transformation.

6) How can the regulator obtained in response to Question 5 be completed to improve the high-frequency input immunity (considering that improving immunity consists of increasing, in contrast with sensitivity)? We will use ω_h ($\omega_h >> \omega_u$) to denote the high transitional frequency introduced by this transformation.

7) In order to reduce the dispersion of the transitional frequencies of the regulator (a dispersion that can prove prejudicial to the stability of the regulator installed numerically), it is appropriate to reduce the gap between the transitional frequencies, ω_b and ω_h, introduced by the complements of the simple phase-lead controller, notably by taking $\omega_b = \omega_u / 7$ and $\omega_h = 7\omega_u$. These complements then lose all their transparency at the frequency, ω_u, which requires the regulator to be synthesized again, taking into account, from the start, the complements obtained in response to Questions 5 and 6. For the same phase margin and the same unit gain frequency in open loop, synthesize the regulator, whose transitional frequencies, ω_1 and ω_2, still satisfy $(\omega_1\omega_2)^{1/2} = \omega_u$ (as the transitional frequencies, ω_b and ω_h, satisfy $(\omega_b\omega_h)^{1/2} = \omega_u$ here).

Solution to Problem 2

1) Phase margin in the absence of phase correction

The phase margin is written, by definition:

$$\Phi_m = \arg \beta(j\omega_u) + \pi \,,$$

that is:

$$\Phi_m = \arg C(j\omega_u) + \arg G(j\omega_u) + \pi ,$$

or, knowing that $\arg C(j\omega_u) = 0$:

$$\Phi_m = \arg G(j\omega_u) + \pi ,$$

or, given the plant phase at the frequency, ω_u:

$$\Phi_m = -\frac{\pi}{2} - \arctan\frac{\omega_u}{\omega_0} + \pi = \frac{\pi}{2} - \arctan\frac{\omega_u}{\omega_0} ,$$

or indeed, numerically:

$$\Phi_m^0 = 90 - \arctan 5 = 90 - 78.69 = 11.31 .$$

2) Regulator phase at the frequency, ω_u

Given that the regulator must make the phase margin change from 11.31° to 51.31°, it must bring a maximum phase lead, φ_m, at the frequency, ω_u, that is, $\varphi_m = \arg C(j\omega_u)$, such that:

$$\varphi_m = \arg C(j\omega_u) = 51.31° - 11.31° = 40° .$$

3) Synthesis of the simple phase-lead controller

The specification (in Question 3) of a maximum phase lead of 40° of the regulator at the frequency, ω_u, simply takes up the answer to Question 2. In other words, it is of no use a priori, but it is of use a posteriori should the response given to Question 2 be incorrect. In fact, such a specification provides the best of help to the student in solving the rest of the problem.

Knowing that the maximum phase lead, φ_m, of the simple phase-lead controller is given by

$$\varphi_m = \arcsin\frac{\alpha - 1}{\alpha + 1} \text{ with } \alpha = \frac{\omega_2}{\omega_1} > 1 ,$$

we can write:

$$\sin\varphi_m = \frac{\alpha-1}{\alpha+1},$$

from which we obtain:

$$\alpha = \frac{1+\sin\varphi_m}{1-\sin\varphi_m},$$

that is, numerically:

$$\alpha = \frac{1+\sin 40°}{1-\sin 40°} = \frac{1+0.6427}{1-0.6427} = 4.597\,.$$

Distributed geometrically with respect to the frequency, ω_u, the transitional frequencies, ω_1 and ω_2, can then be determined in accordance with:

$$\omega_1 = \frac{\omega_u}{\sqrt{\alpha}} = \frac{50}{\sqrt{4.597}} = 23.320 \text{ rd/s}$$

and

$$\omega_2 = \omega_u\sqrt{\alpha} = 50\sqrt{4.597} = 107.203 \text{ rd/s}\,.$$

The static gain, C_0, can be determined from the relation, $|\beta(j\omega_u)| = 1$, through which the frequency, ω_u, is defined. Indeed, this relation enables us to write:

$$|C(j\omega_u)||G(j\omega_u)| = 1\,,$$

or, in this case:

$$C_0\left(\frac{1+\left(\dfrac{\omega_u}{\omega_1}\right)^2}{1+\left(\dfrac{\omega_u}{\omega_2}\right)^2}\right)^{\frac{1}{2}} \frac{\omega_0{}'}{\omega_u\left(1+\left(\dfrac{\omega_u}{\omega_0}\right)^2\right)^{\frac{1}{2}}} = 1\,,$$

from which we obtain:

$$C_0 = \left(\frac{1+\left(\frac{\omega_u}{\omega_2}\right)^2}{1+\left(\frac{\omega_u}{\omega_1}\right)^2} \right)^{\frac{1}{2}} \frac{\omega_u}{\omega_0{}'} \left(1+\left(\frac{\omega_u}{\omega_0}\right)^2\right)^{\frac{1}{2}},$$

that is, numerically:

$$C_0 = \left(\frac{1+\left(\frac{50}{107.203}\right)^2}{1+\left(\frac{50}{23.320}\right)^2} \right)^{\frac{1}{2}} 10(1+25)^{\frac{1}{2}} = 23.761.$$

4) High-frequency input sensitivity

The high-frequency input sensitivity is reduced to the regulator gain at these frequencies, that is:

$$S(\omega)_{HF} = \left|C(j\omega)\right|_{HF} = C_0 \frac{\omega_2}{\omega_1} = C_0 \alpha,$$

or, numerically:

$$S(\omega)_{HF} = (23.761)(4.597) = 109.23.$$

5) First complement of the controller

The first complement of the simple phase-lead controller is as follows:

$$C(s) = PI(s) C_0 \frac{1+\frac{s}{\omega_1}}{1+\frac{s}{\omega_2}},$$

where:

$$PI(s) = \frac{\omega_b}{s} + 1 \text{ with } \omega_b << \omega_u .$$

Thus defined, the first complement, $PI(s)$, is none other than a proportional-integral controller, of transitional frequency, ω_b, transparent at the frequency, ω_u (since $\omega_b << \omega_u$).

6) Second complement of the controller

The second complement of the simple phase-lead controller is as follows:

$$C(s) = PI(s)\, C_0 \frac{1+\dfrac{s}{\omega_1}}{1+\dfrac{s}{\omega_2}}\, PB(s),$$

where:

$$PB(s) = \frac{1}{1+\dfrac{s}{\omega_h}} \text{ with } \omega_h >> \omega_u .$$

Thus defined, the second complement, $PB(s)$, is none other than a first-order low-pass filter, of transitional frequency, ω_h, transparent at the frequency, ω_u (since $\omega_h >> \omega_u$).

7) Synthesis of the complete regulator

The frequency response of the complete regulator is of the form:

$$C(j\omega) = \left(\frac{\omega_b}{j\omega} + 1\right) C_0 \frac{1+j\dfrac{\omega}{\omega_1}}{1+j\dfrac{\omega}{\omega_2}} \frac{1}{1+j\dfrac{\omega}{\omega_h}},$$

that is, at the frequency, ω_u:

$$C\left(j\omega_u\right) - \left(\frac{\omega_b}{j\omega_u}+1\right)C_0\frac{1+j\dfrac{\omega_u}{\omega_1}}{1+j\dfrac{\omega_u}{\omega_2}}\frac{1}{1+j\dfrac{\omega_u}{\omega_h}},$$

with

$$\left(\omega_1\omega_2\right)^{1/2} = \omega_u \text{ and } \left(\omega_b\omega_h\right)^{1/2} = \omega_u .$$

Let us formulate the term relating to the complements introduced in response to Questions 5 and 6, that is:

$$\frac{1+\dfrac{\omega_b}{j\omega_u}}{1+j\dfrac{\omega_u}{\omega_h}} = \frac{1-j\dfrac{\omega_b}{\omega_u}}{1+j\dfrac{\omega_u}{\omega_h}} = \rho\left(\omega_u\right)e^{j\theta\left(\omega_u\right)},$$

with:

$$\rho(\omega_u) = \left(\frac{1+\left(\dfrac{\omega_b}{\omega_u}\right)^2}{1+\left(\dfrac{\omega_u}{\omega_h}\right)^2}\right)^{\frac{1}{2}} \text{ and}$$

$$\theta(\omega_u) = -\arctan\left(\frac{\omega_b}{\omega_u}\right) - \arctan\left(\frac{\omega_u}{\omega_h}\right),$$

or, considering that $\dfrac{\omega_b}{\omega_u} = \dfrac{\omega_u}{\omega_h} = \dfrac{1}{7}$:

$$\rho(\omega_u) = 1$$

and

$$\theta(\omega_u) = -2\arctan\left(\frac{1}{7}\right) = -16.26° .$$

The modulus and argument of $C(j\omega_u)$ can then be written as:

$$\left|C(j\omega_u)\right| = C_0\left(\frac{1+\left(\dfrac{\omega_u}{\omega_1}\right)^2}{1+\left(\dfrac{\omega_u}{\omega_2}\right)^2}\right)^{\frac{1}{2}}$$

and

$$\arg C(j\omega_u) = \varphi_m + \theta(\omega_u) ,$$

with φ_m given by:

$$\varphi_m = arc\sin\frac{\alpha-1}{\alpha+1} \text{ with } \alpha = \frac{\omega_2}{\omega_1} > 1 .$$

The phase margin, Φ_m, which is written, by definition,

$$\Phi_m = \arg \beta(j\omega_u) + \pi ,$$

that is,

$$\Phi_m = \arg C(j\omega_u) + \arg G(j\omega_u) + \pi ,$$

can then be particularized in accordance with:

$$\Phi_m = \varphi_m + \theta(\omega_u) + \arg G(j\omega_u) + \pi ,$$

from which we obtain φ_m:

$$\varphi_m = \left(\Phi_m - \arg G(j\omega_u) - \pi\right) - \theta(\omega_u),$$

that is, numerically:

$$\varphi_m = 40° + 16.26° = 56.26°,$$

a result that expresses that the maximum phase lead, φ_m, brought by the simple phase-lead controller is then greater than that preceding it, namely 40°, to compensate for the phase lags brought at the frequency, ω_u, by the complements, $PI(s)$ and $PB(s)$ (with these lags 16.26° overall).

Knowing that

$$\varphi_m = arc\sin\frac{\alpha - 1}{\alpha + 1} \text{ with } \alpha = \frac{\omega_2}{\omega_1} > 1,$$

it is possible to determine α as a function of φ_m, that is:

$$\alpha = \frac{1 + \sin\varphi_m}{1 - \sin\varphi_m},$$

or, numerically:

$$\alpha = \frac{1 + \sin 56.26°}{1 - \sin 56.26°} = \frac{1 + 0.8315}{1 - 0.8315} = 10.869.$$

Given that the transitional frequencies, ω_1 and ω_2 are always distributed geometrically with respect to the frequency, ω_u, they admit the following expressions, respectively:

$$\omega_1 = \frac{\omega_u}{\sqrt{\alpha}} = \frac{50}{\sqrt{10.869}} = 15.166 \text{ rd/s}$$

and

$$\omega_2 = \omega_u\sqrt{\alpha} = 50\sqrt{10.869} = 164.840 \text{ rd/s}.$$

Concerning the set of transitional frequencies of the regulator, it is appropriate to express the numerical values of the transitional frequencies, ω_b and ω_h, such as presented in Question 7:

$$\omega_b = \frac{\omega_u}{7} = \frac{50}{7} = 7.143 \text{ rd/s}$$

and

$$\omega_h = 7\omega_u = 7(50) = 350 \text{ rd/s}$$

The static gain, C_0, is determined from $|\beta(j\omega_u)| = 1$, in accordance with:

$$|C(j\omega_u)||G(j\omega_u)| = 1,$$

that is, in this case:

$$C_0 \left(\frac{1+\left(\frac{\omega_u}{\omega_1}\right)^2}{1+\left(\frac{\omega_u}{\omega_2}\right)^2} \right)^{\frac{1}{2}} \frac{\omega_0{}'}{\omega_u \left(1+\left(\frac{\omega_u}{\omega_0}\right)^2\right)^{\frac{1}{2}}} = 1,$$

from which we obtain:

$$C_0 = \left(\frac{1+\left(\frac{\omega_u}{\omega_2}\right)^2}{1+\left(\frac{\omega_u}{\omega_1}\right)^2} \right)^{\frac{1}{2}} \frac{\omega_u}{\omega_0{}'} \left(1+\left(\frac{\omega_u}{\omega_0}\right)^2\right)^{\frac{1}{2}},$$

or, numerically:

$$C_0 = \left(\frac{1+\left(\frac{50}{164.840}\right)^2}{1+\left(\frac{50}{15.166}\right)^2} \right)^{\frac{1}{2}} 10(1+25)^{\frac{1}{2}} = 15.466\,.$$

Presentation of Problem 3

Synthesis of a PID Regulator Based on Three Phase-Lead Controller Structures: Comparison and Choice of the Best Structure

The study plant that forms the subject of this problem is achieved by a continuous-current motor, for which the output considered is the angular position. Taking account of its electrical time constant, it is described by a third-order transmittance of the form:

$$G(s) = \frac{1}{\dfrac{s}{\omega_0}\left(1+\dfrac{s}{\omega_m}\right)\left(1+\dfrac{s}{\omega_e}\right)},$$

where ω_0 denotes the transition frequency linked to integration, ω_m denotes the mechanical transitional frequency and ω_e the electrical transitional frequency.

In the absence of a mechanical charge (motor idle), the transmittance parameters admit the following numerical values:

$$\omega_0 = 20\ rd/s\,;\ \omega_m = 50\ rd/s\,;\ \omega_e = 250\ rd/s\,.$$

1) In order for the dynamics of the motor control to be sufficiently fast, the designer sets the unit gain frequency in open loop, ω_u, at a valuc cqualing that of the electrical transitional frequency of the motor, that is, $\omega_u = \omega_e = 250\ rd/s$. Justify this choice.

2) In the absence of phase correction, the regulator then being reduced to a simple proportional, determine the phase margin, ϕ_m, of the control. Would you say its value seems suitable?

3) At the same frequency, ω_u, what phase must be presented by the regulator that is to substitute the proportional in order to ensure a phase margin of 46.31° (this particular value being justified by the resulting simplification in the numerical calculations)?

4) The designer wonders which regulator to choose, hesitating in particular between:

– the structure of a simple phase-lead controller, defined by:

$$C(s) = C_0 \frac{1+\dfrac{s}{\omega_1}}{1+\dfrac{s}{\omega_2}}$$

– that of a double phase-lead controller, defined by:

$$C(s) = C_0 \left(\frac{1+\dfrac{s}{\omega_1}}{1+\dfrac{s}{\omega_2}} \right)^2$$

– finally, that of a triple phase-lead controller, defined by:

$$C(s) = C_0 \left(\frac{1+\dfrac{s}{\omega_1}}{1+\dfrac{s}{\omega_2}} \right)^3 .$$

for each of the three structures considered, determine the parameters of the regulator ensuring a phase margin of 46.31° at the frequency, $\omega_u = 250\ rd/s$.

5) Which structure ensures the lowest dispersion of the regulator's transitional frequencies?

6) Which structure ensures the best precision at low frequency?

7) Which structure ensures the best input immunity at high frequency?

8) In conclusion, given the answers to Questions 5–7, which of the three structures should the designer retain?

9) Determine the static gain in open loop and deduce the step error.

10) Suggest a complexification of the regulator aiming to cancel the ramp error.

11) Suggest a complexification of the regulator so as to ensure that the high-frequency input sensitivity decreases by 12 dB/oct.

Solution to Problem 3

1) Justification of the choice of frequency ω_u

Given that the high transitional frequency, ω_e, is not very large in relation to the low transitional frequency, ω_m (since ω_e has a value of only $5\omega_m$), the effect of ω_e on the cut-off frequency, ω_c, cannot be overlooked. Thus, ω_c is slightly less than ω_m, that is,

$$\omega_c = \omega_m^- = 50^- \ rd/s.$$

Considering that $\omega_u = 250 \ rd/s$,

$$\frac{\omega_u}{\omega_c} = \frac{250}{50^-} = 5^+,$$

a ratio in accordance with the recommendation in the field, that is,

$$5\omega_c \leq \omega_u \leq 20\omega_c.$$

2) Phase margin in the absence of phase correction

As the phase margin is defined by

$$\phi_m = \arg \beta(j\omega_u) + \pi,$$

the nullity of $\arg C(j\omega_u)$ enables ϕ_m to be reduced to:

$$\phi_m = \arg G(j\omega_u) + \pi ,$$

or, given the plant phase:

$$\phi_m = -\frac{\pi}{2} - arctg\frac{\omega_u}{\omega_m} - arctg\frac{\omega_u}{\omega_e} + \pi ,$$

or indeed, numerically:

$$\phi_m = \frac{\pi}{2} - arctg\ 5 - arctg\ 1 ,$$

therefore:

$$\phi_m^{\circ} = 90 - 78.69 - 45 = -33.69 .$$

The phase margin being negative, it is of course unacceptable.

3) Regulator phase at frequency ω_u

Considering that the regulator must make the phase margin change from –33.69° to 46.31°, it must bring a maximum phase lead, φ_m, at the frequency, ω_u, that is, $\varphi_m = \arg C(j\omega_u)$, such that:

$$\varphi_m = \arg C(j\omega_u) = 46.31^{\circ} - (-33.69^{\circ}) = 80^{\circ} .$$

4) Synthesis of the regulator

4.1) *Simple phase-lead controller*

As $\omega_u = (\omega_1\omega_2)^{1/2}$,

$$\arg C(j\omega_u) = \varphi_m = \arcsin\frac{\alpha - 1}{\alpha + 1} \text{ with } \alpha = \frac{\omega_2}{\omega_1} > 1 ,$$

from which we obtain:

$$\sin \varphi_m = \frac{\alpha - 1}{\alpha + 1},$$

therefore:

$$\alpha = \frac{1 + \sin \varphi_m}{1 - \sin \varphi_m} = \frac{1 + \sin 80^\circ}{1 - \sin 80^\circ} = \frac{1 + 0.9848}{1 - 0.9848} = 130.57 .$$

Distributed geometrically with respect to the frequency, ω_u, the transitional frequencies, ω_1 and ω_2, are expressed, respectively, by:

$$\omega_1 = \frac{\omega_u}{\sqrt{\alpha}} = \frac{250}{11.426} = 21.88 \; rd \,/\, s$$

and

$$\omega_2 = \omega_u \sqrt{\alpha} = 250(11.426) = 2856.5 \; rd \,/\, s .$$

As

$$|\beta(j\omega_u)| = 1,$$

we can write:

$$|C(j\omega_u)||G(j\omega_u)| = 1,$$

or, given the literal expression of $C(j\omega)$:

$$C_0 \left(\frac{1 + \left(\dfrac{\omega_u}{\omega_1} \right)^2}{1 + \left(\dfrac{\omega_u}{\omega_2} \right)^2} \right)^{1/2} |G(j\omega_u)| = 1,$$

from which we obtain:

$$C_0 = \frac{1}{|G(j\omega_u)|}\left(\frac{1+\left(\frac{\omega_u}{\omega_2}\right)^2}{1+\left(\frac{\omega_u}{\omega_1}\right)^2}\right)^{1/2},$$

that is, given the numerical values of ω_u, ω_0, ω_m, ω_e, ω_1 and ω_2:

$$C_0 = 7.854\,.$$

4.2) *Double phase-lead controller*

As $\omega_u = (\omega_1\omega_2)^{1/2}$,

$$\arg C(j\omega_u) = \varphi_m = 2\arcsin\frac{\alpha-1}{\alpha+1} \text{ with } \alpha = \frac{\omega_2}{\omega_1} > 1,$$

from which we obtain:

$$\frac{\varphi_m}{2} = \arcsin\frac{\alpha-1}{\alpha+1},$$

or:

$$\sin\left(\frac{\varphi_m}{2}\right) = \frac{\alpha-1}{\alpha+1},$$

therefore:

$$\alpha = \frac{1+\sin\left(\frac{\varphi_m}{2}\right)}{1-\sin\left(\frac{\varphi_m}{2}\right)} = \frac{1+\sin 40°}{1-\sin 40°} = \frac{1+0.6428}{1-0.6428} \simeq 4.6\,.$$

Distributed geometrically with respect to the frequency, ω_u, the frequencies, ω_1 and ω_2, are written, respectively:

$$\omega_1 = \frac{\omega_u}{\sqrt{\alpha}} = \frac{250}{2.144} = 116.60 \; rd/s$$

and

$$\omega_2 = \omega_u \sqrt{\alpha} = 250(2.144) = 536 \; rd/s .$$

As

$$|\beta(j\omega_u)| = 1 ,$$

we can write:

$$|C(j\omega_u)||G(j\omega_u)| = 1 ,$$

or, given the literal expression of $C(j\omega)$:

$$C_0 \frac{1+\left(\dfrac{\omega_u}{\omega_1}\right)^2}{1+\left(\dfrac{\omega_u}{\omega_2}\right)^2} |G(j\omega_u)| = 1 ,$$

from which we obtain:

$$C_0 = \frac{1}{|G(j\omega_u)|} \frac{1+\left(\dfrac{\omega_u}{\omega_2}\right)^2}{1+\left(\dfrac{\omega_u}{\omega_1}\right)^2} ,$$

that is, given the numerical values of ω_u, ω_0, ω_m, ω_e, ω_1 and ω_2:

$$C_0 = 19.584 .$$

4.3) *Triple phase-lead controller*

As $\omega_u = (\omega_1 \omega_2)^{1/2}$,

$$\arg C(j\omega_u) = \varphi_m = 3\arcsin\frac{\alpha - 1}{\alpha + 1} \text{ with } \alpha = \frac{\omega_2}{\omega_1} > 1,$$

from which we obtain:

$$\frac{\varphi_m}{3} = \arcsin\frac{\alpha - 1}{\alpha + 1},$$

or:

$$\sin\left(\frac{\varphi_m}{3}\right) = \frac{\alpha - 1}{\alpha + 1},$$

therefore:

$$\alpha = \frac{1 + \sin\left(\frac{\varphi_m}{3}\right)}{1 - \sin\left(\frac{\varphi_m}{3}\right)} = \frac{1 + \sin 26.66°}{1 - \sin 26.66°} = \frac{1 + 0.4488}{1 - 0.4488} = 2.628 .$$

Distributed geometrically with respect to the frequency, ω_u, the frequencies, ω_1 and ω_2, are written, respectively:

$$\omega_1 = \frac{\omega_u}{\sqrt{\alpha}} = \frac{250}{1.621} = 154.22 \; rd / s$$

and

$$\omega_2 = \omega_u \sqrt{\alpha} = 250(1.621) = 405.25 \; rd / s .$$

As

$$|\beta(j\omega_u)| = 1,$$

we can write:

$$\left|C(j\omega_u)\right|\left|G(j\omega_u)\right| = 1,$$

or, given the literal expression of $C(j\omega)$:

$$C_0\left(\frac{1+\left(\dfrac{\omega_u}{\omega_1}\right)^2}{1+\left(\dfrac{\omega_u}{\omega_2}\right)^2}\right)^{3/2}\left|G(j\omega_u)\right| = 1,$$

from which we obtain:

$$C_0 = \frac{1}{\left|G(j\omega_u)\right|}\left(\frac{1+\left(\dfrac{\omega_u}{\omega_2}\right)^2}{1+\left(\dfrac{\omega_u}{\omega_1}\right)^2}\right)^{3/2},$$

that is, given the numerical values of ω_u, ω_0, ω_m, ω_e, ω_1 and ω_2:

$$C_0 = 21.135.$$

5) Dispersion of the regulator transitional frequencies

Considering that the $\alpha = \omega_2/\omega_1$ ratio is worth, successively, 130.57 for the simple phase-lead controller, 4.6 for the double and 2.628 for the triple, it is therefore the structure of the triple phase-lead controller that ensures the lowest dispersion of the regulator transitional frequencies.

6) Precision at low frequency

Knowing that the low-frequency gain asymptote of the regulator is defined by the gain, C_0, which has a value, successively, of 7.854 for the simple phase-lead controller, 19.584 for the double and 21.135 for the triple, it is the structure of the triple phase-lead controller that ensures the best low-frequency precision of the control.

7) High-frequency input immunity

The high-frequency input sensitivity, that is,

$$S(\omega)_{HF} = |C(j\omega)|_{HF},$$

is expressed, successively, by:

$$S(\omega)_{HF} = C_0 \frac{\omega_2}{\omega_1} = C_0 \alpha = 7.854(130.57) = 1025.5$$

for the first structure;

$$S(\omega)_{HF} = C_0 \left(\frac{\omega_2}{\omega_1}\right)^2 = C_0 \alpha^2 = 19.584(4.6)^2 = 414.4$$

for the second structure;

$$S(\omega)_{HF} = C_0 \left(\frac{\omega_2}{\omega_1}\right)^3 = C_0 \alpha^3 = 21.135(2.628)^3 = 383.6$$

for the third structure.

This set of results shows that the structure of the triple phase-lead controller ensures the lowest input sensitivity at high frequencies, and therefore the best input immunity at these frequencies.

8) Structure to be retained

In light of the answers to Questions 5–7, which each prove to favor the third structure, the designer will need to retain this structure or, otherwise, the second, which gives fairly similar results.

9) Static gain in open loop and step error

The static gain in open loop is given by:

$$|\beta(j0)| = |C(j0)||G(j0)|,$$

that is, considering that:

$$|C(j0)| = C_0 \text{ and } |G(j0)| = \infty:$$

$$|\beta(j0)| = \infty.$$

The step error, inversely proportional to $|1 + \beta(j0)|$, is therefore nil.

10) Canceling the ramp error

A complexification of the regulator aiming to cancel the ramp error consists of integrating into it a proportional-integrator (in cascade), whose transparency at the frequency, ω_u, is ensured by a transitional frequency, ω_b, significantly below ω_u, that is:

$$PI(s) = 1 + \frac{\omega_b}{s} = \frac{\omega_b}{s}\left(1 + \frac{s}{\omega_b}\right) \text{ with } \omega_b << \omega_u.$$

11) Frequency decrease in high-frequency input sensitivity

A complexification of the regulator so as to ensure a decrease in the high-frequency input sensitivity of 12 dB/oct, consists of integrating into it (in cascade) a second-order low-pass filter, whose transparency at the frequency ω_u, is ensured by a transitional frequency, ω_h, significantly above ω_u, that is:

$$PB(s) = \left(\frac{1}{1 + \frac{s}{\omega_h}}\right)^2 \text{ with } \omega_h >> \omega_u.$$

Presentation of Problem 4

Linearizing Control of a Motor Shaft: Linearization by Immersion

This problem relates to linearization by immersion. The study plant is a shaft of torque, $\Gamma(t)$, and moment of inertia, J, which rotates a small mass, m, at a distance, r, from the longitudinal axis, Δ, of the shaft. The small mass is secured to the shaft by means of a rigid, rectilinear arm of length OG and mass that is assumed to be negligible.

- $q(t)$ represents the angular position of the small mass with respect to the vertical.
- I represents the moment of inertia of the small mass with respect to the Δ axis.
- f represents the viscous friction coefficient on the motor shaft.

The nonlinear differential equation describing the dynamic behavior of the plant with respect to the Δ axis is of the form:

$$(J+I)\ddot{q}(t)+f\dot{q}(t)-mgr\sin q(t)=\Gamma(t).$$

Part one: plant linearization

In the case of linearization of the plant by immersion, determine the nonlinear precontroller that ensures the input precorrected plant, $\Gamma'(t)$, takes on the behavior of a second-order linear system, defined by:

$$G'(s) = \frac{Q(s)}{\Gamma'(s)} = \frac{\omega_0'}{s\left(1 + \frac{s}{\omega_0}\right)},$$

and whose ω_0' and ω_0 parameters we will express analytically below.

Part two: linearized plant control

Allocated the appropriate units, the parameters, ω_0' and ω_0, admit, respectively, the following numerical values:

$$\omega_0' = 100 \text{ and } \omega_0 = 10.$$

The precorrected plant, $G'(s)$, is now integrated into an elementary control loop, $(K_1 = K_2 = 1)$, whereby $C(s)$ denotes the regulator.

1) In the absence of phase correction, the regulator then being reduced to a simple proportional, determine the phase margin of the control at the frequency, $\omega_u = 10\omega_0$. Would you say the value obtained seems suitable? Justify your answer.

2) With the designer wishing to produce the regulator by means of a simple phase-lag and phase-lead controller, determine the parameters of the regulator, $C(s)$, which ensures a phase margin of 50° at the same frequency, ω_u, given that the regulator is then described by a transmittance of the form:

$$C(s) = C_0 \frac{1 + s/\omega_1'}{1 + s/\omega_2'} \frac{1 + s/\omega_1}{1 + s/\omega_2},$$

in which:

$$\omega_2' < \omega_1' < \omega_1 < \omega_2,$$

with:

$$\frac{\omega_1'}{\omega_2'} = 100 \text{ and } \frac{\omega_u}{\omega_1'} = 20;\ (\omega_1\omega_2)^{1/2} = \omega_u.$$

3) Aside from the static precision studied in response to Question 5 through the determination of the step error, what can be said about the precision at low frequency when we question the interest of a phase lag and lead in comparison to simply a phase lead?

4) Determine the high-frequency input sensitivity.

5) Determine both the step error and the ramp error, considering that

$$\lim_{t\to\infty} \varepsilon_y(t) = \lim_{s\to 0} s\varepsilon_y(s),$$

$\varepsilon_y(t)$ denoting the output error between the desired value and the real value.

6) In this question, the transitional frequency, ω_2', tends toward zero in accordance with an increase in the static gain, C_0, such that $C_0\omega_2' = C_0'\omega_1'$ (constancy of the frequency gain product), C_0 denoting the proportional asymptotic behavior gain below ω_2' and C_0' the proportional asymptotic behavior gain between ω_1 and ω_1'. The structure of the regulator is then reduced to that of a PID regulator resulting from the cascading of a proportional-integrator and a simple phase-lead controller. Determine the literal expression of the transmittance, $C(s)$, defining this new structure. Show that the ramp error obtained in answer to Question 5 is canceled when ω_2' tends toward zero, considering that C_0' and ω_1' remain finite: comment on the consistency of this result.

7) To extend the comment put forward in the previous question to the non-integer, it is appropriate to conduct a rapid calculation of the ramp error by continuing to apply the end-value theory, but this time from a transmittance in open loop, which identifies with that of a non-integer (real) integrator, that is,

$$\beta(s) = \left(\frac{\omega_u}{s}\right)^n \text{ with } 1 < n < 2,$$

this transmittance resulting from the extension to the entire frequency space of the vertical template of the second-generation CRONE control. After determining the ramp error, comment on the result obtained in order to give a general property concerning the canceling of the ramp error, simultaneously in integer and in non-integer.

Solution to Problem 4

Part one: plant linearization

The differential equation of the study plant, namely

$$(J+I)\ddot{q}(t)+f\dot{q}(t)-mgr\sin q(t)=\Gamma(t),$$

may be rewritten in the form:

$$(J+I)\ddot{q}(t)+f\dot{q}(t)=\Gamma(t)+mgr\sin q(t),$$

that is:

$$(J+I)\ddot{q}(t)+f\dot{q}(t)=\Gamma'(t),$$

by writing:

$$\Gamma'(t)=\Gamma(t)+mgr\sin q(t),$$

from which we obtain:

$$\Gamma(t)=\Gamma'(t)-mgr\sin q(t),$$

a nonlinear equation. conveyed by the functional diagram in Figure P4.1.

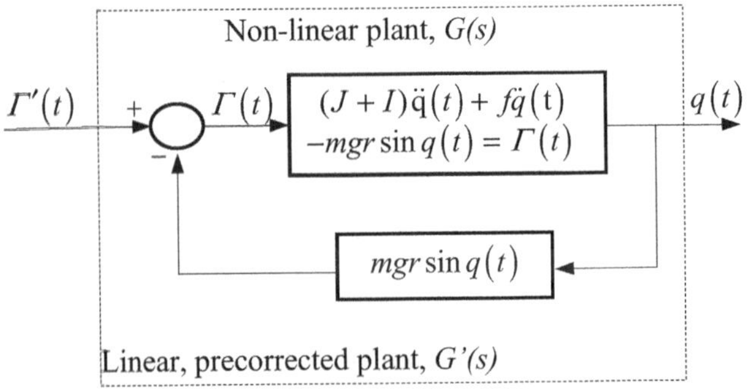

Figure P4.1. *Plant linearization*

The Laplace transform of the precorrected plant differential equation is written as:

$$(J+I)s^2Q(s)+fsQ(s)=\Gamma'(s),$$

or:

$$sQ(s)\left[f+(J+I)s\right]=\Gamma'(s),$$

from which we obtain the transfer function:

$$G'(s)=\frac{Q(s)}{\Gamma'(s)}=\frac{1}{s\left[f+(J+I)s\right]},$$

or:

$$G'(s)=\frac{\frac{1}{f}}{s\left(1+\frac{J+I}{f}s\right)},$$

or indeed:

$$G'(s)=\frac{\omega_0'}{s\left(1+\frac{s}{\omega_0}\right)},$$

by writing:

$$\omega_0'=\frac{1}{f} \text{ and } \omega_0=\frac{f}{J+I}.$$

Part two: linearized plant control

1) Phase margin in the absence of phase correction

The phase margin is written, by definition:

$$\Phi_m=\arg\beta(j\omega_u)+\pi,$$

that is:

$$\Phi_m = \arg C(j\omega_u) + \arg G(j\omega_u) + \pi ,$$

or, considering that $\arg C(j\omega_u) = 0$:

$$\Phi_m = \arg G(j\omega_u) + \pi ,$$

or, given the plant phase at the frequency, ω_u:

$$\Phi_m = -\frac{\pi}{2} - \arctan\frac{\omega_u}{\omega_0} + \pi = \frac{\pi}{2} - \arctan\frac{\omega_u}{\omega_0} ,$$

or indeed, numerically:

$$\Phi_m^{\circ} = 90 - \arctan 10 = 90 - 84.289 = 5.711 .$$

The value thus obtained is not suitable as it is far from satisfying the relevant recommendation, namely:

$$50 \leq \Phi_m^{\circ} \leq 55 .$$

2) Synthesis of the simple phase-lag and phase-lead controller

The phase of the simple phase-lag and phase-lead controller is written, at the frequency, ω_u:

$$\arg C(j\omega_u) = -\arctan\frac{\omega_u}{\omega_2'} + \arctan\frac{\omega_u}{\omega_1'} + \arctan\frac{\omega_u}{\omega_1} - \arctan\frac{\omega_u}{\omega_2} ,$$

or, given the numerical values of $\arg C(j\omega_u)$, and of the ratios, ω_u/ω_2' and ω_u/ω_1', namely

$$\arg C(j\omega_u) = 50^\circ - 5.711^\circ = 44.289^\circ ,$$

$$\omega_u/\omega_2' = (20)(100) = 2{,}000 \text{ and } \omega_u/\omega_1' = 20 :$$

$$44.289° = -\arctan 2{,}000 + \arctan 20 + \arcsin\frac{\alpha-1}{\alpha+1} \text{ with } \alpha = \frac{\omega_2}{\omega_1} > 1,$$

or indeed:

$$44.289° = -89.971° + 87.137° + \arcsin\frac{\alpha-1}{\alpha+1},$$

therefore:

$$\arcsin\frac{\alpha-1}{\alpha+1} = 47.123°,$$

that is:

$$\frac{\alpha-1}{\alpha+1} = \sin(47.123°) = 0.733,$$

from which we obtain:

$$\alpha = \frac{1+0.733}{1-0.733} = 6.490.$$

Distributed geometrically with respect to the frequency, ω_u, the transitional frequencies, ω_1 and ω_2, admit the following expressions:

$$\omega_1 = \frac{\omega_u}{\sqrt{\alpha}} = \frac{100}{\sqrt{6.49}} = 39.253\,rd/s$$

and

$$\omega_2 = \omega_u\sqrt{\alpha} = 100\sqrt{6.49} = 254.755\,rd/s.$$

The transitional frequencies, ω_2' and ω_1', are given by:

$$\omega_2' = \frac{\omega_u}{2{,}000} = \frac{100}{2{,}000} = 0.05\,rd/s$$

and

$$\omega_1' = \frac{\omega_u}{20} = \frac{100}{20} = 5\,rd/s.$$

To determine the static gain, C_0, it is appropriate to write:

$$\left|\beta(j\omega_u)\right| = 1,$$

or:

$$\left|C(j\omega_u)\right|\left|G(j\omega_u)\right| = 1,$$

or, given the literal expressions of $C(j\omega)$ and $G(j\omega)$:

$$C_0\left(\frac{1+\left(\frac{\omega_u}{\omega_1'}\right)^2}{1+\left(\frac{\omega_u}{\omega_2'}\right)^2}\right)^{\frac{1}{2}}\left(\frac{1+\left(\frac{\omega_u}{\omega_1}\right)^2}{1+\left(\frac{\omega_u}{\omega_2}\right)^2}\right)^{\frac{1}{2}}\frac{\omega_0'}{\omega_u}\left(\frac{1}{1+\left(\frac{\omega_u}{\omega_0}\right)^2}\right)^{\frac{1}{2}} = 1,$$

from which we obtain:

$$C_0 = \frac{\omega_u}{\omega_0'}\left(1+\left(\frac{\omega_u}{\omega_0}\right)^2\right)^{\frac{1}{2}}\left(\frac{1+\left(\frac{\omega_u}{\omega_2'}\right)^2}{1+\left(\frac{\omega_u}{\omega_1'}\right)^2}\right)^{\frac{1}{2}}\left(\frac{1+\left(\frac{\omega_u}{\omega_2}\right)^2}{1+\left(\frac{\omega_u}{\omega_1}\right)^2}\right)^{\frac{1}{2}},$$

or, numerically:

$$C_0 = \left(1+10^2\right)^{\frac{1}{2}}\left(\frac{1+\left(\frac{100}{0.05}\right)^2}{1+\left(\frac{100}{5}\right)^2}\right)^{\frac{1}{2}}\left(\frac{1+\left(\frac{100}{254.755}\right)^2}{1+\left(\frac{100}{39.253}\right)^2}\right)^{\frac{1}{2}},$$

therefore:

$$C_0 = (10.05)\frac{2000}{20.025}\frac{1.074}{2.737} = 393.870 .$$

3) Precision at low frequency

With respect to a single phase lead, the presence of a phase lag is of interest in that it brings low-frequency gain via an asymptotic gain increase when the frequency reduces from ω_1' to ω_2'. This supply of gain of the regulator then increases the gain in open loop, and thus improves precision at low frequency (while retaining the same stability degree ensured by the phase lead of the regulator around the frequency, ω_u).

4) High-frequency input sensitivity

The high-frequency input sensitivity is reduced to the regulator gain at these frequencies, that is:

$$S(\omega)_{HF} = |C(j\omega)|_{HF} = C_0 \frac{\omega_2'}{\omega_1'}\frac{\omega_2}{\omega_1} = \frac{C_0}{100}\alpha ,$$

or, given the numerical values of C_0 and α :

$$S(\omega)_{HF} = 25.562 .$$

5) Step and ramp errors

According to the lesson, the step and ramp errors in tracking are given, respectively, by:

$$\lim_{t\to\infty} \varepsilon_y(t) = \lim_{s\to 0} \frac{1}{1+\beta(s)}$$

and

$$\lim_{t\to\infty} \varepsilon_y(t) = \lim_{s\to 0} \frac{1}{s(1+\beta(s))},$$

with the general expression of the transmittance in open loop, $\beta(s)$, being particularized here by:

$$\beta(s) = C(s)G'(s) = C_0 \frac{1+\frac{s}{\omega_1'}}{1+\frac{s}{\omega_2'}} \frac{1+\frac{s}{\omega_1}}{1+\frac{s}{\omega_2}} \frac{\omega_0'}{s\left(1+\frac{s}{\omega_0}\right)}.$$

The denominators, $1+\beta(s)$ and $s\,(1+\beta(s))$, become:

$$1+\beta(s) = 1+\frac{1+\frac{s}{\omega_1'}}{1+\frac{s}{\omega_2'}} \frac{1+\frac{s}{\omega_1}}{1+\frac{s}{\omega_2}} \frac{C_0\omega_0'}{s\left(1+\frac{s}{\omega_0}\right)}$$

and

$$s(1+\beta(s)) = s+\frac{1+\frac{s}{\omega_1'}}{1+\frac{s}{\omega_2'}} \frac{1+\frac{s}{\omega_1}}{1+\frac{s}{\omega_2}} \frac{C_0\omega_0'}{1+\frac{s}{\omega_0}},$$

that is, for s tending toward zero:

$$\lim_{s\to 0}(1+\beta(s)) = 1+\frac{C_0\omega_0'}{0} = \infty$$

and

$$\lim_{s\to 0} s(1+\beta(s)) = 0 + C_0\omega_0' = C_0\omega_0',$$

making it possible to write the following, for the step error:

$$\lim_{t\to\infty} \varepsilon_y(t) = \frac{1}{\infty} = 0,$$

and for the ramp error:

$$\lim_{t \to \infty} \varepsilon_y(t) = \frac{1}{C_0 \omega_0'}.$$

6) Toward low-frequency integration of the regulator and canceling of the ramp error

Starting from the transmittance, $C(s)$, of the regulator, that is,

$$C(s) = C_0 \frac{1+\dfrac{s}{\omega_1'}}{1+\dfrac{s}{\omega_2'}} \frac{1+\dfrac{s}{\omega_1}}{1+\dfrac{s}{\omega_2}},$$

if ω_2' tends toward zero, s/ω_2' then being very large with respect to 1, $C(s)$ can be reduced in accordance with:

$$C(s) = C_0 \frac{1+\dfrac{s}{\omega_1'}}{\dfrac{s}{\omega_2'}} \frac{1+\dfrac{s}{\omega_1}}{1+\dfrac{s}{\omega_2}},$$

or:

$$C(s) = C_0 \omega_2' \left(\frac{1}{s} + \frac{1}{\omega_1'} \right) \frac{1+\dfrac{s}{\omega_1}}{1+\dfrac{s}{\omega_2}},$$

or, given that

$C_0 \omega_2' = C_0' \omega_1'$:

$$C(s) = C_0' \omega_1' \left(\frac{1}{s} + \frac{1}{\omega_1'} \right) \frac{1+\dfrac{s}{\omega_1}}{1+\dfrac{s}{\omega_2}},$$

or indeed:

$$C(s) = \left(1+\frac{\omega_1'}{s}\right) C_0' \frac{1+\frac{s}{\omega_1}}{1+\frac{s}{\omega_2}},$$

a literal expression of transmittance, $C(s)$, which clearly defines the structure of a PID regulator resulting from the cascading of a proportional-integrator,

$$1+\frac{\omega_1'}{s},$$

and a simple phase-lead controller,

$$C_0' \frac{1+\frac{s}{\omega_1}}{1+\frac{s}{\omega_2}} \text{ with } \omega_2 > \omega_1 .$$

Moreover, the relation expressing the constancy of the frequency gain product in the case of an asymptotic gain decrease at a rate of $6\,dB\,/\,oct$, that is,

$$C_0 \omega_2' = C_0' \omega_1',$$

enables C_0 to be expressed in accordance with:

$$C_0 = C_0' \frac{\omega_1'}{\omega_2'},$$

with this result showing that, for finite C_0' and ω_1' values, C_0 tends toward infinity when ω_2' tends toward zero.

The ramp error obtained in response to Question 5, that is,

$$\lim_{t \to \infty} \varepsilon_y(t) = \frac{1}{C_0 \omega_0'},$$

then tends toward zero when ω_2' tends toward zero. This result is consistent as it clearly confirms the property according to which the ramp error is canceled for at least two low-frequency integrations in open loop. Here, two integrations of this type are then present that were brought by the plant and that are now brought by the regulator.

7) Integer and non-integer ramp error

To extend the comment put forward in Question 6 to the non-integer, it is appropriate to apply the general expression of the ramp error in tracking given in the answer to Question 5, that is,

$$\lim_{t\to\infty} \varepsilon_y(t) = \lim_{s\to 0} \frac{1}{s\left(1+\beta(s)\right)},$$

then to particularize it by the open-loop transmittance proposed in Question 7, that is,

$$\beta(s) = \left(\frac{\omega_u}{s}\right)^n \text{ with } 1 < n < 2 .$$

The denominator, $s\left(1+\beta(s)\right)$, then becomes:

$$s\left(1+\beta(s)\right) = s + s\beta(s) = s + s\left(\frac{\omega_u}{s}\right)^n = s + \frac{\omega_u^n}{s^{n-1}} \text{ with } 0 < n-1 < 1,$$

that is, for s tending toward zero:

$$\lim_{s\to 0} s\left(1+\beta(s)\right) = 0 + \frac{\omega_u^n}{0} = \infty,$$

making it possible to write the following (for $1 < n < 2$):

$$\lim_{t\to\infty} \varepsilon_y(t) = \frac{1}{\infty} = 0 .$$

This result, which expresses that the ramp error is nil for an integration order greater than one, leads to the formulation of a more general property than that established for integer systems, namely a property relating to the integer and non-integer, according to which the ramp error is canceled for more than one low-frequency integration in open loop (that is, for at least two integrations in the integer case).

Presentation of Problem 5

Linearizing Control of a Motor Shaft: Linearization by Disturbance Rejection

This problem relates to linearization by disturbance rejection, with the nonlinear study plant subject to a linearizing control being that considered in the lesson, namely, a motor shaft rotating a small mass, m, whose linear part is characterized by:

$$G'(s) = \frac{1}{k + fs + (J+I)s^2} = \frac{G_0}{1 + 2\zeta\, s/\omega_n + s^2/\omega_n^2},$$

the parameters being defined in accordance with the lesson. The interpretation of its nonlinear part in terms of input disturbance is conveyed by:

$$p'(t) = mgr \sin q(t),$$

$q(t)$ denoting the angular position of the motor shaft.

In the case of a drive at constant angular speed, that is, $q(t) = \omega_0 t + q_0$, the plant input disturbance is a sinusoidal vibration such that $p'(t) = mgr \sin(\omega_0 t + q_0)$.

1) Determine the amplitude, p_m, of the output disturbance, $p(t)$, of the plant.

2) For a unit gain frequency in open loop, ω_u, such that $\omega_u = 100\omega_0$, determine, at ω_0, the general form of the frequency response in regulation (as a function of the frequency response in open loop).

3) For $\omega_u = 100\omega_0$ and $\omega_u = 10\omega_n$, propose, on the frequency axis, the relative distribution of characteristic frequencies of the control, namely ω_0, ω_n, ω_1 and ω_2, considering that ω_1 and ω_2 denote the transitional frequencies of the simple phase-lead controller,

$$C(s) = C_0 \frac{1 + s/\omega_1}{1 + s/\omega_2} \text{ with } \frac{\omega_2}{\omega_1} = \alpha > 1,$$

which ensures a phase margin of 45° at the frequency, ω_u (the textbook case considered acceptable here by the designer). The quantifying of this relative distribution of characteristic frequencies will be conducted taking into consideration that the damping ratio, ζ, is low in relation to one.

4) Using the legitimate approximations dictated by the ratios between characteristic frequencies, determine the amplitude, A, of the control output disturbance as a function of m, g, r and C_0. After modifying as appropriate the expression of A as determined, show how the amplitude, A, varies as a function of the frequency, ω_u.

5) Express the high-frequency input sensitivity as a function of m, g, r, α and A.

6) What happens to this sensitivity when the regulator is defined by:

$$C(s) = C_0 \left(\frac{1 + s/\omega_1}{1 + s/\omega_2} \right)^2 \text{ with } \frac{\omega_2}{\omega_1} = \alpha > 1?$$

Solution to Problem 5

On the characterization of the linear part of the plant

The transmittance characterizing the linear part of the plant, that is,

$$G'(s) = \frac{1}{k + fs + (J + I)s^2},$$

can be rewritten as:

$$G'(s) = \frac{1/k}{1 + \frac{f}{k}s + \frac{J+I}{k}s^2},$$

an expression that admits the canonical form:

$$G'(s) = \frac{G_0}{1 + 2\zeta\frac{s}{\omega_n} + \frac{s^2}{\omega_n^2}},$$

by writing:

$$G_0 = \frac{1}{k},$$

$$2\frac{\zeta}{\omega_n} = \frac{f}{k}, \text{ that is, } \zeta = \frac{\omega_n}{2}\frac{f}{k},$$

and

$$\frac{1}{\omega_n^2} = \frac{J+I}{k}, \text{ that is, } \omega_n = \left(\frac{k}{J+I}\right)^{\frac{1}{2}},$$

ζ then becoming:

$$\zeta = \frac{1}{2}\frac{f}{k}\left(\frac{k}{J+I}\right)^{\frac{1}{2}} = \frac{1}{2}\frac{f}{\left(k(J+I)\right)^{1/2}}.$$

1) Amplitude of the plant output disturbance

Given that $q(t) = \omega_0 t + q_0$, the plant input disturbance, $p'(t)$, is written as:

$$p'(t) = mgr\sin\left(\omega_0 t + q_0\right).$$

We associate with the concrete function, $p'(t)$, the complex quantity, $\overline{p'(t)}$, whose imaginary part is none other than $p'(t)$, that is:

$$\overline{p'(t)} = mgr\, e^{j(\omega_0 t + q_0)} .$$

The complex quantity associated with the plant output disturbance, $p(t)$, is then written as:

$$\overline{p(t)} = G'(j\omega_0)\overline{p'(t)},$$

from which we obtain the amplitude, p_m, of $p(t)$:

$$p_m = \left|\overline{p(t)}\right| = \left|G'(j\omega_0)\right|\left|\overline{p'(t)}\right|,$$

that is:

$$p_m = \frac{G_0}{\left[\left(1-\left(\frac{\omega_0}{\omega_n}\right)^2\right)^2 + 4\zeta^2\left(\frac{\omega_0}{\omega_n}\right)^2\right]^{1/2}} mgr .$$

2) $F_r(j\omega_0)$ as a function of $\beta(j\omega_0)$

The frequency response in regulation, $F_r(j\omega)$, also known as the sensitivity, and notated $S(j\omega)$, is written, at the frequency, ω_0:

$$F_r(j\omega_0) = S(j\omega_0) = \frac{1}{1+\beta(j\omega_0)} .$$

Given that the frequency, ω_0, is low with respect to ω_u since $\omega_0 = \omega_u/100$, $|\beta(j\omega_0)|$ is high in relation to one, that is, $|\beta(j\omega_0)| >> 1$, thus enabling $F_r(j\omega_0)$ to be reduced to:

$$F_r(j\omega_0) = \frac{1}{\beta(j\omega_0)} .$$

3) Relative distribution of characteristic frequencies

Since $\omega_u = 100\omega_0$ and $\omega_u = 10\omega_n$, we can write:

$$10\omega_n = 100\omega_0,$$

from which we obtain:

$$\omega_n = 10\omega_0.$$

We thus have the expressions of ω_u and ω_n as a function of ω_0, these expressions defining the relative location of these frequencies, as presented graphically here.

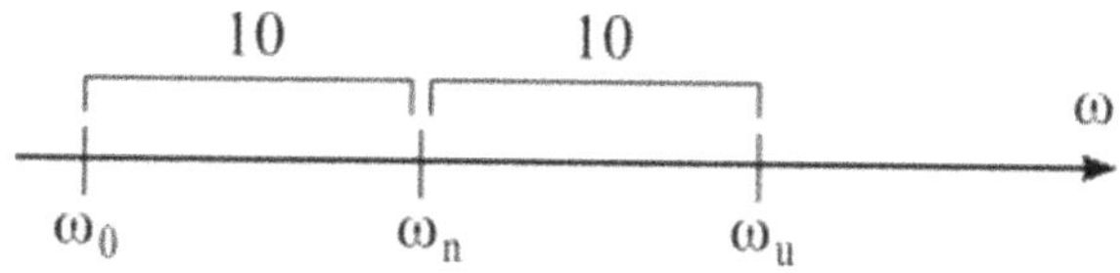

All that remains is to locate the transitional frequencies, ω_1 and ω_2, of the regulator with respect to the frequency, ω_u, it being understood that ω_1 and ω_2 are distributed geometrically with respect to ω_u.

Considering that a low damping ratio is accompanied by a rapid variation in phase around the transitional frequency, we can consider that the plant phase reaches the asymptote, –180°, beyond the frequency, ω_n, and notably at the frequency, $\omega_u = 10\omega_n$, that is, $\arg G'(j\omega_u) = -180°$.

Thus, the phase margin defined by

$$\Phi_m = \arg \beta(j\omega_u) + 180°,$$

that is,

$$\Phi_m = \arg G'(j\omega_u) + \arg C(j\omega_u) + 180°,$$

becomes, in this case:

$$\Phi_m = -180° + \arg C(j\omega_u) + 180° = \arg C(j\omega_u),$$

expressing that the phase margin, Φ_m, of the control is then reduced to the maximum phase lead, φ_m, of the regulator at the frequency, ω_u, therefore, for $\Phi_m = 45°$:

$$\varphi_m = \arg C(j\omega_u) = \Phi_m = 45° = arc\sin\frac{\alpha-1}{\alpha+1},$$

from which we obtain:

$$\alpha = \frac{1+\sin\varphi_m}{1-\sin\varphi_m} = \frac{1+\sin 45°}{1-\sin 45°} = 5.825,$$

the ratio between the frequencies, ω_2 and ω_1, which, distributed geometrically with respect to ω_u, are expressed, respectively, by:

$$\omega_1 = \frac{\omega_u}{\sqrt{\alpha}} = \frac{\omega_u}{2.413}$$

and

$$\omega_2 = \omega_u\sqrt{\alpha} = 2.413\omega_u.$$

Thus, the relative distribution of the charasteristic control frequencies is represented here on the frequency axis.

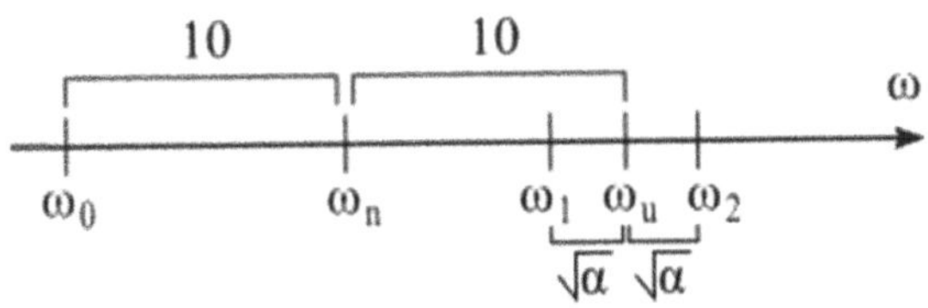

4) Amplitude of the control output disturbance

Considering that

$$\omega_0 << \omega_n \text{ and } \omega_0 << \omega_1,$$

$G'(j\omega_0)$ and $C(j\omega_0)$ are reduced, respectively, to:

$$G'(j\omega_0) = G_0 \text{ and } C(j\omega_0) = C_0.$$

The amplitude, p_m, of the plant output disturbance and the modulus of the frequency response in regulation (gain in regulation) at ω_0 thus become:

$$p_m = G_0 mgr$$

and

$$|F_r(j\omega_0)| = \frac{1}{|\beta(j\omega_0)|} = \frac{1}{|C(j\omega_0)||G'(j\omega_0)|} = \frac{1}{C_0 G_0}.$$

The amplitude, A, of the control output disturbance then admits an immediate expression:

$$A = |F_r(j\omega_0)|\, p_m = \frac{mgr}{C_0},$$

this result expressing that the amplitude, A, is inversely proportional to the static gain, C_0, of the regulator, and from which we can obtain, to answer the other questions of the problem:

$$C_0 = \frac{mgr}{A}.$$

Considering the nature of such a result, it is appropriate to make the best use of it, notably to show how the amplitude, A, varies as a function of the frequency, ω_u.

The asymptote gain diagram of the regulator, as represented here, enables us to write:

$$C_0 = |C(j\omega_u)| \frac{\omega_1}{\omega_u},$$

that is:

$$C_0 = \frac{|C(j\omega_u)|}{\sqrt{\alpha}},$$

or, considering that $|C(j\omega_u)||G'(j\omega_u)| = 1$:

$$C_0 = \frac{1}{\sqrt{\alpha}|G'(j\omega_u)|},$$

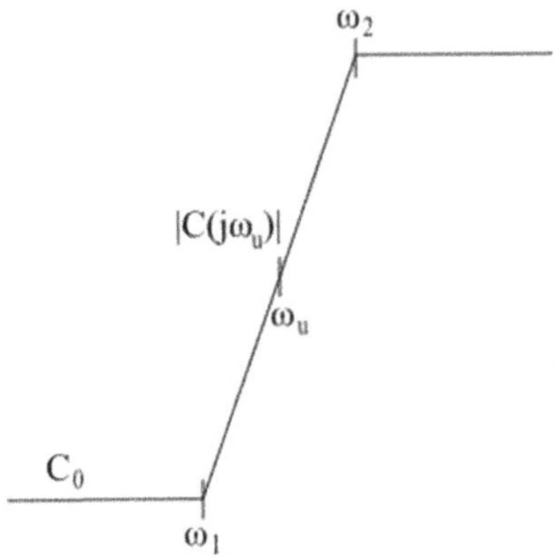

or indeed, given that $|G'(j\omega_u)|$ admits the reduction,

$$|G'(j\omega_u)| = \frac{G_0}{\left(\frac{\omega_u}{\omega_n}\right)^2} = G_0\left(\frac{\omega_n}{\omega_u}\right)^2:$$

$$C_0 = \frac{1}{\sqrt{\alpha}G_0}\left(\frac{\omega_u}{\omega_n}\right)^2,$$

the amplitude, A, then becomes:

$$A = mgr\sqrt{\alpha}G_0\left(\frac{\omega_n}{\omega_u}\right)^2,$$

with this result expressing that the amplitude, A, is inversely proportional to the square of the frequency, ω_u.

5) High-frequency input sensitivity

$$S(\omega)_{HF} = |C(j\omega)|_{HF} = C_0 \frac{\omega_2}{\omega_1} = C_0 \alpha = \frac{mgr}{A} \alpha ,$$

whereby it is to be recalled that the ratio, $\alpha = \omega_2 / \omega_1$, is such that

$$\varphi_m = \arg C(j\omega_u) = 45° = arc \sin \frac{\alpha - 1}{\alpha + 1}.$$

6) High-frequency input sensitivity

$$S(\omega)_{HF} = |C(j\omega)|_{HF} = C_0 \left(\frac{\omega_2}{\omega_1}\right)^2 = C_0 \alpha^2 = \frac{mgr}{A} \alpha^2 ,$$

whereby it is to be specified that the ratio, $\alpha = \omega_2 / \omega_1$, is such that

$$\varphi_m = \arg C(j\omega_u) = 45° = 2 arc \sin \frac{\alpha - 1}{\alpha + 1},$$

α then being given by

$$\alpha = \frac{1 + \sin(\phi_m / 2)}{1 - \sin(\phi_m / 2)}.$$

Presentation of Problem 6

High-Gain Control: Characterization in Tracking and Regulation

The principle of high-gain control forms the subject of this problem, which aims to illustrate such a principle through the particularization of the general transmittances, $G(s)$ and $H^{-1}(s)$, of the plant and the reaction chain.

This particularization is compliant with the transmittances:

$$G(s) = \frac{G_0}{\left(1+\frac{s}{\omega_0}\right)^2} \text{ and } H^{-1}(s) = 1+\frac{s}{\omega_0},$$

in which the G_0 and ω_0 parameters admit the following numerical values:

$$G_0 = 10 \text{ and } \omega_0 = 1\ rd/s.$$

After highlighting an elementary control loop by restructuring the control scheme as presented in the lesson, we obtain a study configuration that better complies with the results that are widely known in the field. This is therefore the study framework proposed for solving this problem.

1) Determine the expression of the open-loop transmittance of the elementary control loop obtained from said restructuring.

2) For a sufficiently high gain, K, that leads to a high open-loop transition frequency, ω_u, in relation to the transitional frequency, ω_0, of the plant, plot the

curve of the open-loop Nichols locus and determine the corresponding phase margin.

3) Determine the expression of the transmittance in tracking the high-gain control. We will demonstrate that this expression defines the transmittance of two low-pass filters in cascade, for which we will determine, for $KG_0 >> 1$, the static gain and the transitional frequencies. Let us use $\omega_0^{'}$ to denote the highest transitional frequency.

4) Determine the expression of the transmittance in regulating the high-gain control. We will show that this expression defines the transmittance of a simple phase-lead controller, for which we will determine, for $KG_0 >> 1$, the static gain and the transitional frequencies, the highest of which will be denoted by $\omega_0^{'}$. We will also determine the high-frequency gain and plot the asymptotic gain diagram that clearly illustrates the cut-off frequency, $\omega_0^{'}$, in regulation.

5) Using the results obtained in the lesson on the elementary control loop, it is easy to determine the input sensitivities in tracking and regulating the high-gain control. Determine, therefore, the general expressions of these sensitivities, taking care not to express on $\beta(j\omega)$ in order to facilitate the reduction of these general expressions at low and high frequency ($\omega << \omega_u$ and $\omega >> \omega_u$). Next, particularize the reduced expressions thus obtained by the respective expressions of $G(j\omega)$ and $H^{-1}(j\omega)$.

Solution to Problem 6

As a preamble, in order to clearly specify the study framework proposed in order to solve the problem, it is convenient to remember the restructuring of the control scheme, from which the elementary control loop arises, the controller transmittance, $C(s)$, which is none other than the product, $KH^{-1}(s)$ (Figure P6.1).

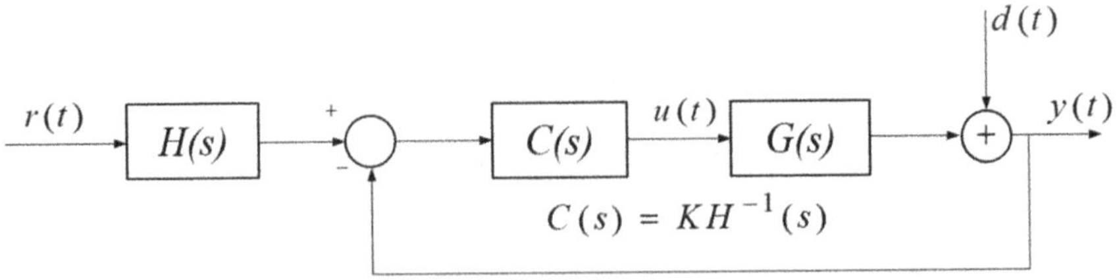

Figure P6.1. *Highlighting of an elementary control loop through restructuring of the control scheme*

1) Open-loop transmittance

If the general expression of the transmittance in open loop is given in this case by

$$\beta(s) = KH^{-1}(s)G(s),$$

the particularized expression is given by:

$$\beta(s) = K(1 + \frac{s}{\omega_0})\frac{G_0}{(1 + \frac{s}{\omega_0})^2} = \frac{KG_0}{1 + \frac{s}{\omega_0}}.$$

2) Open-loop Nichols locus and phase margin

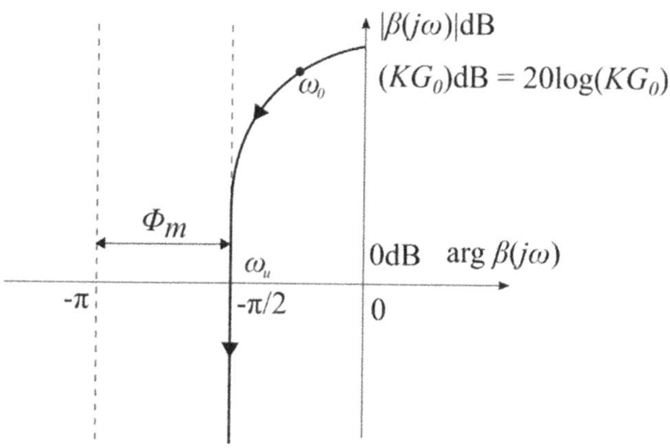

Figure P6.2. *Curve of the open-loop Nichols locus*

Given that the frequency, ω_u, is high with respect to the transitional frequency, ω_0 (simultaneously that of the plant and in open loop), the open-loop phase has more than reached its asymptotic behavior at the frequency, ω_u (Figure P6.2), therefore

$$\arg \beta(j\omega_u) = -\frac{\pi}{2},$$

and, consequently:

$$\phi_m = \arg \beta(j\omega_u) + \pi = \frac{\pi}{2}, \text{ that is, } \phi_m^{\circ} = 90 .$$

3) Transmittance in tracking

3.1) *General expression*

$$F_r(s) = \left[\frac{Y(s)}{R(s)}\right]_{D(s)=0} = H(s)\frac{\beta(s)}{1+\beta(s)} .$$

3.2) *Particularized expression*

As

$$H(s) = \frac{1}{1+\dfrac{s}{\omega_0}} \text{ and } \beta(s) = \frac{KG_0}{1+\dfrac{s}{\omega_0}},$$

$F_r(s)$ becomes:

$$F_r(s) = \frac{1}{1+\dfrac{s}{\omega_0}} \frac{\dfrac{KG_0}{1+\dfrac{s}{\omega_0}}}{1+\dfrac{KG_0}{1+\dfrac{s}{\omega_0}}},$$

or:

$$F_r(s) = \frac{1}{1+\dfrac{s}{\omega_0}} \frac{KG_0}{1+\dfrac{s}{\omega_0}+KG_0},$$

or indeed:

$$F_r(s) = \frac{KG_0}{1+KG_0} \frac{1}{1+\dfrac{s}{\omega_0}} \frac{1}{1+\dfrac{s}{(1+KG_0)\omega_0}},$$

that is, for $KG_0 >> 1$:

$$F_r(s) = \frac{1}{1+\frac{s}{\omega_0}} \frac{1}{1+\frac{s}{\omega_0^{'}}} \text{ with } \omega_0^{'} = KG_0\omega_0 .$$

The expression thus obtained defines the transmittance of two first-order low-pass filters in cascade, the static gain of which is unitary, that is,

$$|F_r(j0)| = 1,$$

and the transitional frequencies of which are

$$\omega_0 \text{ and } \omega_0^{'} = KG_0\omega_0 >> \omega_0 .$$

The transitional frequency, $\omega_0^{'}$, which is very high in relation to the transitional frequency, ω_0, defines the cut-off frequency at –3 dB (or the bandwidth) in tracking the elementary control loop, thus expressing the significant dynamic rapidity in tracking this loop.

4) Transmittance in regulation

4.1) *General expression*

$$F_d(s) = \left[\frac{Y(s)}{D(s)}\right]_{R(s)=0} = \frac{1}{1+\beta(s)} .$$

4.2) *Particularized expression*

As

$$\beta(s) = \frac{KG_0}{1+\frac{s}{\omega_0}},$$

$F_d(s)$ becomes:

$$F_d(s) = \frac{1}{1 + \dfrac{KG_0}{1 + \dfrac{s}{\omega_0}}},$$

or:

$$F_d(s) = \frac{1 + \dfrac{s}{\omega_0}}{1 + \dfrac{s}{\omega_0} + KG_0},$$

or indeed:

$$F_d(s) = \frac{1}{1 + KG_0} \frac{1 + \dfrac{s}{\omega_0}}{1 + \dfrac{s}{(1 + KG_0)\omega_0}},$$

that is, for $KG_0 >> 1$:

$$F_d(s) = \frac{1}{KG_0} \frac{1 + \dfrac{s}{\omega_0}}{1 + \dfrac{s}{\omega_0'}} \text{ with } \omega_0' = KG_0\omega_0 .$$

This expression defines the transmittance of a simple phase-lead controller, whose static gain is given by

$$|F_d(j0)| = \frac{1}{KG_0},$$

and the transitional frequencies of which are

$$\omega_0 \text{ and } \omega_0' = KG_0\omega_0 >> \omega_0 .$$

The transitional frequency, ω_0', which is very high in relation to the transitional frequency, ω_0, defines the cut-off frequency (at –3 dB) in the regulation of the elementary control loop, thus expressing the significant dynamic rapidity in the regulation of this loop (in fact the same as in tracking, as expected).

To help plot the asymptotic gain diagram of $F_d(j\omega)$, which clearly illustrates the regulation cut-off frequency, ω_0' (Figure P6.3), it is convenient to express the high-frequency gain, that is:

$$\left|F_d(j\omega)\right|_{HF} = \frac{1}{KG_0}\left|\frac{1+j\frac{\omega}{\omega_0}}{1+j\frac{\omega}{\omega_0'}}\right|_{HF} = \frac{1}{KG_0}\frac{\omega_0'}{\omega} = \frac{1}{KG_0}KG_0 = 1,$$

or, in decibels:

$$\left|F_d(j\omega)\right|_{HF} dB = 20\log 1 = 0.$$

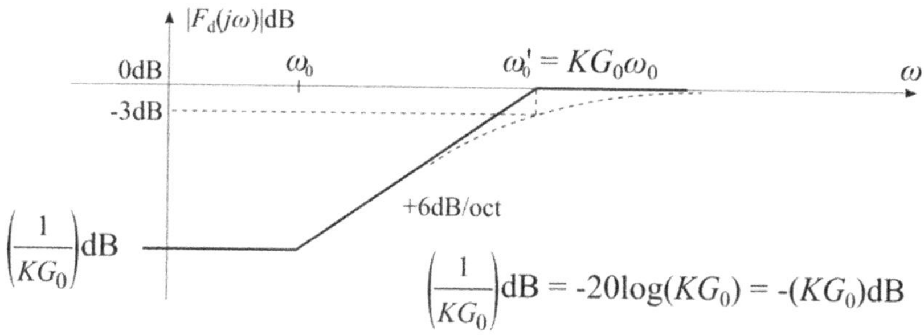

Figure P6.3. *Asymptotic gain diagram of the frequency response in regulation*

5) Input sensitivities in tracking and regulation

In tracking $(D(j\omega) = 0)$, the input sensitivity is defined by:

$$S_r(\omega) = \left|\frac{U(j\omega)}{R(j\omega)}\right|_{D(j\omega)=0},$$

that is:

$$S_r(\omega) = \text{[gain between } u(t) \text{ and the input, (+), of the comparator]}$$

$$\times \text{[gain between the input, (+), of the comparator and } r(t)\text{]},$$

therefore:

$$S_r(\omega) = \left|\frac{C(j\omega)}{1+\beta(j\omega)}\right| \left|H(j\omega)\right|,$$

the first term of the product expressing the input sensitivity in tracking and regulation of the elementary control loop.

Given that

$$C(j\omega) = KH^{-1}(j\omega),$$

$S_r(\omega)$ becomes:

$$S_r(\omega) = \left|\frac{KH^{-1}(j\omega)}{1+\beta(j\omega)}\right| \left|H(j\omega)\right| = \left|\frac{K}{1+\beta(j\omega)}\right|.$$

In regulation $(R(j\omega) = 0)$, the input sensitivity is defined by:

$$S_d(\omega) = \left|\frac{U(j\omega)}{D(j\omega)}\right|_{R(j\omega)=0},$$

that is, since the input, (+), of the comparator is nil:

$$S_d(\omega) = \left|\frac{C(j\omega)}{1+\beta(j\omega)}\right|,$$

this expression being none other than that of the input sensitivity in the tracking and regulation of the elementary control loop.

Given that

$$C(j\omega) = KH^{-1}(j\omega),$$

$S_d(\omega)$ becomes:

$$S_d(\omega) = \left|\frac{KH^{-1}(j\omega)}{1+\beta(j\omega)}\right|.$$

5.1) *Reduction of* $S_r(\omega)$ *and* $S_d(\omega)$ *at low and high frequency*

At low frequency ($\omega \ll \omega_u$ and therefore $|\beta(j\omega)| \gg 1$), $S_r(\omega)$ and $S_d(\omega)$ admit the following reduced expressions, respectively:

$$S_r(\omega)_{BF} = \left|\frac{K}{\beta(j\omega)}\right| = \left|\frac{\mathrm{K}}{\mathrm{KH}^{-1}(j\omega)G(j\omega)}\right| = \left|\frac{1}{H^{-1}(j\omega)G(j\omega)}\right|$$

and

$$S_d(\omega)_{BF} = \left|\frac{\mathrm{KH}^{-1}(j\omega)}{\beta(j\omega)}\right| = \left|\frac{\mathrm{KH}^{-1}(j\omega)}{\mathrm{KH}^{-1}(j\omega)G(j\omega)}\right| = \left|\frac{1}{G(j\omega)}\right|.$$

At high frequency ($\omega \gg \omega_u$ and therefore $|\beta(j\omega)| \ll 1$), $S_r(\omega)$ and $S_d(\omega)$ admit the following reduced expressions, respectively:

$$S_r(\omega)_{HF} = \mathrm{K}$$

and

$$S_d(\omega)_{HF} = K\left|\mathrm{H}^{-1}(j\omega)\right|.$$

5.2) *Particularization of the reduced expressions of* $S_r(\omega)$ *and* $S_d(\omega)$

For

$$G(j\omega)=\frac{G_0}{\left(1+j\frac{\omega}{\omega_0}\right)^2} \text{ and } H^{-1}(j\omega)=1+j\frac{\omega}{\omega_0},$$

the product, $H^{-1}(j\omega)G(j\omega)$, then becoming

$$H^{-1}(j\omega)G(j\omega)=\frac{G_0}{1+j\frac{\omega}{\omega_0}},$$

the reduced expressions of $S_r(\omega)$ and $S_d(\omega)$ become, respectively:

$$S_r(\omega)_{BF}=\left|\frac{1+j\frac{\omega}{\omega_0}}{G_0}\right|=\frac{1}{G_0}\left[1+\left(\frac{\omega}{\omega_0}\right)^2\right]^{\frac{1}{2}};$$

$$S_d(\omega)_{BF}=\left|\frac{\left(1+j\frac{\omega}{\omega_0}\right)^2}{G_0}\right|=\frac{1}{G_0}\left[1+\left(\frac{\omega}{\omega_0}\right)^2\right];$$

$$S_r(\omega)_{HF}=K;$$

$$S_d(\omega)_{HF}=K\left|1+j\frac{\omega}{\omega_0}\right|=K\left[1+\left(\frac{\omega}{\omega_0}\right)^2\right]^{\frac{1}{2}}=K\frac{\omega}{\omega_0} \text{ since } \omega \gg \omega_u \gg \omega_0.$$

Presentation of Problem 7

Feedforward Control: Characterization in Tracking and Regulation by a Direct Approach and by an Indirect Approach via a Reference Prefilter

In this problem relating to feedforward control, the plant is considered to be perfectly modeled, that is, $\tilde{G}(s) = G(s)$, in order to directly obtain the main results characterizing this control.

1) Given the functional diagram of the feedforward control as presented in the lesson, show that the particularization, $\tilde{G}(s) = G(s)$, ensures, in tracking, a direct cascading of the inverse model of the plant, and of the plant itself, through the canceling of the signal, $u_1(t)$. Next determine the transmittance in tracking.

2) Still based on the same functional diagram, determine the general expression of the transmittance in regulation.

3) Show that feedforward comes down to filtering the reference input, $r(t)$, with a *prefilter*, whose transmittance is none other than the inverse of the general expression of transmittance in tracking of the elementary control loop. Can this result not be predicted to ensure a unitary transmittance in tracking? Moreover, is it compatible with the general expression of transmittance in regulation, determined in response to Question 2?

4) Based on the functional diagram illustrating the change to the elementary control loop via a reference prefilter, determine the general expressions of the input sensitivities in tracking and regulation, without necessarily expressing $\beta(j\omega)$ in

order to facilitate the reduction of these general expressions at low and high frequencies ($\omega << \omega_u$ and $\omega >> \omega_u$).

5) What are the general expressions obtained in Question 4 reduced to at low and high frequencies?

6) Finally, particularize the reduced expressions resulting from Question 5 by the plant of the problem on high-gain control, and by a simple phase-lead controller, that is,

$$G(s) = \frac{G_0}{(1+s/\omega_0)^2} \text{ and } C(s) = C_0 \frac{1+s/\omega_1}{1+s/\omega_2}, \quad \text{with} \quad \omega_1 < \omega_2 .$$

Solution to Problem 7

1) Behavior in tracking

In tracking $(D(s)=0)$, the functional diagram of the feedforward control enables us to write:

$$U_1(s) = C(s)\left[R(s) - G(s)\left(G^{-1}(s)R(s) + U_1(s)\right)\right],$$

or:

$$U_1(s) = C(s)\left[R(s) - R(s) - G(s)U_1(s)\right],$$

or:

$$U_1(s)\left(1 + C(s)G(s)\right) = 0 ,$$

or indeed:

$$U_1(s) = 0, \text{ therefore } u_1(t) = 0 ,$$

thus expressing a direct cascading of the inverse model of the plant, $G^{-1}(s)$, and of the plant itself, $G(s)$.

Determination of transmittance in tracking, $F_r(s)$, is then immediate, since $Y(s)$ is written as:

$$Y(s) = G(s)G^{-1}(s)R(s),$$

that is,

$$Y(s) = R(s),$$

we immediately obtain:

$$F_r(s) = \left[\frac{Y(s)}{R(s)}\right]_{D(s)=0} = 1.$$

2) Behavior in regulation

In regulation $(R(s) = 0)$, the functional diagram of the feedforward control enables us to write:

$$u_2(t) = 0, \text{ therefore } U_2(s) = 0,$$

with the transmittance in regulation, $F_d(s)$, thus being reduced to that of the elementary control loop, of which $\beta(s) = C(s)G(s)$ is the transmittance in open loop, that is:

$$F_d(s) = \left[\frac{Y(s)}{D(s)}\right]_{R(s)=0} = \frac{1}{1+\beta(s)}.$$

3) From feedforward to a reference prefilter

The functional diagram of the elementary control loop enables us to write:

$$U(s) = \frac{1}{G(s)}R(s) + C(s)\big(R(s) - Y(s)\big),$$

that is, by multiplying and dividing by $C(s)$:

$$U(s) = C(s)\left[\frac{1}{C(s)G(s)}R(s) + R(s) - Y(s)\right],$$

or, knowing that $\beta(s) = C(s)G(s)$:

$$U(s) = C(s)\left[\frac{R(s)}{\beta(s)} + R(s) - Y(s)\right],$$

or:

$$U(s) = C(s)\left[\left(\frac{1}{\beta(s)} + 1\right)R(s) - Y(s)\right],$$

or indeed:

$$U(s) = C(s)\left[\frac{1+\beta(s)}{\beta(s)}R(s) - Y(s)\right].$$

The symbolic equation thus obtained is conveyed by the functional diagram (Figure P7.1), which illustrates the change to the elementary control loop via a *reference prefilter* of transmittance,

$$\frac{1+\beta(s)}{\beta(s)},$$

which is none other than the inverse of the *general expression of transmittance in tracking of the elementary control loop*, namely

$$\frac{\beta(s)}{1+\beta(s)}.$$

A result such as this can certainly be predicted to ensure a unitary transmittance in tracking, since the product of the transmittances of the reference prefilter and the elementary control loop in tracking $(D(s) = 0)$ is expressed by:

$$F_r(s) = \left[\frac{Y(s)}{R(s)}\right]_{D(s)=0} = \frac{1+\beta(s)}{\beta(s)}\frac{\beta(s)}{1+\beta(s)} = 1.$$

Moreover, this result is compatible with the general expression of transmittance in regulation determined in response to Question 2. Indeed, with the reference prefilter imposing a nil reference input of the elementary control loop for $R(s) = 0$, the transmittance in regulation is reduced to that of the elementary control loop (of transmittance in open loop, $\beta(s) = C(s)G(s)$), that is:

$$F_d(s) = \left[\frac{Y(s)}{D(s)}\right]_{R(s)=0} = \frac{1}{1+\beta(s)}.$$

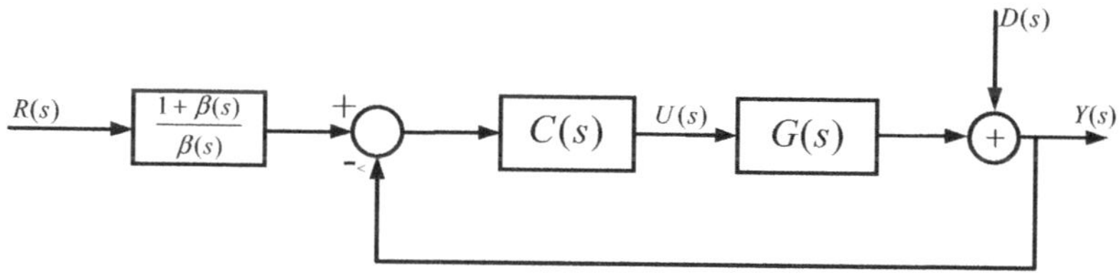

Figure P7.1. *Change to the elementary control loop via a reference prefilter*

4) Input sensitivities in tracking and regulation

The functional diagram illustrating the change to the elementary control loop via a reference prefilter enables us to write, in tracking $(D(j\omega) = 0)$:

$$S_r(\omega) = \left|\frac{U(j\omega)}{R(j\omega)}\right|_{D(j\omega)=0} = \left|\frac{C(j\omega)}{1+\beta(j\omega)}\right|\left|\frac{1+\beta(j\omega)}{\beta(j\omega)}\right|$$

$$= \left|\frac{C(j\omega)}{\beta(j\omega)}\right| = \left|\frac{C(j\omega)}{C(j\omega)G(j\omega)}\right| = \left|\frac{1}{G(j\omega)}\right|,$$

and, in regulation $(R(j\omega) = 0)$:

$$S_d(\omega) = \left|\frac{U(j\omega)}{D(j\omega)}\right|_{R(j\omega)=0} = \left|\frac{C(j\omega)}{1+\beta(j\omega)}\right|.$$

5) Reduction of $S_r(\omega)$ and $S_d(\omega)$ at low and high frequencies

At low frequencies ($\omega << \omega_u$ and therefore $|\beta(j\omega)| >> 1$), $S_r(\omega)$ and $S_d(\omega)$ admit the following reduced expressions, respectively:

$$S_r(\omega)_{BF} = \left|\frac{1}{G(j\omega)}\right|$$

and

$$S_d(\omega)_{BF} = \left|\frac{C(j\omega)}{\beta(j\omega)}\right| = \left|\frac{C(j\omega)}{C(j\omega)G(j\omega)}\right| = \left|\frac{1}{G(j\omega)}\right|.$$

At high frequencies ($\omega >> \omega_u$ and therefore $|\beta(j\omega)| << 1$), $S_r(\omega)$ and $S_d(\omega)$ admit the following reduced expressions, respectively:

$$S_r(\omega)_{HF} = \left|\frac{1}{G(j\omega)}\right|$$

and

$$S_d(\omega)_{HF} = |C(j\omega)|.$$

6) Particularization of reduced expressions

At low frequencies, the reduced expressions are particularized in accordance with:

$$S_r(\omega)_{BF} = S_d(\omega)_{BF} = \left|\frac{1}{G(j\omega)}\right| = \left|\frac{\left(1+j\frac{\omega}{\omega_0}\right)^2}{G_0}\right|$$

$$= \frac{1+\left(\frac{\omega}{\omega_0}\right)^2}{G_0} = \frac{1}{G_0} \quad \text{for} \quad \omega \ll \omega_0.$$

At high frequencies, the reduced expressions are particularized in accordance with:

$$S_r(\omega)_{HF} = \left|\frac{1}{G(j\omega)}\right| = \left|\frac{\left(1+j\frac{\omega}{\omega_0}\right)^2}{G_0}\right|$$

$$= \frac{1+\left(\frac{\omega}{\omega_0}\right)^2}{G_0} = \frac{1}{G_0}\left(\frac{\omega}{\omega_0}\right)^2 \quad \text{for} \quad \omega >> \omega_0$$

and

$$S_d(\omega)_{HF} = |C(j\omega)| = C_0 \left| \frac{1 + j\dfrac{\omega}{\omega_1}}{1 + j\dfrac{\omega}{\omega_2}} \right|$$

$$= C_0 \left(\frac{1 + \left(\dfrac{\omega}{\omega_1}\right)^2}{1 + \left(\dfrac{\omega}{\omega_2}\right)^2} \right)^{\frac{1}{2}} = C_0 \frac{\omega_2}{\omega_1} \quad \text{for} \quad \omega >> \omega_2.$$

Presentation of Problem 8

Synthesis of an Internal-Model Control Using a PID Controller of the Equivalent Elementary Control Loop

By developing a synthesis strategy of an internal model control, based on an optimal PID regulator structure, this problem is a significant, original contribution provided in terms of both internal model control and PID regulator.

The study plant that forms the subject of this problem is again achieved by a continuous-current motor (but parameterized differently), for which the output considered is the angular speed. Taking into account its electrical time constant, it is described by a second-order transmittance of the form:

$$G(s) = \frac{G_0}{\left(1+\dfrac{s}{\omega_m}\right)\left(1+\dfrac{s}{\omega_e}\right)},$$

in which ω_m represents the mechanical transitional frequency and ω_e the electrical transitional frequency, with G_0 denoting the static gain. The numerical values of the parameters are as follows:

$$G_0 = 100,\ \omega_m = 1\,rd\,/\,s \text{ and } \omega_e = 20\,rd\,/\,s\,.$$

The plant thus defined is subject to an internal model control, the principle of which is illustrated by the functional diagram in Figure P8.1.

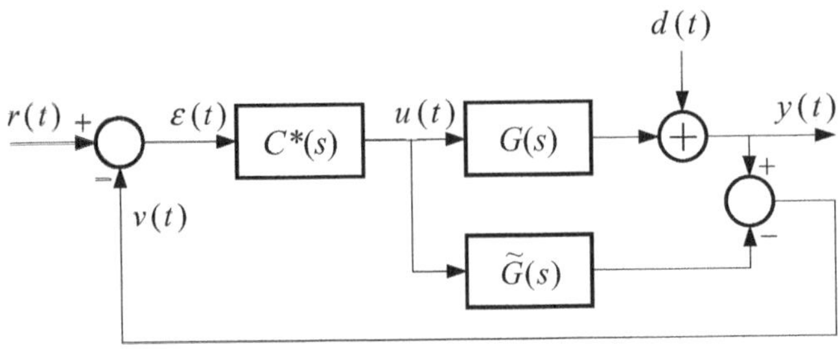

Figure P8.1. *Block diagram of internal-model control*

1) Determine the general form of the transmittance of the regulator, $C(s)$, which is cascaded with the plant in the equivalent elementary control loop.

2) In order for the dynamics of the motor control to be sufficiently fast, the designer sets the unit gain frequency in open loop, ω_u, at a value equaling that of the electrical transitional frequency of the motor, that is, $\omega_u = \omega_e = 20\,rd\,/\,s$. Justify this choice.

3) In the absence of phase correction, the regulator, $C(s)$, then being reduced to a simple proportional, determine the phase margin, ϕ_m, of the control. Would you say its value seems suitable?

4) At the same frequency, ω_u, what phase must be presented by the regulator that is to substitute the proportional in order to ensure a phase margin of 60°?

5) Within the framework of this problem, the regulator chosen by the designer is a PID regulator that is convenient to present in comparison with that of Problem 1, for which the synthesis is performed by a progressive complexification of a simple phase-lead controller: this involves the introduction of low and high transitional frequencies, ω_b and ω_h, distanced from the frequency, ω_u, ($\omega_b << \omega_u << \omega_h$) so as not to affect the results obtained beforehand at the frequency, ω_u, with the simple phase-lead controller.

Here, the regulator considered is a PID regulator, whose low and high transitional frequencies, ω_b and ω_h, of the proportional integral (at low frequency) and the low-pass (at high frequency), are no longer far from the frequency, ω_u : the first, ω_b, is no longer very small in relation to ω_u, and the second, ω_h, is no longer very large in relation to ω_u. This frequency configuration, which has the advantage of reducing the dispersion of the regulator transitional frequencies, is accompanied, in this problem, by a refinement consisting of equaling ω_h and ω_2: this refinement presents the advantage of reducing the regulator structure to a minimum structure with four parameters, which nevertheless meets the precision requirements at low frequency and the input immunity requirements at high frequency.

As such, the PID regulator as presented is defined by a transmittance of the form:

$$C(s) = K\left(1+\frac{\omega_b}{s}\right)\frac{1+s/\omega_1}{\left(1+s/\omega_2\right)^2},$$

with

$$\omega_b < \omega_1 < \omega_2 .$$

K is the gain of the proportional asymptotic behavior between ω_b and ω_1; ω_b is the transitional frequency of the proportional-integral action; ω_1 and ω_2 are the extremal transitional frequencies of the differential asymptotic behavior around the frequency, ω_u, such that

$$\omega_u = \left(\omega_1\omega_2\right)^{1/2} .$$

Based on the structure of the regulator of Problem 1, the term to the power of 2 is explained by the identity of the transitional frequency, ω_h, of the low-pass filter of order 1 and the high transitional frequency, ω_2, of the simple phase-lead controller.

In fact, the PID regulator behaves *asymptotically* here as:

– an *integrator* for $\omega < \omega_b$, in order to improve precision at low frequency (by increasing the regulator gain);

– a *proportional* for $\omega_b < \omega < \omega_1$ (which introduces an adjustment parameter);

– a *differentiator* for $\omega_1 < \omega < \omega_2$, in order to improve the stability degree (by increasing the angular distance (or phase margin) at the frequency, $\omega_u = \left(\omega_1\omega_2\right)^{1/2}$);

– an *integrator* for $\omega > \omega_2$ in order to improve input immunity at high frequency (by decreasing the regulator gain).

For a phase margin of 60° at the frequency, $\omega_u = 20\,rd/s$, determine the parameters of the regulator, considering that the ratio between ω_u and ω_b is set at 5 by the designer.

6) Determine the static gain in open loop. What error is it able to cancel?

7) Determine the low-frequency input sensitivity.

8) Determine the input sensitivity, first, at high frequency as a function of ω, and, second, at infinite frequencies.

9) Finally, determine the literal expression of the transmittance of the regulator, $C^*(s)$, as is presented in the functional diagram of the internal-model control.

Solution to Problem 8

1) Regulator, $C(s)$, of the equivalent elementary control loop

The block diagram of the internal-model control enables us to write:

$$U(s) = C^*(s)\left[R(s) - \left(Y(s) - \tilde{G}(s)U(s)\right)\right],$$

or:

$$U(s) = C^*(s)\left[R(s) - Y(s) + \tilde{G}(s)U(s)\right],$$

or:

$$U(s) = C^*(s)\left(R(s) - Y(s)\right) + C^*(s)\tilde{G}(s)U(s),$$

or indeed:

$$U(s)\left(1 - C^*(s)\tilde{G}(s)\right) = C^*(s)\left(R(s) - Y(s)\right),$$

from which we obtain:

$$U(s) = \frac{C^*(s)}{1 - C^*(s)\tilde{G}(s)}\left(R(s) - Y(s)\right),$$

that is:

$$U(s) = C(s)\big(R(s) - Y(s)\big),$$

by writing:

$$C(s) = \frac{C^*(s)}{1 - C^*(s)\tilde{G}(s)},$$

this expression being none other than that of the regulator transmittance, $C(s)$, of the equivalent elementary control loop.

2) Justification of the choice of frequency ω_u

Considering that the high transitional frequency, ω_e, (of the plant) is not large enough in relation to the low transitional frequency, ω_m, to enable its effect on the cut-off frequency, ω_c, to be totally overlooked, ω_c remains less than ω_m, (but by very little), that is:

$$\omega_c = \omega_m^- = 1^- \, rd\,/\,s\,.$$

Since $\omega_u = 20\, rd\,/\,s$,

$$\frac{\omega_u}{\omega_c} = \frac{20}{1^-} = 20^+\,.$$

Given the low overshoot of 20, this ratio remains acceptable, taking into account the relevant recommendation (which is, moreover, only given as an indication):

$$5\omega_c \le \omega_u \le 20\omega_c\,.$$

3) Phase margin in the absence of phase correction

As

$$\phi_m = \arg \beta(j\omega_u) + \pi\,,$$

the nullity of $\arg C(j\omega_u)$ reduces ϕ_m to:

$$\phi_m = \arg G(j\omega_u) + \pi ,$$

or, given the plant phase:

$$\phi_m = -\arctan\frac{\omega_u}{\omega_m} - \arctan\frac{\omega_u}{\omega_e} + \pi ,$$

or indeed, numerically:

$$\phi_m = -\arctan 20 - \arctan 1 + \pi ,$$

therefore:

$$\phi_m^{\circ} = -87.137 - 45 + 180 = 47.86 .$$

While this phase margin is far from aberrant, it nevertheless does not meet the designer's requirement, formulated by a 60° phase margin in Question 4.

4) Regulator phase at the frequency, ω_u

Considering that the regulator must make the phase margin change from 47.86° to 60°, at the frequency, ω_u, it must present a phase such that:

$$\arg C(j\omega_u) = 60^{\circ} - 47.86^{\circ} = 12.14^{\circ} .$$

5) Synthesis of the regulator, $C(s)$, (in this case a PID)

Figure P8.2 proposes Bode asymptotic diagrams of the PID regulator, $C(s)$, as defined by Question 5 of the presentation.

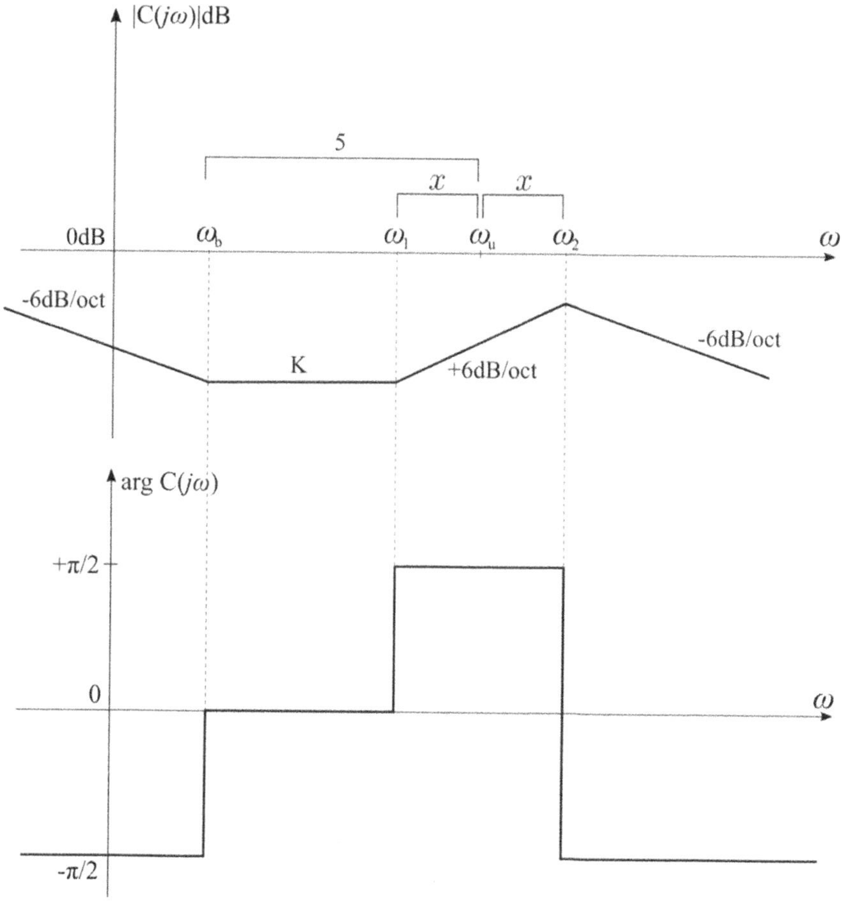

Figure P8.2. *Bode asymptotic diagrams of the PID regulator*

By rewriting $C(s)$ in the form:

$$C(s) = K\frac{\omega_b}{s}\left(1+\frac{s}{\omega_b}\right)\frac{1+s/\omega_1}{\left(1+s/\omega_2\right)^2},$$

it is easy to express the phase of the regulator, $C(s)$, that is:

$$\arg C(j\omega) = -\frac{\pi}{2} + \arctan\frac{\omega}{\omega_b} + \arctan\frac{\omega}{\omega_1} - 2\arctan\frac{\omega}{\omega_2},$$

or, at the frequency, ω_u:

$$\arg C(j\omega_u) = -\frac{\pi}{2} + \arctan\frac{\omega_u}{\omega_b} + \arctan\frac{\omega_u}{\omega_1} - 2\arctan\frac{\omega_u}{\omega_2},$$

or, given the numerical value of the ω_u/ω_b ratio and by writing

$$\frac{\omega_u}{\omega_1} = \frac{\omega_2}{\omega_u} = x:$$

$$\arg C(j\omega_u) = -\frac{\pi}{2} + \arctan 5 + \arctan x - 2\arctan\frac{1}{x},$$

or indeed, knowing that

$$\arctan x + \arctan\frac{1}{x} = \frac{\pi}{2}:$$

$$\arg C(j\omega_u) = -\frac{\pi}{2} + \arctan 5 + \arctan x - 2\left(\frac{\pi}{2} - \arctan x\right),$$

therefore:

$$\arg C(j\omega_u) = -3\frac{\pi}{2} + \arctan 5 + 3\arctan x,$$

from which we obtain:

$$\arctan x = \frac{\pi}{2} + \frac{1}{3}\left(\arg C(j\omega_u) - \arctan 5\right),$$

that is:

$$x = \tan\left[\frac{\pi}{2} + \frac{1}{3}\left(\arg C(j\omega_u) - \arctan 5\right)\right],$$

the numerical value of $\arg C(j\omega_u)$, 12.14°, particularizing this result in compliance with:

$$x = \tan\left[90^\circ + \frac{1}{3}(12.14^\circ - 78.69^\circ)\right] = 2.452\,.$$

Distributed geometrically with respect to the frequency, ω_u, the transitional frequencies, ω_1 and ω_2, admit the following expressions:

$$\omega_1 = \frac{\omega_u}{x} = \frac{20}{2.452} = 8.15\,rd\,/\,s$$

and

$$\omega_2 = \omega_u x = 20(2.452) = 49.04\,rd\,/\,s\,.$$

Concerning the transitional frequency, ω_b, it is given by:

$$\omega_b = \frac{\omega_u}{5} = \frac{20}{5} = 4\,rd\,/\,s\,.$$

With regard to the gain, K, it satisfies the relation

$$|\beta(j\omega_u)| = 1\,,$$

that is:

$$|C(j\omega_u)||G(j\omega_u)| = 1\,,$$

or, given the expressions of $C(j\omega)$ and $G(j\omega)$:

$$K\left(1+\left(\frac{\omega_b}{\omega_u}\right)^2\right)^{\frac{1}{2}} \frac{\left(1+\left(\frac{\omega_u}{\omega_1}\right)^2\right)^{\frac{1}{2}}}{1+\left(\frac{\omega_u}{\omega_2}\right)^2} \frac{G_0}{\left(1+\left(\frac{\omega_u}{\omega_m}\right)^2\right)^{\frac{1}{2}}\left(1+\left(\frac{\omega_u}{\omega_e}\right)^2\right)^{\frac{1}{2}}} = 1\,,$$

from which we obtain:

$$K = \frac{1+\left(\dfrac{\omega_u}{\omega_2}\right)^2}{\left(1+\left(\dfrac{\omega_b}{\omega_u}\right)^2\right)^{\frac{1}{2}}\left(1+\left(\dfrac{\omega_u}{\omega_1}\right)^2\right)^{\frac{1}{2}}} \frac{\left(1+\left(\dfrac{\omega_u}{\omega_m}\right)^2\right)^{\frac{1}{2}}\left(1+\left(\dfrac{\omega_u}{\omega_e}\right)^2\right)^{\frac{1}{2}}}{G_0},$$

or, numerically:

$$K = 0.122 .$$

6) Static gain in open loop

Determined at zero frequency, the static gain in the open loop is given by:

$$|\beta(j0)| = |C(j0)||G(j0)| = \infty . G_0 = \infty .$$

The infinity of this gain enables the step error to be canceled, thus ensuring an excellent static precision.

7) Low-frequency input sensitivity

The low-frequency input sensitivity is given by:

$$S(\omega)_{BF} = \frac{1}{|G(j\omega)|_{BF}} = \frac{1}{G_0} = 10^{-2} .$$

8) High-frequency input sensitivity

The high-frequency input sensitivity is given by:

$$S(\omega)_{HF} = |C(j\omega)|_{HF} = K\frac{\omega_2^2}{\omega_1}\frac{1}{\omega},$$

and, at infinite frequency, by:

$$S(\infty) = |C(j\infty)| = 0 .$$

9) Returning to regulator, $C^*(s)$, of internal-model control

Given the response to the first question, that is

$$C(s) = \frac{C^*(s)}{1 - C^*(s)\tilde{G}(s)},$$

it is easy to express $C^*(s)$ as a function of $C(s)$:

$$C^*(s) = \frac{C(s)}{1 + C(s)\tilde{G}(s)},$$

or, given the respective expressions of $C(s)$ and $\tilde{G}(s)$, that is,

$$C(s) = K\left(1 + \frac{\omega_b}{s}\right)\frac{1 + s/\omega_1}{\left(1 + s/\omega_2\right)^2}$$

and

$$\tilde{G}(s) = \frac{G_0}{\left(1 + \dfrac{s}{\omega_m}\right)\left(1 + \dfrac{s}{\omega_e}\right)}:$$

$$C^*(s) = \frac{K\left(1 + \dfrac{\omega_b}{s}\right)\left(1 + \dfrac{s}{\omega_1}\right)\left(1 + \dfrac{s}{\omega_m}\right)\left(1 + \dfrac{s}{\omega_e}\right)}{\left(1 + \dfrac{s}{\omega_2}\right)^2\left(1 + \dfrac{s}{\omega_m}\right)\left(1 + \dfrac{s}{\omega_e}\right) + KG_0\left(1 + \dfrac{\omega_b}{s}\right)\left(1 + \dfrac{s}{\omega_1}\right)}.$$

Presentation of Problem 9

Quadratic-Criterion Control

The study plant considered is a fundamental first-order system (the fundamental nature implying a constant-numerator transfer), with this system thus described by a transmittance of the form:

$$G(s) = \frac{G_0}{1 + \dfrac{s}{\omega_0}},$$

with $G_0 = 100$ and $\omega_0 = 10$ rd/s.

The plant thus defined forms the subject of a quadratic-criterion control.

1) Represent the plant by expressing its output, $y(t)$, as a function of $u(t)$ (its input) and $\dot{y}(t)$, then a_1, where $a_1 = 1/\omega_0$.

2) The designer needs the output, $y(t)$, of the plant to respond to the reference input, $r(t)$, of the control, with the given dynamics set by a fundamental second-order reference model, unit static gain, damping ratio, $\zeta_0' = 0.5$, and natural frequency without damping (or transitional) $\omega_0' = 10\omega_0$. Justify the choice of ζ_0' and ω_0' comparatively to the performances of an elementary control loop in closed loop, whose phase margin is $45°$ at the unit gain frequency in open loop, $\omega_u = 10\omega_0$ (a frequency often retained, given that ω_0 is the plant cut-off frequency here): for the damping ratio, it is advisable to use a second generation CRONE control.

3) Synthesize the control law, ensuring that the plant output, $y(t)$, is as close as possible to the output, $y_M(t)$, of the reference model, in accordance with the minimization of the quadratic criterion.

$$J(t) = \left[y(t) - y_M(t)\right]^2 + \delta G_0 u^2(t),$$

with δ being a control weighting factor. Establish the functional diagram of the control, integrating both the reference model already defined, and the control law thus synthesized.

4) The functional diagram established in response to Question 3 highlights a control loop, with $Y_M(s)$ as the reference input and $Y(s)$ as the output. Reduce this control loop to an elementary control loop, whose regulator is remarkably simple.

5) Based on the general expression of the input sensitivity in tracking, determine this sensitivity at high frequency before and after changing to an elementary control loop. Start by determining the sensitivity for the control loop and for the equivalent elementary control loop, which arise from the functional representation of the quadratic-criterion control. After comparing the results obtained, determine the sensitivity for the control by taking into account the frequency response of the reference model. What happens when the weighting factor, δ, is adjusted to offer the control the same dynamic range as the reference model?

Solution to Problem 9

1) Plant representation adapted to the quadratic-criterion control

The plant transmittance,

$$G(s) = \frac{Y(s)}{U(s)} = \frac{G_0}{1 + \dfrac{s}{\omega_0}},$$

enables us to write the symbolic equation:

$$\frac{s}{\omega_0} Y(s) + Y(s) = G_0 U(s),$$

from which we obtain the differential equation:

$$\frac{1}{\omega_0}\dot{y}(t) + y(t) = G_0 u(t),$$

which may be rewritten in the form:

$$y(t) = G_0 u(t) - \frac{1}{\omega_0}\dot{y}(t),$$

that is, by writing

$$a_1 = \frac{1}{\omega_0}:$$

$$y(t) = G_0 u(t) - a_1 \dot{y}(t).$$

2) Justification of the reference model

To justify the choice of $\zeta_0' = 0.5$ and $\omega_0' = 10\omega_0$, in comparison with the closed-loop performances of an elementary control loop, whereby $\Phi_m = 45°$ and $\omega_u = 10\omega_0$, it is necessary to determine these performances and then show that they correspond to the values allocated to ζ_0' and ω_0'.

For a phase margin, Φ_m, of $45°$, in closed loop, the damping ratio, ζ, can be estimated by that of a second-generation CRONE control with the same phase margin, which then implies an order, $n = 3/2$, and, consequently, a damping ratio, $\zeta = -\cos(\pi/n) = 0.5$, with this result complying with $\zeta_0' = 0.5$, thus justifying this choice.

For an open-loop unit gain (or transition) frequency, ω_u, such that $\omega_u = 10\omega_0$, in closed loop, the transitional frequency, ω_t, which equals (as a first approximation) ω_u, is then expressed by $\omega_t = \omega_u = 10\omega_0$, a result that clearly corresponds to $\omega_0' = 10\omega_0$, thus justifying this choice.

3) Control law synthesis

Given the expression of $y(t)$ determined in response to Question 1, that is,

$$y(t) = G_0 u(t) - a_1 \dot{y}(t),$$

the criterion, $J(t)$, becomes:

$$J(t) = \left[G_0 u(t) - a_1 \dot{y}(t) - y_M(t)\right]^2 + \delta G_0 u^2(t).$$

Its minimization with respect to $u(t)$ is conveyed by:

$$\frac{\partial J(t)}{\partial u(t)} = 0,$$

that is:

$$2G_0\left[G_0 u(t) - a_1 \dot{y}(t) - y_M(t)\right] + 2\delta G_0 u(t) = 0,$$

or:

$$2G_0\left[G_0 u(t) - a_1 \dot{y}(t) - y_M(t) + \delta u(t)\right] = 0,$$

or:

$$2G_0\left[\left(G_0 + \delta\right) u(t) - a_1 \dot{y}(t) - y_M(t)\right] = 0,$$

or indeed, since G_0 is not nil:

$$\left(G_0 + \delta\right) u(t) - a_1 \dot{y}(t) - y_M(t) = 0,$$

from which we obtain the control equation, $u(t)$, of the plant:

$$u(t) = \frac{1}{G_0 + \delta}\left[y_M(t) + a_1 \dot{y}(t)\right],$$

or, in Laplace transforms:

$$U(s) = \frac{1}{G_0 + \delta}\left[Y_M(s) + a_1 s Y(s)\right].$$

The symbolic equation thus obtained is conveyed by the functional diagram of the control, incorporating the reference model already defined and the control law synthesized here (see Figure P9.1).

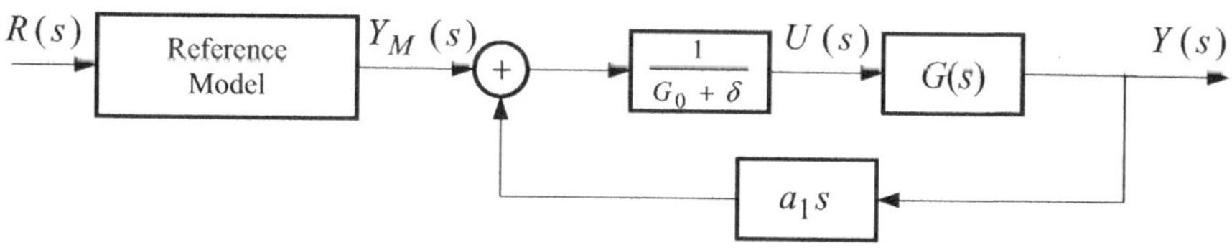

Figure P9.1. *Control scheme incorporating the reference model and the control law through the cascading of the reference model and of a control loop*

4) Change to an elementary control loop

Let us return to the control loop highlighted in the functional diagram drawn up in response to Question 3, and make it more conventional: first, by replacing the summator with a comparator and second, by allocating the minus sign to the transmittance of the reaction chain, in order to compensate for the minus sign introduced by the comparator at its inverting input.

The general expression of transmittance in tracking then enables us to determine a transmittance in the closed loop of the form:

$$\frac{Y(s)}{Y_M(s)} = \frac{1}{-a_1 s}\,\frac{-\dfrac{a_1 s}{G_0 + \delta}\dfrac{G_0}{1 + s/\omega_0}}{1 - \dfrac{a_1 s}{G_0 + \delta}\dfrac{G_0}{1 + s/\omega_0}},$$

or, since $a_1 = 1/\omega_0$:

$$\frac{Y(s)}{Y_M(s)} = \frac{\dfrac{1}{G_0 + \delta}\dfrac{G_0}{1 + a_1 s}}{1 - \dfrac{a_1 s}{G_0 + \delta}\dfrac{G_0}{1 + a_1 s}},$$

or:

$$\frac{Y(s)}{Y_M(s)} = \frac{G_0}{(G_0 + \delta)(1 + a_1 s) - a_1 s G_0},$$

or indeed:

$$\frac{Y(s)}{Y_M(s)} = \frac{G_0}{G_0 + \delta + (G_0 + \delta) a_1 s - a_1 s G_0},$$

that is:

$$\frac{Y(s)}{Y_M(s)} = \frac{G_0}{G_0 + \delta + \delta a_1 s} = \frac{G_0}{G_0 + \delta(1 + a_1 s)},$$

or, by displaying the plant transmittance:

$$\frac{Y(s)}{Y_M(s)} = \frac{\dfrac{G_0}{\delta(1 + a_1 s)}}{1 + \dfrac{G_0}{\delta(1 + a_1 s)}},$$

therefore, ultimately:

$$\frac{Y(s)}{Y_M(s)} = \frac{\dfrac{1}{\delta} G(s)}{1 + \dfrac{1}{\delta} G(s)},$$

this transmittance being none other than that of an elementary control loop, whose regulator is reduced to a proportional of gain $1/\delta$ (see Figure P9.2).

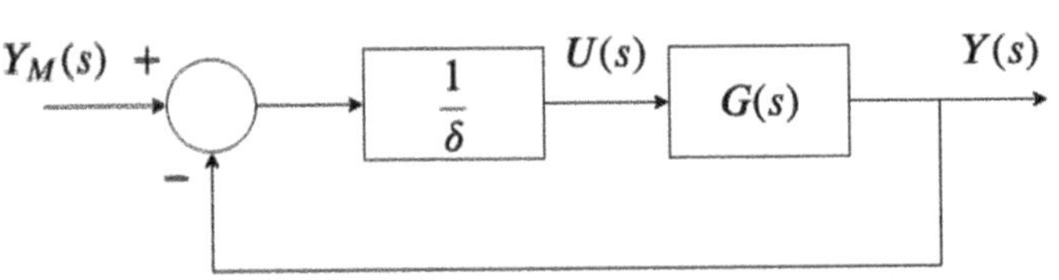

Figure P9.2. *Elementary control loop equivalent to the control loop in Figure P9.1*

5) High-frequency input sensitivity in tracking

Given the general expression of the input sensitivity in tracking, that is,

$$\left|\frac{K_1 C(j\omega)}{1+\beta(j\omega)}\right|,$$

this sensitivity is particularized here in accordance with what is discussed below.

For the control loop, where

$$K_1 = 1, \quad C(j\omega) = \frac{1}{G_0 + \delta}$$

and

$$\beta(j\omega) = \frac{1}{G_0 + \delta}\frac{G_0}{1 + a_1 j\omega}(-a_1 j\omega),$$

the sensitivity becomes:

$$\left|\frac{U(j\omega)}{Y_M(j\omega)}\right| = \left|\frac{\dfrac{1}{G_0 + \delta}}{1 - \dfrac{G_0}{G_0 + \delta}\dfrac{a_1 j\omega}{1 + a_1 j\omega}}\right|,$$

that is, at high frequency (and considering that δ is positive):

$$\left|\frac{U(j\omega)}{Y_M(j\omega)}\right|_{HF} = \left|\frac{\dfrac{1}{G_0 + \delta}}{1 - \dfrac{G_0}{G_0 + \delta}}\right| = \left|\frac{1}{G_0 + \delta - G_0}\right| = \left|\frac{1}{\delta}\right| = \frac{1}{\delta}.$$

For the equivalent elementary control loop, where

$$K_1 = 1, \quad C(j\omega) = \frac{1}{\delta}$$

and

$$\beta(j\omega) = \frac{1}{\delta}\frac{G_0}{1+a_1 j\omega},$$

the sensitivity becomes:

$$\left|\frac{U(j\omega)}{Y_M(j\omega)}\right| = \left|\frac{\dfrac{1}{\delta}}{1+\dfrac{1}{\delta}\dfrac{G_0}{1+a_1 j\omega}}\right|,$$

that is, at high frequency (and considering that δ is positive):

$$\left|\frac{U(j\omega)}{Y_M(j\omega)}\right|_{HF} = \left|\frac{1}{\delta}\right| = \frac{1}{\delta}.$$

In terms of comparison, the results obtained are identical, namely $1/\delta$, thus expressing that the input sensitivity in tracking at high frequency is indeed the same for the control loop and the equivalent elementary control loop (as was of course to be expected).

Consequently, the high-frequency input sensitivity in tracking of the quadratic-criterion control, which is written as:

$$\left|\frac{U(j\omega)}{R(j\omega)}\right|_{HF} = \left|\frac{U(j\omega)}{Y_M(j\omega)}\right|_{HF}\left|\frac{Y_M(j\omega)}{R(j\omega)}\right|_{HF},$$

then becomes:

$$\left|\frac{U(j\omega)}{R(j\omega)}\right|_{HF} = \frac{1}{\delta}\left|\frac{1}{1+2\zeta_0'\left(j\dfrac{\omega}{\omega_0'}\right)+\left(j\dfrac{\omega}{\omega_0'}\right)^2}\right|_{HF},$$

that is:

$$\left|\frac{U(j\omega)}{R(j\omega)}\right|_{HF} = \frac{1}{\delta}\left(\frac{\omega_0'}{\omega}\right)^2.$$

When, through the action of the weighting factor, δ, the control dynamic range identifies with that of the reference model, the dynamic range of the control loop or the equivalent elementary control loop is deleted and is then infinitely fast. This implies an infinite bandwidth for each of the loops, and therefore a weighting factor, δ, tending toward zero, causing the input sensitivity to tend toward infinity, as determined by the last relation. We will show how the bandwidth of the loops tends toward infinity when δ tends toward zero.

The functional diagram of the elementary control loop (see Figure P9.2) expresses the closed-loop frequency response:

$$\frac{Y(j\omega)}{Y_M(j\omega)} = \frac{\beta(j\omega)}{1+\beta(j\omega)} = \frac{\frac{1}{\delta}G(j\omega)}{1+\frac{1}{\delta}G(j\omega)},$$

that is, given the expression of $G(j\omega)$ (in which $a_1 = 1/\omega_0$):

$$\frac{Y(j\omega)}{Y_M(j\omega)} = \frac{\frac{1}{\delta}\frac{G_0}{1+ja_1\omega}}{1+\frac{1}{\delta}\frac{G_0}{1+ja_1\omega}},$$

or:

$$\frac{Y(j\omega)}{Y_M(j\omega)} = \frac{G_0}{\delta + j\delta a_1\omega + G_0} = \frac{G_0}{G_0+\delta+j\delta a_1\omega},$$

or:

$$\frac{Y(j\omega)}{Y_M(j\omega)} = \frac{\frac{G_0}{G_0+\delta}}{1+j\frac{\delta a_1}{G_0+\delta}\omega} = \frac{\frac{G_0}{G_0+\delta}}{1+j\frac{\delta}{G_0+\delta}\frac{\omega}{\omega_0}},$$

or indeed:

$$\frac{Y(j\omega)}{Y_M(j\omega)} = \frac{G_0}{G_0+\delta}\frac{1}{1+j\frac{\omega}{\omega_0''}},$$

with

$$\omega_0'' = \left(1 + \frac{G_0}{\delta}\right)\omega_0 \,,$$

with these expressions enabling the calculation of the limits:

$$\lim_{\delta \to 0} \frac{G_0}{G_0 + \delta} = 1$$

and

$$\lim_{\delta \to 0} \omega_0'' = \infty \,,$$

with the latter limit indeed expressing that the cut-off frequency at $-3\,\text{dB}$, ω_0'', which defines the loop bandwidth, tends toward infinity when the weighting factor, δ, tends toward zero.

Presentation of Problem 10

Synthesis of a Constant Phase-Lead CRONE Controller: The Essential Stage in the Synthesis of the Fractional PID Whose (Integer) Integration at Low Frequency Simply Results from a Cascade Proportional–Integral

This problem is related to the frequency control of a continuous-dye laser designed and achieved in 1975 at Bordeaux 1 University in order to conduct high-resolution molecular spectroscopic experiments.

1) Considering that its transmittance in tracking of order $3/2$ admits an expression of the form

$$F_r(s) = \frac{1}{1+\left(\dfrac{s}{\omega_0}\right)^{\frac{3}{2}}},$$

determine the expression of the transmittance in open loop presenting an infinite static gain.

2) Using n to denote the second-generation CRONE control, the reduced first overshoot $D_1^{'}$ (expressed as a percentage) of the step response in closed loop is given by a polynomial approximation defined by the equation:

$$D_1^{'} = an^2 + bn + c\,,$$

with $a = 79.195$, $b = -138.507$ and $c = 59.528$.

Thus, determine the numerical value of $D_1^{'}$ in the case of the problem posed.

3) Determine the control phase margin.

a) Is it acceptable?

b) Is it optimal?

4) In order to increase the phase margin and thus decrease the first overshoot, the designer suggests incorporating a constant-phase CRONE controller providing a 10° phase lead into the current regulator.

Determine the new phase margin at the unit gain frequency in open loop, $\omega_u = \omega_0$, as well as the new reduced first overshoot.

5) The constant-phase CRONE controller is defined:

– where the ideal version is concerned, by non-integer transmittance,

$$C_m(s) = C_0 \left(\frac{1 + \dfrac{s}{\omega_b}}{1 + \dfrac{s}{\omega_h}} \right)^m$$

– for the real version, by integer transmittance,

$$C_N(s) = C_0 \prod_{i=1}^{N} \frac{1 + \dfrac{s}{\omega_i^{'}}}{1 + \dfrac{s}{\omega_i}}\,,$$

with

$$\frac{\omega_{i+1}^{'}}{\omega_i^{'}} = \frac{\omega_{i+1}}{\omega_i} = \alpha\eta > 1 \text{ and } \frac{\omega_i}{\omega_i^{'}} = \alpha > 0\,,\ \frac{\omega_{i+1}^{'}}{\omega_i} = \eta > 0\,.$$

5.1) Determine the parameters of the ideal version, $C_m(s)$, that ensure, in open loop, a vertical template extending across two and a half decades and centered geometrically on the frequency, ω_u, such that $f_u = 20\,KHz$.

5.2) Determine the parameters of the real version, $C_N(s)$ based on five zeros and five poles.

6) Determine the control resonance frequency and ratio in tracking before and after correction.

7) Determine the natural frequency and the damping ratio of the control before and after correction.

8) Using the legitimate approximations dictated by the nature of the presentation of the problem, determine the extremal values of the interval in which ω_0 can vary without any degradation of the phase margin.

Solution to Problem 10

1) Transmittance in open loop of infinite static gain

The general expression of the transmittance in tracking, that is,

$$F_r(s) = K\frac{\beta(s)}{1+\beta(s)},$$

enables us to write:

$$K\beta(s) = F_r(s) + F_r(s)\beta(s),$$

that is:

$$\beta(s)\left[K - F_r(s)\right] = F_r(s),$$

from which we draw the general expression of the transmittance in open loop:

$$\beta(s) = \frac{F_r(s)}{K - F_r(s)}.$$

This general expression can be particularized in accordance with:

$$\beta(s) = \frac{\dfrac{1}{1+\left(\dfrac{s}{\omega_0}\right)^{\frac{3}{2}}}}{K - \dfrac{1}{1+\left(\dfrac{s}{\omega_0}\right)^{\frac{3}{2}}}},$$

or:

$$\beta(s) = \frac{1}{K + K\left(\dfrac{s}{\omega_0}\right)^{\frac{3}{2}} - 1},$$

or:

$$\beta(s) = \frac{1}{K - 1 + K\left(\dfrac{s}{\omega_0}\right)^{\frac{3}{2}}},$$

or indeed, for $\omega = 0$, that is, $s = j\omega = 0$:

$$\beta(0) = \frac{1}{K-1},$$

static gain in open loop tending toward infinity when K tends toward one, in which case $\beta(s)$ becomes:

$$\beta(s) = \left(\frac{\omega_0}{s}\right)^{\frac{3}{2}}.$$

2) Reduced first overshoot of the step response in closed loop

For $n = 3/2$, the reduced first overshoot (expressed as a percentage) of the step response in closed loop admits the following numerical value:

$$D_1^{'}(\%) = 79.195(1.5)^2 - 138.507(1.5) + 59.528,$$

that is:

$$D_1^{'} = 29.956\ \%.$$

3) Control phase margin

By definition, the phase margin is given by:

$$\Phi_m = \arg \beta(j\omega) - (-\pi),$$

that is:

$$\Phi_m = \arg \beta(j\omega) + \pi,$$

or, in this case, considering that

$$\arg \beta(j\omega_u) = \arg\left(\frac{\omega_0}{j\omega_u}\right)^{\frac{3}{2}} = \frac{3}{2}\left(-\frac{\pi}{2}\right) = -\frac{3\pi}{4}:$$

$$\Phi_m = -\frac{3\pi}{4} + \pi = -\frac{3\pi}{4} + \frac{4\pi}{4} = \frac{\pi}{4},$$

therefore:

$$\Phi_m = \frac{\pi}{4} rd = 45°.$$

If this phase margin can be deemed acceptable as a case study, it is nevertheless not optimal considering the recommended interval $[50°, 55°]$.

4) New phase margin and new reduced first overshoot

Given that the constant-phase CRONE controller brings a phase lead of 10°, that is, a constant phase lead of 10° around the frequency, $\omega_u = \omega_0$, it is possible to determine the new phase margin, Φ_m, that is:

$$\Phi_m^{\circ} = 45 + 10 = 55 .$$

With regard to the determination of the new reduced first overshoot, $D_1^{'}$, it is appropriate to begin by determining the new n order of the control in accordance with:

$$-n\ 90° = -135° + 10° = -125° ,$$

from which we obtain:

$$n = \frac{125}{90} = 1.388 ,$$

the new reduced first overshoot (expressed as a percentage) then admitting the following numerical value:

$$D_1^{'}(\%) = 79.195(1.388)^2 - 138.507(1.388) + 59.528 ,$$

that is:

$$D_1^{'} = 19.853\% .$$

5) Constant-phase CRONE controller synthesis

5.1) *Ideal version*

To determine the order, m, of the controller, it is sufficient to write that

$$m\ 90° = 10° ,$$

from which we obtain:

$$m = \frac{10}{90} = 0.111 .$$

The extremal frequencies, ω_A and ω_B, of the vertical template of the second-generation CRONE control are given by:

$$\omega_A = \frac{\omega_u}{\sqrt{10^{2.5}}} = \frac{2\pi(20.10^3)}{\sqrt{10^{2.5}}} = 7066.606\, rd\ /\ s$$

and

$$\omega_B = \omega_u\sqrt{10^{2.5}} = 2\pi(20.10^3)\sqrt{10^{2.5}} = 2234.657\ 10^3\, rd\ /\ s\,.$$

With regard to the transitional frequencies, ω_b and ω_h, of the controller, these are deduced from ω_A and ω_B in accordance with the relations (used by default):

$$\omega_b = \frac{\omega_A}{10} = 706.660\ rd\ /\ s$$

and

$$\omega_h = 10\omega_B = 2234.657\ 10^4\ rd\ /\ s\,.$$

To determine the static gain, C_0, of the controller, it is sufficient to express that the gain in open loop after correction is unitary at the frequency, ω_u, that is, in this case:

$$C_0\left(\frac{1+\left(\dfrac{\omega_u}{\omega_b}\right)^2}{1+\left(\dfrac{\omega_u}{\omega_h}\right)^2}\right)^{\frac{m}{2}}\left(\frac{\omega_0}{\omega_u}\right)^{\frac{3}{2}} = 1,$$

this relationship clearly showing that the controller gain is also unitary at the frequency, ω_u, since $\omega_u = \omega_0$, a relationship from which we obtain:

$$C_0 = \left(\frac{1+\left(\dfrac{\omega_u}{\omega_h}\right)^2}{1+\left(\dfrac{\omega_u}{\omega_b}\right)^2}\right)^{\frac{m}{2}},$$

or, numerically:

$$C_0 = 0.562.$$

5.2) *Real version*

Since

$$m = \frac{\log \alpha}{\log(\alpha\eta)},$$

from which we obtain:

$$\alpha = (\alpha\eta)^m \text{ and } \eta = (\alpha\eta)^{1-m},$$

and given that

$$(\alpha\eta)^N = \frac{\omega_h}{\omega_b},$$

from which we obtain:

$$\alpha\eta = \left(\frac{\omega_h}{\omega_b}\right)^{\frac{1}{N}},$$

α and η are written as:

$$\alpha = \left(\frac{\omega_h}{\omega_b}\right)^{\frac{m}{N}} \text{ and } \eta = \left(\frac{\omega_h}{\omega_b}\right)^{\frac{1-m}{N}}.$$

Given that

$$\frac{\omega_B}{\omega_A} = 10^{2.5} \text{ (given in the presentation)}$$

and

$$\frac{\omega_A}{\omega_b} = \frac{\omega_h}{\omega_B} = 10 \text{ (taken by default)},$$

it is possible to express the ω_h/ω_b ratio, that is,

$$\frac{\omega_h}{\omega_b} = 10^2(10^{2.5}) = 10^{4.5},$$

then to particularize α and η in accordance with:

$$\alpha = \left(\frac{\omega_h}{\omega_b}\right)^{\frac{m}{N}} = \left(10^{4.5}\right)^{\frac{0.111}{5}} = 1.258$$

and

$$\eta = \left(\frac{\omega_h}{\omega_b}\right)^{\frac{1-m}{N}} = \left(10^{4.5}\right)^{\frac{1-0.111}{5}} = 6.311,$$

the product $\alpha\eta$ (always greater than one) is then:

$$\alpha\eta = 7.939 .$$

It is now possible to determine the transitional frequencies, $\omega_i^{'}$ and ω_i, of the controller according to the following relations:

$$\omega_1^{'} = \sqrt{\eta}\,\omega_b = 2.512(706.660) = 1,775.250\, rd\ /\ s$$

$$\omega_1 = \alpha\omega_1^{'} = 1.258(1775.250) = 2,233.264\, rd\ /\ s$$

$$\omega_2^{'} = (\alpha\eta)\omega_1^{'} = (7.939)1775.250 = 14,093.710\, rd\ /\ s$$

$$\omega_2 = (\alpha\eta)\omega_1 = (7.939)2233.264 = 17,729.883\, rd\ /\ s$$

$$\omega_3^{'} = (\alpha\eta)^2\,\omega_1^{'} = (63.028)1775.250 = 1,118.904\ 10^2\ rd\ /\ s$$

$$\omega_3 = (\alpha\eta)^2\,\omega_1 = (63.028)2233.264 = 1,407.581\ 10^2\ rd\ /\ s$$

$$\omega_4^{'} = (\alpha\eta)^3\,\omega_1^{'} = (500.377)1775.250 = 888.294\ 10^3\ rd\ /\ s$$

$$\omega_4 = (\alpha\eta)^3 \omega_1 = (500.377)2233.264 = 1{,}117.474\ 10^3\ rd\ /\ s$$

$$\omega_5^{'} = (\alpha\eta)^4 \omega_1^{'} = (3972.493)1775.250 = 705.2168\ 10^4\ rd\ /\ s$$

$$\omega_5 = (\alpha\eta)^4 \omega_1 = (3972.493)2233.264 = 887.1625\ 10^4\ rd\ /\ s\ .$$

NOTE.– An appendix dedicated to the first frequency relation,

$$\omega_1^{'} = \omega_b \eta^{\frac{1}{2}},$$

is provided at the end of the correction of the current problem, where the validity of this relation is demonstrated in graphical form.

6) Resonance frequency and ratio in tracking before and after correction

Given by

$$\omega_r = \left(-\cos n\frac{\pi}{2}\right)^{\frac{1}{n}} \omega_u \text{ and } Q = \frac{1}{\sin n\frac{\pi}{2}},$$

the control resonance frequency and ratio in tracking admit the following values:

before correction $(n = 3/2 = 1.5)$,

$$\omega_r = 99.739\ 10^3\ rd\ /\ s \text{ and } Q = \sqrt{2} = 1.414\ ;$$

after correction $(n = 1.388)$,

$$\omega_r = 84.073\ 10^3\ rd\ /\ s \text{ and } Q = 1.219\ .$$

7) Natural frequency and damping ratio before and after correction

Given by

$$\omega_p = \omega_u \sin\frac{\pi}{4} \text{ and } \zeta = -\cos\frac{\pi}{n},$$

the natural frequency and the damping ratio of the control admit the following numerical values:

before correction ($n = 3/2 = 1.5$),

$$\omega_p = 108.828\ 10^3\ rd/s \text{ and } \zeta = 0.5;$$

after correction ($n = 1.388$),

$$\omega_p = 96.709\ 10^3\ rd/s \text{ and } \zeta = 0.638.$$

8) Admissible variation interval of ω_0

Transmittance in open loop (after correction) admits the expression:

$$\beta(s) = \left(\frac{\omega_0}{s}\right)^n C_0 \left(\frac{1+\dfrac{s}{\omega_b}}{1+\dfrac{s}{\omega_h}}\right)^m \quad \text{with } n = 1.5 \text{ and } m = 0.111,$$

or, for $\omega_A \le \omega \le \omega_B$, therefore for $\omega_b << \omega << \omega_h$:

$$\beta(s) = \left(\frac{\omega_0}{s}\right)^n C_0 \left(\frac{\dfrac{s}{\omega_b}}{1}\right)^m,$$

or:

$$\beta(s) = \left(\frac{\omega_0}{s}\right)^n C_0 \left(\frac{\omega_b}{s}\right)^{-m},$$

or indeed:

$$\beta(s) = \frac{C_0 \omega_0^n \omega_b^{-m}}{s^{n-m}} = \frac{\left[\left(C_0 \omega_0^n \omega_b^{-m}\right)^{\frac{1}{n-m}}\right]^{n-m}}{s^{n-m}} = \left(\frac{\omega_u}{s}\right)^{n-m},$$

by writing:

$$\omega_u = \left(C_0 \omega_0^n \omega_b^{-m}\right)^{\frac{1}{n-m}}.$$

The variations of ω_u consecutive to the variations of ω_0 do not affect the phase margin as long as ω_u remains within the interval $[\omega_A, \omega_B]$, that is:

$$\omega_A \leq \omega_u \leq \omega_B,$$

therefore:

$$\omega_A \leq \left(C_0 \omega_0^n \omega_b^{-m}\right)^{\frac{1}{n-m}} \leq \omega_B,$$

or:

$$\omega_A^{n-m} \leq C_0 \omega_0^n \omega_b^{-m} \leq \omega_B^{n-m},$$

that is, ultimately:

$$\left(\frac{\omega_A^{n-m}}{C_0 \omega_b^{-m}}\right)^{\frac{1}{n}} \leq \omega_0 \leq \left(\frac{\omega_B^{n-m}}{C_0 \omega_b^{-m}}\right)^{\frac{1}{n}}.$$

Numerical application:

$$\left(\frac{\omega_A^{n-m}}{C_0 \omega_b^{-m}}\right)^{\frac{1}{n}} = 8,671.840,$$

$$\left(\frac{\omega_B^{n-m}}{C_0 \omega_b^{-m}}\right)^{\frac{1}{n}} = 17,815.42 \times 10^2,$$

therefore:

$$8,671.840 \leq \omega_0 \leq 17,815.42 \times 10^2,$$

that is, a ratio of 205.44 between the extremal values of ω_0, a ratio that we can compare to the ratio of 316.22 between the extremal values, ω_B and ω_A, of ω_u.

Appendix

In order to determine the ratio between the first transitional frequency, ω_1', of the real version of the constant-phase CRONE controller and the low transitional frequency, ω_b, of its ideal version, it is appropriate to analyze the asymptotic gain diagrams of the ideal version and the real version.

The asymptotic gain diagram of the ideal version comprises, successively (according to the increasing frequencies):

– a horizontal gain asymptote, C_0, for $\omega < \omega_b$;

– an asymptote of slope, $6\, m\, dB / oct$, for $\omega_b < \omega < \omega_h$;

– a horizontal gain asymptote, $C_0\, \omega_h / \omega_b$, for $\omega > \omega_h$.

The asymptotic gain diagram of the real version comprises, successively (again according to the increasing frequencies):

– a horizontal gain asymptote, C_0, for $\omega < \omega_1'$;

– an alternating sequence of ramps at $6\, dB / oct$ and stages (at $0\, dB / oct$), for $\omega_1' < \omega < \omega_N$;

– a horizontal gain asymptote, $C_0\, \omega_h / \omega_b$, for $\omega > \omega_N$.

It transpires that the gain smoothing straight-line of the alternating sequence of ramps and steps (which cuts through the stages halfway along, that is, at $(1/2)\log\eta = \log\eta^{1/2}$) identifies with the slope asymptote, $6\, m\, dB / oct$, of the ideal version between ω_b and ω_h. As a result, by immediately reading graphically:

$$\omega_b = \frac{\omega_1'}{\eta^{1/2}} \text{ and } \omega_h = \omega_N \eta^{1/2},$$

that is:

$$\omega_1' = \omega_b \eta^{1/2} \text{ and } \omega_N = \frac{\omega_h}{\eta^{1/2}},$$

remarkably simple results making it possible to express the first and last transitional frequency of the real version ($\omega_1^{'}$ and ω_N) from the low and high transitional frequencies of the ideal version (ω_b and ω_h).

Presentation of Problem 11

Synthesis of a Constant Phase-Lead CRONE Controller with Successively Symmetrical and Asymmetrical Frequency Placement

The study plant that forms the subject of a control (integer then non-integer) within the framework of this problem is carried out by a continuous-current CEM-AXEM, F9M4-type motor, for which the output considered is the angular position of the shaft.

Its mechanical and electrical characteristics, provided by the manufacturer, allow it to be defined by a second-order transmittance of the form (with the electrical transitional frequency not being taken into account):

$$G(s) = \frac{1}{\dfrac{s}{\omega_0'}\left(1+\dfrac{s}{\omega_0}\right)},$$

in which the nominal parameters are given by:

– $\omega_0' = 16.89\, rd/s$, the transition frequency (linked to the integration between speed and angular position);

– $\omega_0 = 50\ rd/s$, the mechanical transitional frequency.

The variation of the parametric state of the plant around its nominal parametric state is such that the transitional frequency, ω_0, initially equaling $50 rd/s$ (idle motor), can decrease to $5 rd / s$ (motor under load, its inertia on the shaft can reach 10 times that of the idle motor).

The plant thus defined is subject to a position control, ensured:

– *by a simple phase-lead controller, defined by the integer transmittance*

$$C(s) = C_0 \frac{1+s/\omega_1}{1+s/\omega_2}$$

or by a constant-phase CRONE controller, defined

– *where the ideal version is concerned, by non-integer transmittance*

$$C_m(s) = C_0' \left(\frac{1+s/\omega_b}{1+s/\omega_h} \right)^m$$

– *and for the real version, by integer transmittance*

$$C_N(s) = C_0' \prod_{i=1}^{N} \frac{1+s/\omega_i'}{1+s/\omega_i}.$$

The simple phase-lead controller and then the constant-phase CRONE controller provide the same phase lead of 50° at the frequency, ω_u, set by the designer, at 10 times the transitional frequency, ω_0, of the idle motor, that is, $\omega_u = 10\omega_0 = 500 rd / s$.

1) Determine the control phase margin. Does it seem suitable?

2) In the case of the simple phase-lead controller whose frequency placement is (geometrically) symmetrical to the frequency, ω_u, that is, $(\omega_1\omega_2)^{1/2} = \omega_u$, determine its set of parameters.

3) In the case of the constant-phase CRONE controller whose frequency placement is (geometrically) symmetrical to the frequency, ω_u, that is, $(\omega_b\omega_h)^{1/2} = \omega_u$, determine the parameters of its ideal version in the event that it presents a constant phase over 1.3 decades (frequency ratio of 20, for simplicity). To this end, we will use $\omega_A = 10\omega_b$ and $\omega_B = \omega_h/10$ to denote the extremal

frequencies of the phase locking range, which is equivalent to admitting the approximation considering that the phase reaches the phase locking (or asymptotic behavior) at these frequencies.

4) Determine the input sensitivity (of the plant) at high frequency with the simple phase-lead controller and then with the constant-phase CRONE controller. Conclude which controller provides the best input immunity.

5) In order to improve the input immunity with the CRONE, the designer takes advantage of the fact that the unit gain frequency in open loop decreases when the inertia increases, to place the controller asymmetrically to the frequency, ω_u, in accordance with $\omega_B = \omega_u$. Then determine the new parameters of its ideal version, and the parameters of its real version sized by five zeros and five poles.

6) Determine the new high-frequency input sensitivity and conclude with the improvement of the corresponding immunity.

Solution to Problem 11

1) Phase margin

By definition, the phase margin is written as:

$$\Phi_m^\circ = \arg \beta(j\omega_u) - (-180),$$

or:

$$\Phi_m^\circ = \arg C(j\omega_u) + \arg G(j\omega_u) + 180,$$

or, given the controller and the plant:

$$\Phi_m^\circ = 50 - 90 - \arctan\frac{\omega_u}{\omega_0} + 180,$$

or indeed, numerically:

$$\Phi_m^\circ = 140 - \arctan 10 = 140 - 84.29,$$

therefore:

$$\Phi_m^\circ = 55.71.$$

This phase margin is suitable considering the recommendation in this field.

2) Simple phase-lead controller

2.1) Transitional frequencies, ω_1 and ω_2

The maximum phase lead is given by:

$$\varphi_m = \arcsin\frac{\alpha - 1}{\alpha + 1} \text{ with } \alpha = \frac{\omega_2}{\omega_1} > 1,$$

from which we obtain:

$$\sin\varphi_m = \frac{\alpha - 1}{\alpha + 1},$$

that is:

$$\alpha - 1 = (\alpha + 1)\sin\varphi_m,$$

or:

$$\alpha(1 - \sin\varphi_m) = 1 + \sin\varphi_m,$$

or indeed:

$$\alpha = \frac{1 + \sin\varphi_m}{1 - \sin\varphi_m} = \frac{1 + \sin 50}{1 - \sin 50} = \frac{1 + 0.766}{1 - 0.766} = 7.55.$$

The transitional frequencies, ω_1 and ω_2, then admit the following expressions:

$$\omega_1 = \frac{\omega_u}{\sqrt{\alpha}} = \frac{500}{\sqrt{7.55}} = 181.96\, rd\,/\,s$$

and

$$\omega_2 = \omega_u \sqrt{\alpha} = 500\sqrt{7.55} = 1{,}373.86\, rd\,/\,s.$$

2.2) Static gain C_0

As

$$|\beta(j\omega_u)| = 1,$$

we have:

$$|C(j\omega_u)||G(j\omega_u)| = 1\,,$$

therefore:

$$C_0 \left(\frac{1 + \left(\dfrac{\omega_u}{\omega_1} \right)^2}{1 + \left(\dfrac{\omega_u}{\omega_2} \right)^2} \right)^{\frac{1}{2}} \frac{1}{\dfrac{\omega_u}{\omega_0'} \left(1 + \left(\dfrac{\omega_u}{\omega_0} \right)^2 \right)^{\frac{1}{2}}} = 1,$$

from which we obtain C_0:

$$C_0 = \frac{\omega_u}{\omega_0'} \left(1 + \left(\frac{\omega_u}{\omega_0} \right)^2 \right)^{\frac{1}{2}} \left(\frac{1 + \left(\dfrac{\omega_u}{\omega_2} \right)^2}{1 + \left(\dfrac{\omega_u}{\omega_1} \right)^2} \right)^{\frac{1}{2}},$$

or, numerically:

$$C_0 = \frac{500}{16.89} \left(1 + \left(\frac{500}{50} \right)^2 \right)^{\frac{1}{2}} \left(\frac{1 + \left(\dfrac{500}{1{,}373.86} \right)^2}{1 + \left(\dfrac{500}{181.96} \right)^2} \right)^{\frac{1}{2}},$$

that is, ultimately:

$$C_0 = (29.60)(10.05)(0.363) = 107.98.$$

3) Constant-phase CRONE controller

Figure P11.1 presents the frequency configuration corresponding to a symmetrical frequency placement with respect to the frequency ω_u; this placement is defined by $(\omega_b \omega_h)^{1/2} = \omega_u$.

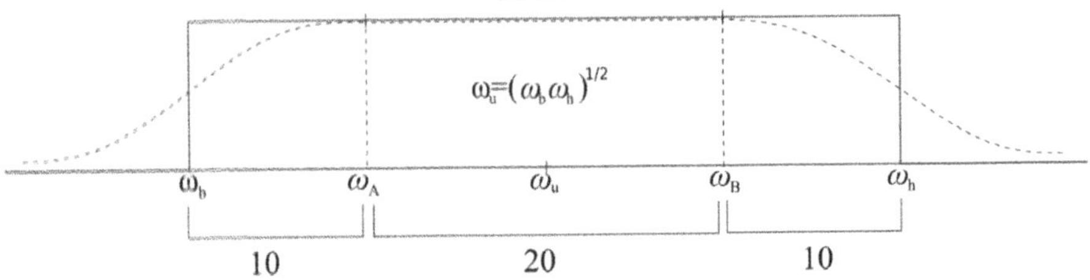

Figure P11.1. *Symmetrical frequency placement of the controller*

3.1) *Transitional frequencies,* ω_b *and* ω_h

The extremal frequencies, ω_A and ω_B, such that

$$\frac{\omega_B}{\omega_A} = 10^{1.3} = 19.95 \simeq 20,$$

are given by:

$$\omega_A = \frac{\omega_u}{\sqrt{20}} \text{ and } \omega_B = \omega_u \sqrt{20}.$$

The transitional frequencies, ω_b and ω_h, meanwhile, are given by:

$$\omega_b = \frac{\omega_A}{10} = \frac{\omega_u}{10\sqrt{20}} = \frac{500}{10\sqrt{20}} = 11.18 rd / s$$

and

$$\omega_h = 10\omega_B = \omega_u 10\sqrt{20} = 500\left(10\sqrt{20}\right) = 22360.68\, rd / s.$$

3.2) *m order*

The expression of the controller phase lead makes it possible to write:

$$m90° = 50°,$$

from which we obtain:

$$m = \frac{50°}{90°} = 0.55\,.$$

3.3) Static gain C_0'

As

$$\left|\beta(j\omega_u)\right| = 1,$$

we have:

$$\left|C_m(j\omega_u)\right|\left|G(j\omega_u)\right| = 1,$$

therefore:

$$C_0'\left(\frac{1+\left(\frac{\omega_u}{\omega_b}\right)^2}{1+\left(\frac{\omega_u}{\omega_h}\right)^2}\right)^{\frac{m}{2}} \frac{1}{\frac{\omega_u}{\omega_0'}\left(1+\left(\frac{\omega_u}{\omega_0}\right)^2\right)^{\frac{1}{2}}} = 1\,,$$

from which we obtain C_0':

$$C_0' = \frac{\omega_u}{\omega_0'}\left(1+\left(\frac{\omega_u}{\omega_0}\right)^2\right)^{\frac{1}{2}}\left(\frac{1+\left(\frac{\omega_u}{\omega_h}\right)^2}{1+\left(\frac{\omega_u}{\omega_b}\right)^2}\right)^{\frac{m}{2}},$$

or, numerically:

$$C_0' = \frac{500}{16.89}\left(1+\left(\frac{500}{50}\right)^2\right)^{\frac{1}{2}}\left(\frac{1+\left(\frac{500}{22{,}360.68}\right)^2}{1+\left(\frac{500}{11.18}\right)^2}\right)^{0.27},$$

that is, ultimately:

$$C_0' = (29.60)(10.05)(0.123) = 36.59.$$

4) Input sensitivity in HF

As the high-frequency input sensitivity is reduced to the regulator gain at these frequencies, we can write:

$$S(\omega)_{HF} = \left|C(j\omega)\right|_{HF} = \begin{cases} C_0\dfrac{\omega_2}{\omega_1} & \text{for } C(s) \\ C_0'\left(\dfrac{\omega_h}{\omega_b}\right)^m & \text{for } C_m(s) \end{cases},$$

or, numerically:

$$C_0\frac{\omega_2}{\omega_1} = C_0\alpha = (107.98)(7.55) = 815.25$$

and

$$C_0'\left(\frac{\omega_h}{\omega_b}\right)^m = 36.59\left(\frac{22,360.68}{11.18}\right)^{0.55} = 36.59(2,000)^{0.55} = 2,392.92.$$

The simple phase-lead controller ensures the best input immunity since $1/815.25 > 1/2,392.92$.

5) Asymmetrical placement of constant-phase CRONE controller

Figure P11.2 illustrates an asymmetrical frequency placement of the constant-phase CRONE controller, which is defined in this case by $\omega_B = \omega_u$.

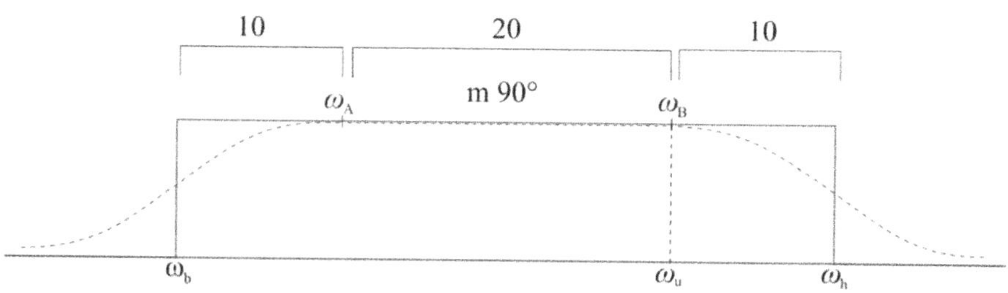

Figure P11.2. *Asymmetrical frequency placement of the controller*

5.1) *Ideal version*

5.1.1) Transitional frequencies, ω_b and ω_h

The extremal frequencies, ω_A and ω_B, such that

$$\frac{\omega_B}{\omega_A} = 10^{1.3} = 19.95 \simeq 20,$$

are given here by:

$$\omega_A = \frac{\omega_u}{20} \text{ and } \omega_B = \omega_u.$$

The transitional frequencies, ω_b and ω_h, meanwhile, are given by:

$$\omega_b = \frac{\omega_A}{10} = \frac{\omega_u}{200} = \frac{500}{200} = 2.5\,rd\,/\,s$$

and

$$\omega_h = 10\omega_B = 10\omega_u = (10)(500) = 5{,}000\,\text{rd/s}\,.$$

5.1.2) *m* order

The expression of the controller phase lead makes it possible to write:

$$m90° = 50°,$$

from which we obtain:

$$m = \frac{50}{90} = 0.55\,.$$

5.1.3) Static gain C_0'

As the literal expression of C_0' is the same as previously mentioned, that is,

$$C_0' = \frac{\omega_u}{\omega_0'}\left(1+\left(\frac{\omega_u}{\omega_0}\right)^2\right)^{\frac{1}{2}}\left(\frac{1+\left(\dfrac{\omega_u}{\omega_h}\right)^2}{1+\left(\dfrac{\omega_u}{\omega_b}\right)^2}\right)^{\frac{m}{2}},$$

it is particularized here in accordance with the new numerical values of the parameters:

$$C_0' = (29.60)(10.05)\left(\frac{1+\left(\dfrac{500}{5{,}000}\right)^2}{1+\left(\dfrac{500}{2.5}\right)^2}\right)^{0.275},$$

that is, ultimately:

$$C_0' = (29.60)(10.05)\left(\frac{1.01}{40,001}\right)^{0.275} = 16.18.$$

5.2) *Real version*

As static gain C_0' is the same as for the ideal version, all that remains is to determine the transitional frequencies, ω_i' and ω_i. To this end, the recursive factors, α and η, are numerically particularized, in accordance with:

$$\alpha = \left(\frac{\omega_h}{\omega_b}\right)^{\frac{m}{N}} = 2,000^{\frac{0.55}{5}} = 2.30$$

and

$$\eta = \left(\frac{\omega_h}{\omega_b}\right)^{\frac{1-m}{N}} = 2,000^{\frac{0.45}{5}} = 1.98,$$

the product $\alpha\eta$ (always greater than one) is then worth:

$$\alpha\eta = 4.554,$$

and the transitional frequencies deduced therefrom are:

$$\omega_1' = \sqrt{\eta}\,\omega_b = \sqrt{1.98}\,(2.5) = 3.52\,rd\,/\,s$$

$$\omega_1 = \alpha\omega_1' = 2.30(3.52) = 8.09\,rd\,/\,s$$

$$\omega_2' = (\alpha\eta)\,\omega_1' = 4.554(3.52) = 16.03\,rd\,/\,s$$

$$\omega_2 = (\alpha\eta)\,\omega_1 = 4.554(8.09) = 36.84\,rd\,/\,s$$

$$\omega_3' = (\alpha\eta)^2\,\omega_1' = (4.554)^2\,(3.52) = 73.00\,rd\,/\,s$$

$$\omega_3 = (\alpha\eta)^2 \omega_1 = (4.554)^2 (8.09) = 167.77\, rd\ /\ s$$

$$\omega_4' = (\alpha\eta)^3 \omega_1' = (4.554)^3 (3.52) = 332.44\, rd\ /\ s$$

$$\omega_4 = (\alpha\eta)^3 \omega_1 = (4.554)^3 (8.09) = 764.06\, rd\ /\ s$$

$$\omega_5' = (\alpha\eta)^4 \omega_1' = (4.554)^4 (3.52) = 1513.96\, rd\ /\ s$$

$$\omega_5 = (\alpha\eta)^4 \omega_1 = (4.554)^4 (8.09) = 3479.53\, rd\ /\ s.$$

6) New input sensitivity in HF

Given the new parameters, the high-frequency input sensitivity becomes:

$$S(\omega)_{HF} = C_0' \left(\frac{\omega_h}{\omega_b} \right)^m = (16.18)2{,}000^{0.55} = 1{,}058.14\,.$$

Changing from 2,392.92 to 1,058.14, the input sensitivity has substantially decreased, even approaching that obtained with the simple phase-lead controller, namely 815.25. The input immunity has therefore clearly improved.

Presentation of Problem 12

Synthesis of a Constant Phase-Lag CRONE Controller and Synthesis Parameters of the Third-Generation CRONE Control

Let us consider a plant defined by a transmittance of the form:

$$G(s) = G_0 \frac{1 + \dfrac{\omega_0}{s}}{1 + \dfrac{s}{\omega_0'}},$$

in which:

$$G_0 = 10$$
$$\omega_0 = 1 \text{ rd/s}$$
$$\omega_0' = 5 \text{ rd/s}.$$

The plant thus defined behaves asymptotically like:

– an integrator for $\omega < \omega_0$;

– a proportional for $\omega_0 < \omega < \omega_0'$;

– an integrator for $\omega > \omega_0'$.

The plant is integrated into an elementary control loop.

1) In the absence of phase correction, with the regulator then being reduced to a simple proportional, determine the phase margin of the control at the unit gain frequency in open loop, ω_u, set by the designer at 50 times the transitional frequency, ω_0', of the plant, that is, $\omega_u = 50\omega_0'$.

Discuss, first, the phase margin thus obtained and, second, the choice of frequency, ω_u.

2) Considering that the reduced first overshoot, D_1' (as a percentage) of the step response in closed loop is given by a polynomial approximation defined by the equation

$$D_1' = an^2 + bn + c,$$

with $a = 79.195$, $b = -138.507$ and $c = 59.528$, determine the second-generation CRONE control order that ensures a value of D_1' deemed suitable by the designer, in this case, 10%.

3) It transpires that the regulator enabling the vertical template of the second-generation CRONE control to be obtained is, within the study context, a constant-phase CRONE regulator:

– the ideal version of which is defined by:

$$C_m(s) = C_0 \left(\frac{1 + \frac{s}{\omega_b}}{1 + \frac{s}{\omega_h}} \right)^m$$

– the real version of which is defined by:

$$C_N(s) = C_0 \prod_{i=1}^{N} \frac{1 + \frac{s}{\omega_i'}}{1 + \frac{s}{\omega_i}}.$$

3.1) Determine the parameters of the ideal version, $C_m(s)$, that ensure, in open loop, a vertical template extending across one and a half decades centered geometrically on $\omega_u = 50\omega_0'$.

3.2) Determine the parameters of the real version, $C_N(s)$, based on four zeros and four poles.

4) Determine the resonance frequencies and the resonance ratio in tracking and regulation.

5) Determine the natural frequency and the damping ratio in tracking and regulation.

6) In view of the results thus obtained, conclude as to the robustness of the dynamic performances of the control.

7) Using the legitimate approximations dictated by the nature of the presentation of the problem, determine the extremal values of the interval in which G_0 can vary without any degrading of the phase margin.

8) Let us now look to the general form of the open-loop transmittance of the third-generation CRONE control.

The open-loop transition frequency, ω_u, determines the rapidity of the closed-loop dynamics.

The gain, K, is an adjustment parameter enabling the frequency, ω_u, to be set at the value chosen by the user.

The four parameters, a and b, real and imaginary (non-integer) integration orders, as well as ω_b and ω_h, extremal transitional frequencies of the generalized template, are determined by optimization after the parameters, ω_u, n_b and n_h, have been set.

The parameters, n_b and n_h, are determined by the user:

– the power, n_b, is the (integer) order of the open-loop behavior at low frequency;

– the power, n_h, is the (integer) order of the open-loop behavior at high frequency.

8.1) For a second-order plant, determine n_b and n_h such that the step error is nil and the high-frequency input sensitivity remains constant.

8.2) Same question for a nil ramp error and for a high-frequency input sensitivity that decreases by 6dB/oct.

Solution to Problem 12

As a preamble, it is convenient to rapidly plot (or sketch) the Bode asymptotic diagrams of the plant by presenting the frequency, ω_u (Figure P12.1).

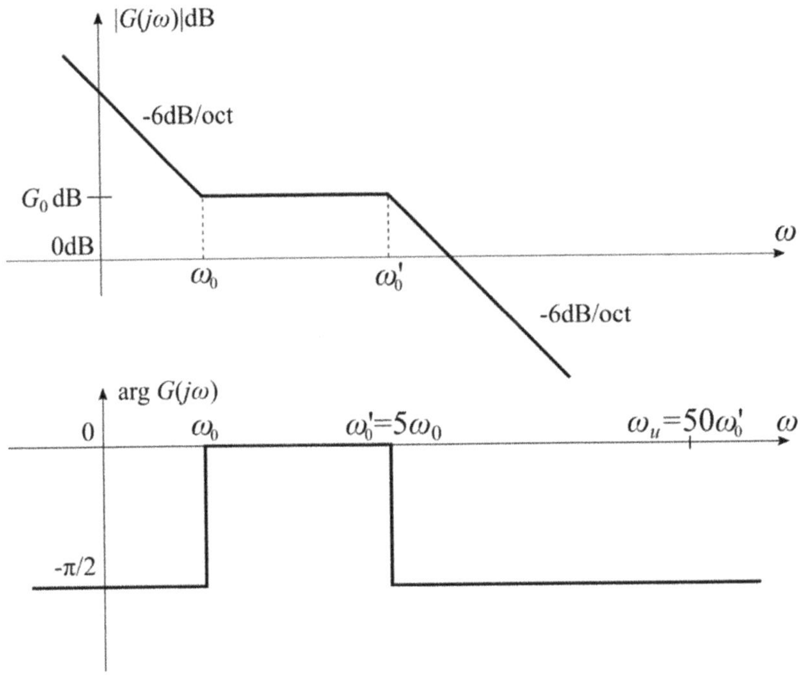

Figure P12.1. *Shape of plant Bode asymptotic diagrams*

1) Phase margin

Given the asymptotic phase diagram of the plant and the significant ratio between ω_u and ω_0', we can write, as a first approximation:

$$\arg G(j\omega_u) = -90° .$$

The phase margin, Φ_m, which, for $\arg C(j\omega) = 0$, is reduced to

$$\Phi_m = \arg G(j\omega_u) + \pi ,$$

can then be particularized in accordance with:

$$\Phi_m = -90° + 180° = 90°.$$

The phase margin thus obtained is too large as it is significantly greater than the recommended interval, $[50°, 55°]$.

Considering that the shape of the asymptotic gain diagram does not help to define a cut-off frequency, ω_c, it proves difficult to seek to justify the choice of frequency, ω_u, with respect to the frequency, ω_c. There is therefore no further comment to be made on this matter, other than that the choice of frequency, ω_u, such that $\omega_u >> \omega_0'$, ensures phase locking of the plant at $-\pi/2$ around the frequency, ω_u.

2) Second-generation CRONE control order

The polynomial equation as proposed in the presentation of the problem, that is,

$$D_1' = an^2 + bn + c,$$

may be rewritten in the form:

$$an^2 + bn + c - D_1' = 0,$$

or:

$$an^2 + bn + c' = 0 \text{ with } c' = c - D_1' = 49.528.$$

The roots of this second-degree equation are given by:

$$n = \frac{-b \pm \sqrt{\Delta}}{2a},$$

where

$$\Delta = b^2 - 4ac' = 3{,}494.709 \text{ and } \sqrt{\Delta} = 59.116,$$

therefore:

$$n = \frac{138.507 \pm 59.116}{2(79.195)},$$

from which we obtain the two roots, n_+ and n_-, such that:

$$n_+ = 1.247 \text{ and } n_- = 0.501.$$

Considering that the order, n, of the second-generation CRONE control is between 1 and 2, that is, $1 < n < 2$, only the first root, n_+, is suitable, therefore:

$$n = n_+ = 1.247\,.$$

3.1) *Parameters of the ideal version of the regulator*

The Nichols loci, around the frequency, ω_u, of the frequency responses in open loop in the absence of phase correction and in the presence of a constant-phase CRONE regulator, illustrate the phase that needs to be brought by the regulator, $C_m(j\omega)$, around the frequency, ω_u, that is $m\pi/2$ (Figure P12.2).

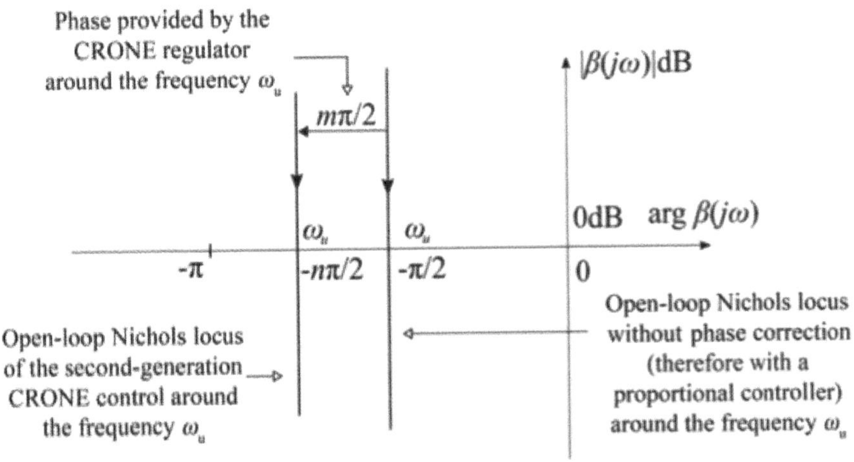

Figure P12.2. *Open-loop Nichols loci (around the frequency ω_u) with a proportional controller and a CRONE controller*

This figure shows that the regulator phase brought, $m\pi/2$, is expressed by the difference:

$$m\frac{\pi}{2} = -n\frac{\pi}{2} - \left(-\frac{\pi}{2}\right) = -n\frac{\pi}{2} + \frac{\pi}{2} = \frac{\pi}{2}(1-n),$$

from which we obtain the m order of the regulator:

$$m = 1 - n = 1 - 1.247 = -0.247 \cdot$$

The extremal frequencies, ω_A and ω_B, of the vertical template of the second-generation CRONE control are given by:

$$\omega_A = \frac{\omega_u}{\sqrt{10^{1.5}}} = \frac{250}{\sqrt{10^{1.5}}} = 44.45 \text{ rd/s}$$

and

$$\omega_B = \omega_u \sqrt{10^{1.5}} = 250\sqrt{10^{1.5}} = 1{,}405.85 \text{ rd/s} \cdot$$

With regard to the transitional frequencies, ω_b and ω_h, of the regulator, these are deduced from ω_A and ω_B in accordance with the relations:

$$\omega_b = \frac{\omega_A}{10} = 4.445 \text{ rd/s}$$

and

$$\omega_h = 10\omega_B = 14{,}058.5 \text{ rd/s} \cdot$$

To determine the static gain, C_0, of the regulator, it is sufficient to use relation, $|\beta(j\omega_u)| = 1$, which enables us to write:

$$|C(j\omega_u)||G(j\omega_u)| = 1,$$

that is, in this case:

$$C_0\left(\frac{1+\left(\dfrac{\omega_u}{\omega_b}\right)^2}{1+\left(\dfrac{\omega_u}{\omega_h}\right)^2}\right)^{\frac{m}{2}} G_0\left(\frac{1+\left(\dfrac{\omega_0}{\omega_u}\right)^2}{1+\left(\dfrac{\omega_u}{\omega_0'}\right)^2}\right)^{\frac{1}{2}} = 1,$$

from which we obtain:

$$C_0 = \frac{1}{G_0}\left(\frac{1+\left(\dfrac{\omega_u}{\omega_h}\right)^2}{1+\left(\dfrac{\omega_u}{\omega_b}\right)^2}\right)^{\frac{m}{2}}\left(\frac{1+\left(\dfrac{\omega_u}{\omega_0'}\right)^2}{1+\left(\dfrac{\omega_0}{\omega_u}\right)^2}\right)^{\frac{1}{2}},$$

or, numerically:

$$C_0 = \frac{1}{10}\left(\frac{1.0003}{3164.27}\right)^{-0.1235}(2501)^{\frac{1}{2}},$$

therefore:

$$C_0 = 13.53.$$

3.2) *Parameters of the real version of the regulator*

While the relations

$$\alpha = \left(\frac{\omega_h}{\omega_b}\right)^{\frac{m}{N}} \text{ and } \eta = \left(\frac{\omega_h}{\omega_b}\right)^{\frac{1-m}{N}},$$

are generally used for positive m, and for α and η greater than 1, they remain valid for negative m, and for values of α and η that are both not greater than 1 (but are always greater than zero). This is why they are used here to determine α and η:

$$\alpha = \left(\frac{14,058.5}{4.445}\right)^{-\frac{0.247}{4}} = 0.608$$

and

$$\eta = \left(\frac{14,058.5}{4.445} \right)^{\frac{1.247}{4}} = 12.335,$$

the product, $\alpha\eta$ (always greater than one) is then worth:

$$\alpha\eta = 7.5.$$

It is now possible to determine the transitional frequencies, ω_i' and ω_i, of the regulator in accordance with the following relations:

$$\omega_1' = \sqrt{\eta}\,\omega_b = \sqrt{12.335}\,(4.445) = 15.611 \text{ rd/s}$$
$$\omega_1 = \alpha\omega_1' = 0.608(15.611) = 9.491 \text{ rd/s}$$
$$\omega_2' = (\alpha\eta)\omega_1' = 7.5(15.611) = 117.082 \text{ rd/s}$$
$$\omega_2 = (\alpha\eta)\omega_1 = 7.5(9.491) = 71.182 \text{ rd/s}$$
$$\omega_3' = (\alpha\eta)^2\,\omega_1' = (7.5)^2\,(15.611) = 878.118 \text{ rd/s}$$
$$\omega_3 = (\alpha\eta)^2\,\omega_1 = (7.5)^2\,(9.491) = 533.868 \text{ rd/s}$$
$$\omega_4' = (\alpha\eta)^3\,\omega_1' = (7.5)^3\,(15.611) = 6,585.890 \text{ rd/s}$$
$$\omega_4 = (\alpha\eta)^3\,\omega_1 = (7.5)^3\,(9.491) = 4,004.015 \text{ rd/s}.$$

4) Resonance frequencies and ratio

The resonance frequencies and ratio of the control are given by:

$$\omega_{r_r} = \left(-\cos n\frac{\pi}{2} \right)^{\frac{1}{n}} \omega_u = 144.66 \text{ rd/s},$$

$$\omega_{r_d} = \left(-\cos n\frac{\pi}{2} \right)^{-\frac{1}{n}} \omega_u = 545.08 \text{ rd/s}$$

and

$$Q = \frac{1}{\sin n\frac{\pi}{2}} = 1.08\,,$$

the numerical values of the resonance frequencies clearly satisfy the relation $\left(\omega_{r_r}\omega_{r_d}\right)^{1/2} = \omega_u$.

5) Natural frequency and damping ratio

The natural frequency and the damping ratio of the control are given by:

$$\omega_p = \omega_u \sin\frac{\pi}{n} = 145.72 \text{ rd/s}$$

and

$$\zeta = -\cos\frac{\pi}{n} = 0.81\,.$$

6) Control robustness

The resonance ratio and the damping ratio are exclusively linked to the n order of the control, enabling the introduction of the concepts of *robust resonance* and *robust oscillatory mode*, and a conclusion thus to be drawn as to the robustness of the dynamic performances of the control.

7) Admissible variation interval of G_0

The transmittance in open loop admits the expression:

$$\beta(s) = C(s)G(s)\,,$$

that is:

$$\beta(s) = C_0\left(\frac{1+\dfrac{s}{\omega_b}}{1+\dfrac{s}{\omega_h}}\right)^m G_0\frac{1+\dfrac{\omega_0}{s}}{1+\dfrac{s}{\omega_0'}}\,,$$

or, for $\omega_A \le \omega \le \omega_B$, therefore for $\omega_b << \omega << \omega_h$, and for $\omega >> \omega_0$ and ω_0':

$$\beta(s) = C_0 G_0 \left(\frac{s}{\omega_b} \right)^m \frac{\omega_0'}{s} = C_0 G_0 \frac{\omega_0'}{\omega_b{}^m} \frac{1}{s^{1-m}},$$

or:

$$\beta(s) = \left[\left(C_0 G_0 \frac{\omega_0'}{\omega_b{}^m} \right)^{\frac{1}{1-m}} \right]^{1-m} \frac{1}{s^{1-m}},$$

or indeed:

$$\beta(s) = \left(\frac{\omega_u}{s} \right)^{1-m},$$

by writing:

$$\omega_u = \left(C_0 G_0 \frac{\omega_0'}{\omega_b{}^m} \right)^{\frac{1}{1-m}},$$

or, considering that $m = 1 - n$:

$$\omega_u = \left(C_0 G_0 \frac{\omega_0'}{\omega_b{}^m} \right)^{\frac{1}{n}}.$$

The variations of ω_u consecutive to the variations of G_0 do not affect the phase margin as long as ω_u remains within the interval $[\omega_A, \omega_B]$, that is:

$$\omega_A \le \omega_u \le \omega_B,$$

therefore:

$$\omega_A \le \left(C_0 G_0 \frac{\omega_0'}{\omega_b{}^m} \right)^{\frac{1}{n}} \le \omega_B,$$

or:

$$\omega_A{}^n \leq C_0 G_0 \frac{\omega_0'}{\omega_b{}^m} \leq \omega_B{}^n ,$$

that is, ultimately:

$$\frac{\omega_b{}^m}{\omega_0'} \frac{\omega_A{}^n}{C_0} \leq G_0 \leq \frac{\omega_b{}^m}{\omega_0'} \frac{\omega_B{}^n}{C_0} .$$

Numerical application:

$$\frac{\omega_b{}^m}{\omega_0'} \frac{\omega_A{}^n}{C_0} = \frac{4.445^{-0.247}}{5} \frac{44.45^{1.247}}{13.53} = 1.157,$$

$$\frac{\omega_b{}^m}{\omega_0'} \frac{\omega_B{}^n}{C_0} = \frac{4.445^{-0.247}}{5} \frac{1,405.85^{1.247}}{13.53} = 85.917,$$

therefore:

$$1.157 \leq G_0 \leq 85.917 \cdot$$

8) Third-generation CRONE control

As a preamble to the third-generation CRONE control, it is appropriate to recall, or indeed to specify, the structure and parameters of the transfer in open loop, which integrates the generalized template.

The generalized template is described in principle by the *restriction in the operational plane*, $\mathbb{C}_j$, of the transmittance of a complex, non-integer integrator, that is:

$$\beta^*(s) = \left[\left(\frac{\omega_u}{s} \right)^n \right]_{\mathbb{C}_j} ,$$

with $s \in \mathbb{C}_j$ $(s = \sigma + j\omega)$ and $n \in \mathbb{C}_i$ $(n = a + ib)$, or, since $n = a + ib$:

$$\beta^*(s) = \left[\left(\frac{\omega_u}{s} \right)^{a+ib} \right]_{\mathbb{C}_j} = \left[\left(\frac{\omega_u}{s} \right)^a \left(\frac{\omega_u}{s} \right)^{ib} \right]_{\mathbb{C}_j} ,$$

or:

$$\beta^*(s) = \left(\frac{\omega_u}{s}\right)^a \left[\left(\frac{\omega_u}{s}\right)^{ib}\right]_{\mathbb{C}_j}$$

$$= \left(\frac{\omega_u}{s}\right)^a \left[\cos\left(b\ln\left(\frac{\omega_u}{s}\right)\right) + i\sin\left(b\ln\left(\frac{\omega_u}{s}\right)\right)\right]_{\mathbb{C}_j}$$

$$= \left(\frac{\omega_u}{s}\right)^a \cos\left(b\ln\left(\frac{\omega_u}{s}\right)\right).$$

In fact, given the developments conducted in [OUS 95a], we can show that the generalized template is subsequently described by a more complex transmittance of the form:

$$\beta^*(s) = \left(\cosh\left(b\frac{\pi}{2}\right)\right)^{sign(b)} \left(\frac{\omega_u}{s}\right)^a \left(\left[\left(\frac{\omega_u}{s}\right)^{ib}\right]_{\mathbb{C}_j}\right)^{-sign(b)},$$

or, by a truncation of this transmittance in accordance with that of the generalized template (this truncation introducing the low and high transitional frequencies, ω_b and ω_h):

$$\beta^*(s) = C^{sign(b)} \left(C_0 \frac{1+\dfrac{s}{\omega_h}}{1+\dfrac{s}{\omega_b}}\right)^a \left(\left[\left(C_0 \frac{1+\dfrac{s}{\omega_h}}{1+\dfrac{s}{\omega_b}}\right)^{ib}\right]_{\mathbb{C}_j}\right)^{-sign(b)},$$

with:

$$C_0 - \left(\frac{1+\left(\dfrac{\omega_u}{\omega_b}\right)^2}{1+\left(\dfrac{\omega_u}{\omega_h}\right)^2}\right)^{1/2} = \left[\frac{\omega_u}{\omega_b}\right]_{\omega_u=(\omega_b\omega_h)^{1/2}}$$

and

$$C = \cosh\left[b\left(\arctan\frac{\omega_u}{\omega_b} - \arctan\frac{\omega_u}{\omega_h}\right)\right],$$

C ensuring a unit gain to $\beta^*(s)$ at the frequency, ω_u, and being iteratively calculated during optimization.

Thus, the open-loop transmittance taking into account the generalized template admits a general expression of the form:

$$\beta(s) = \beta_b(s)\beta^*(s)\beta_h(s),$$

in which:

$$\beta_b(s) = C_b\left(\frac{\omega_b}{s} + 1\right)^{n_b}$$

and

$$\beta_h(s) = \frac{C_h}{\left(1 + \frac{s}{\omega_h}\right)^{n_h}}.$$

C_b and C_h ensure a unit gain to $\beta_b(s)$ and $\beta_h(s)$ for the frequency, ω_u. They are calculated iteratively during optimization in accordance with the following relations, deduced from $|\beta_b(j\omega_u)| = 1$ and $|\beta_h(j\omega_u)| = 1$:

$$C_b = \frac{1}{\left(1 + \left(\frac{\omega_b}{\omega_u}\right)^2\right)^{\frac{n_b}{2}}}$$

and

$$C_h = \left(1 + \left(\frac{\omega_u}{\omega_h}\right)^2\right)^{\frac{n_h}{2}}.$$

These expressions are worth comparing to those of C_0 and C, in the sense that all of these expressions enable (transparent) calculation of the parameters, C_0, C, C_b and C_h, as a function of the frequencies, ω_u, ω_b and ω_h, and of the n_b and n_h orders

8.1) *Step error and input sensitivity*

For the step error to be nil, the open-loop behavior must present at least one integration at low frequency, therefore:

$$n_b = 1 \cdot$$

With regard to the n_h parameter, we can write:

n_h (order in open loop at HF) =
order of the plant at HF + order of the regulator at HF,

that is, since

$$|C(j\omega)|_{\text{HF}} = S(\omega)_{\text{HF}}:$$

n_h = order of the plant at HF (i.e. 2)
+ order corresponding to the sensitivity
of the input at HF (in this case, 0),

therefore:

$$n_h = 2 \cdot$$

8.2) *Ramp error and input sensitivity*

For the ramp error to be nil, the open-loop behavior must present at least one double integration at low frequency, therefore:

$$n_b = 2 \cdot$$

With regard to the n_h parameter, we can write:

n_h (order in open loop at HF) =
order of the plant at HF + order of the regulator at HF,

that is, since

$$\left|C(j\omega)\right|_{\mathrm{HF}} = S(\omega)_{\mathrm{HF}}:$$

n_h = order of the plant at HF (i.e. 2)
+ order corresponding to the sensitivity
of the input at HF (in this case, 1),

therefore:

$$n_h = 3.$$

Presentation of Problem 13

Synthesis of a Variable-Phase CRONE Controller for the Synthesis of a Narrow-Band (Vertical and Generalized) Template

This problem is dedicated to determining a variable-phase CRONE regulator, required by the synthesis of both a vertical template for the second generation and a generalized template for the third generation, when the unit gain frequency in open loop, ω_u, does not belong to an asymptotic behavior of the plant. The study framework belongs to a wider context: that of the *synthesis of a narrow-band template*. This synthesis is conducted in a narrow band of frequencies geometrically centered on the frequency, ω_u. Taking such a frequency band into consideration results in an analytical synthesis of the regulator, whose calculation difficulty is brought down to a level required for a practical task.

Let there be a second-order plant described by a transmittance of the form:

$$G(s) = \frac{1}{\dfrac{s}{\omega_0'}\left(1+\dfrac{s}{\omega_0}\right)},$$

with:

$$\omega_0' = 10\sqrt{2}\ \text{rd/s and}\ \omega_0 = 10\ \text{rd/s},$$

ω_0' being the transition frequency linked to the integration and ω_0 the transitional frequency.

1) With the plant integrated into an elementary control loop, show that in the absence of correction, the unit gain frequency in open loop, ω_u, is equal to the transitional frequency, ω_0, of the plant, then deduce therefrom the phase margin, Φ_m.

2) It transpires that the frequency, ω_u, thus determined satisfies the specifications imposed by the designer concerning the rapidity of the control dynamics. In considering a regulator in cascade with the plant, what should the value of its gain be at this frequency?

3) We wish for the corrected open-loop frequency response locus to simultaneously tangent the 3dB amplitude contours of the Nichols chart and its inverse. To this end, we introduce a regulator (aimed at compensating locally for the plant phase variation), whose frequency response is of the form:

$$C(j\omega) = C_0 \left(\frac{\left(1 + j\dfrac{\omega}{\omega_1'}\right)\left(1 + j\dfrac{\omega}{\omega_2'}\right)}{\left(1 + j\dfrac{\omega}{\omega_1}\right)\left(1 + j\dfrac{\omega}{\omega_2}\right)} \right)^m \quad \text{with } m \in \mathbb{R},$$

the transitional frequency distribution being defined as follows:

$$(\omega_1'\omega_2')^{1/2} = (\omega_1\omega_2)^{1/2} = \omega_u;$$

$$\frac{\omega_1'}{\omega_1} = \frac{\omega_2}{\omega_2'} = \alpha > 1 \text{ and } \frac{\omega_2'}{\omega_1'} = \eta > 1.$$

3.1) For $m > 0$, then $m < 0$, plot the asymptotic phase diagram as well as the shape of the phase diagram of $C(j\omega)$, considering that the phase presents excellent linearity around the frequency, ω_u.

3.2) By likening the open-loop behavior around the frequency, ω_u, to the vertical template (in the Nichols plane) of the second-generation CRONE control, it is possible to analytically determine the resonance ratio in tracking and regulation by the formula:

$$Q = \frac{1}{\sin n\dfrac{\pi}{2}},$$

in which n denotes the order of the control, with n being comprised between 1 and 2.

For $Q\text{dB} = 20\log Q = 3$, determine the n order, then the corresponding phase margin.

Deduce therefrom the phase that the regulator must present at the frequency, ω_u. Is it compliant with that featured on the plot given in response to Question 3.1?

4) Determine the phase derivative of $C(j\omega)$ at the frequency, ω_u, with respect to the decimal logarithm of ω, that is:

$$\left[\frac{d}{d\log\omega}\arg C(j\omega)\right]_{\omega=\omega_u},$$

considering that $d\omega = 2.3\omega\, d\log\omega$. We will take care to express the result as a function of α and η: to this end, it is sufficient to express the ratios between ω_u and the different transitional frequencies of $C(j\omega)$ featuring on the phase diagrams obtained in response to Question 3.1.

5) Determine the phase derivative of $G(j\omega)$ at the same frequency, again with respect to the decimal logarithm of ω, that is:

$$\left[\frac{d}{d\log\omega}\arg G(j\omega)\right]_{\omega=\omega_u},$$

which we will write as equal to p_u.

6) It then becomes possible to establish an obvious equation ensuring phase locking in open loop around the frequency, ω_u: write this as a function of p_u.

7) The phase locking resulting from the verification of the equation thus obtained can be ensured by two dual strategies:

- the first consists of setting the values of α and η (notably $\alpha = \eta = 10$), then determining the appropriate value of m;

- the second consists of setting the values of m and η (notably $m = 1$ and $\eta = 10$), then determining the appropriate value of a.

7.1) Graphically illustrate each of these strategies by highlighting, on the phase diagram of $C(j\omega)$, the influence, first of m and second of α, on the phase slope around the frequency, ω_u.

7.2) In calculation terms, which of these two strategies would you say is the simplest?

7.3) In terms of feasibility, which of the two strategies would you say is the simplest?

8) With regard to the first strategy, determine the literal expression, then the numerical value of m for $\alpha = \eta = 10$.

9) With regard to the second strategy, introducing the function

$$g(p_u, \eta, m) = \frac{1}{1+\eta} + \frac{p_u}{4.6m\,\eta^{1/2}},$$

enables a second-degree equation in α to be established, from which it is possible to obtain the value of α, by setting those of m and η, considering that the value of p_u constitutes a datum resulting from knowledge of the plant: establish this equation and determine α for $m = 1$ and $\eta = 10$.

10) It transpires that the reduced first overshoot, D_1', of the step response in tracking or in regulation is too large, taking into account of the designer's requirements.

Considering that D_1' (as a percentage) is given analytically by a polynomial approximation defined by the equation,

$$D_1' = an^2 + bn + c,$$

with $a = 79.195$,

$b = -138.507$

and $c = 59.528$.

determine the n control order that ensures a value of D_1' deemed suitable by the designer, in this case, 10%.

11) Determine the parameters of the constant-phase controller, $C'(j\omega)$, around ω_u, which, placed in cascade with the regulator, $C(j\omega)$, allows us to ensure such an overshoot, considering that the structure of $C'(j\omega)$ is defined by:

$$C'(j\omega) = C_0' \frac{(1 + j\omega/\Omega_1')(1 + j\omega/\Omega_2')}{(1 + j\omega/\Omega_1)(1 + j\omega/\Omega_2)},$$

with:

$$(\Omega_1 \Omega_2')^{1/2} = (\Omega_1' \Omega_2)^{1/2} = \omega_u;$$

$$\frac{\Omega_1}{\Omega_1'} = \frac{\Omega_2}{\Omega_2'} = \alpha' > 1 \quad \text{and} \quad \frac{\Omega_2'}{\Omega_1} = \eta' > 1.$$

For a given ratio, η' (in this case, $\eta' = 5$), we will determine the ratio, α', then the transitional frequencies and lastly the static gain, C_0'.

12) So far, the questions have related to the synthesis of a narrow-band vertical template, through phase locking in open loop around the frequency, ω_u. Now, the aim is to focus on the synthesis of a narrow-band generalized template. Given the regulator, $C(j\omega)$, whose structure is retained, how can its new function be qualified, so as to ensure an open-loop phase variation around the frequency, ω_u, that complies with such a template? Next develop the strategy enabling this new function to be satisfied.

Solution to Problem 13

1) Given the definition of the frequency, ω_u, it is possible to write:

$$|\beta(j\omega_u)| = \frac{1}{\dfrac{\omega_u}{\omega_0'}\left[1 + \left(\dfrac{\omega_u}{\omega_0}\right)^2\right]^{1/2}} = 1,$$

therefore:

$$\frac{\omega_u}{\omega_0'}\left[1+\left(\frac{\omega_u}{\omega_0}\right)^2\right]^{1/2}=1,$$

or indeed:

$$\left(\frac{\omega_u}{\omega_0'}\right)^2\left[1+\left(\frac{\omega_u}{\omega_0}\right)^2\right]=1,$$

from which we obtain, by developing then multiplying each of the members by $\omega_0'^2\omega_0^2$:

$$\omega_u^{\ 4}+\omega_0^{\ 2}\omega_u^{\ 2}-\omega_0'^2\omega_0^{\ 2}=0,$$

that is, by writing

$$X=\omega_u^{\ 2}:$$

$$X^2+\omega_0^{\ 2}X-\omega_0'^2\omega_0^{\ 2}=0,$$

a second-degree equation in X, whose discriminant is given by:

$$\Delta=\omega_0^{\ 4}\left(1+4\frac{\omega_0'^2}{\omega_0^{\ 2}}\right)$$

and with the following roots:

$$X=\frac{\omega_0^{\ 2}}{2}\left[-1\pm\sqrt{1+4\left(\frac{\omega_0'}{\omega_0}\right)^2}\right].$$

Considering that X needs to be positive, it is convenient to only retain

$$X=\frac{\omega_0^{\ 2}}{2}\left[-1+\sqrt{1+4\left(\frac{\omega_0'}{\omega_0}\right)^2}\right],$$

or, for

$$\omega_0'/\omega_0 = \sqrt{2}:$$

$$X = \omega_0^2,$$

from which we ultimately obtain (since $X = \omega_u^2$):

$$\omega_u = \omega_0 = 10 \text{ rd/s}.$$

By definition, the phase margin is written as:

$$\Phi_m = \arg\beta(j\omega_u) + \pi,$$

that is, in the absence of any correction:

$$\Phi_m = \arg G(j\omega_u) + \pi,$$

or, given that $\omega_u = \omega_0$:

$$\Phi_m = \arg G(j\omega_0) + \pi = -3\pi/4 + \pi,$$

therefore:

$$\Phi_m = \pi/4 \text{ rd} = 45°.$$

2) Before correction (study case for Question 1):

$$|\beta(j\omega_u)| = |G(j\omega_u)| = 1.$$

After correction:

$$|\beta(j\omega_u)| = |C(j\omega_u)||G(j\omega_u)| = 1.$$

Therefore:

$$|C(j\omega_u)| = 1.$$

3.1) Figure P13.1 presents the phase diagrams of the regulator, $C(j\omega)$, for $m > 0$ (Figure P13.1(a)), then $m < 0$ (Figure P13.1(b)), both asymptotically and with respect to the curve shape.

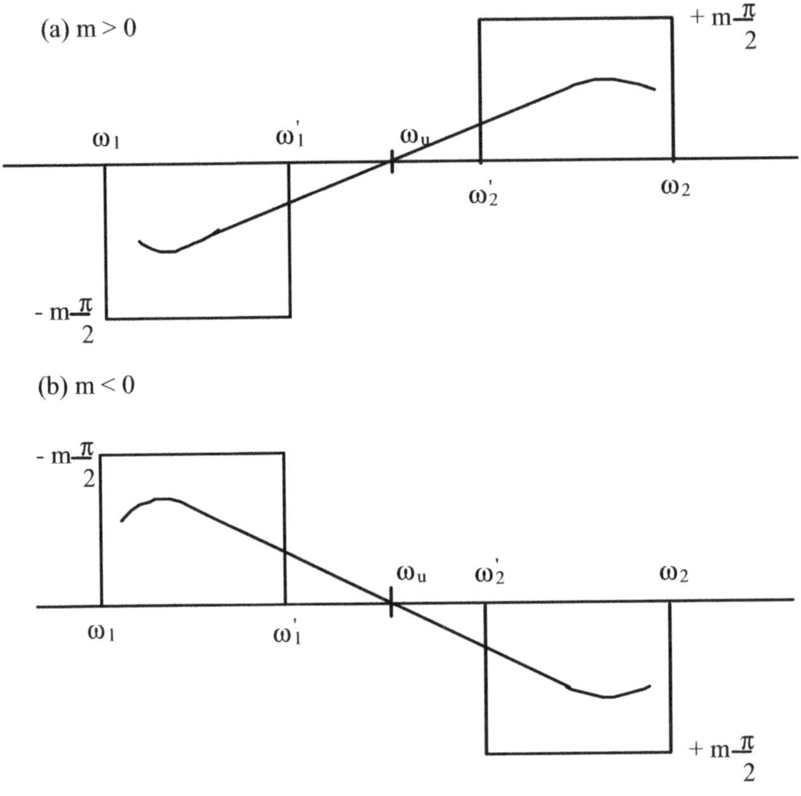

Figure P13.1. *Asymptotic phase diagram and shape of the phase diagram of* $C(j\omega)$

3.2) From $Q\text{dB} = 20\log Q = 3$, we obtain:

$$\log Q = 0.15,$$

therefore:

$$Q = 1.4125.$$

From the literal expression of Q, namely $Q = 1/\sin(n\pi/2)$, we obtain:

$$n\frac{\pi}{2} = \arcsin\frac{1}{Q},$$

with

$$\frac{\pi}{2} < \arcsin\frac{1}{Q} < \pi \text{ since } 1 < n < 2,$$

n then being given by:

$$n = \frac{2}{\pi}\arcsin\frac{1}{Q},$$

that is, for $Q = 1.4125$:

$$n = \frac{2}{\pi}\left(\pi - \frac{\pi}{4}\right) = 1.5.$$

The phase margin, meanwhile, is written as:

$$\Phi_m = \arg\beta(j\omega_u) + \pi,$$

that is:

$$\Phi_m = -n\frac{\pi}{2} + \pi = -\frac{3\pi}{4} + \pi,$$

therefore:

$$\Phi_m = \pi/4 \text{ rd} = 45°.$$

Considering that, first, the desired phase margin is $45°$ (that is, $\Phi_m = 45°$) and, second, the plant presents a phase of $-135°$ at the desired frequency, ω_u (i.e. $\arg G(j\omega_u) = -135°$), the phase that the regulator must provide at this frequency satisfies the equation:

$$\Phi_m = \arg C(j\omega_u) + \arg G(j\omega_u) + 180°,$$

from which we obtain:

$$\arg C(j\omega_u) = \Phi_m - \arg G(j\omega_u) - 180°,$$

therefore:

$$\arg C(j\omega_u) = 45° + 135° - 180° = 0,$$

with this phase complying with that of the plot given in response to Question 3.1.

4) To determine the slope of the regulator phase at the frequency, ω_u, it is sufficient to express the phase, that is,

$$\arg C(j\omega) = m\left[\arctan\frac{\omega}{\omega_1'} + \arctan\frac{\omega}{\omega_2'} - \arctan\frac{\omega}{\omega_1} - \arctan\frac{\omega}{\omega_2}\right],$$

then to calculate the derivative with respect to ω, to ultimately express the derivative with respect to $\log\omega$, using the relation $d\omega = 2.3\omega d\log\omega$ (obtained from $d\mathrm{Log}\omega = d\omega/\omega$), that is:

$$\frac{d}{d\log\omega}\arg C(j\omega) = 2.3\omega\, m\left[\frac{1/\omega_1'}{1+(\omega/\omega_1')^2} + \frac{1/\omega_2'}{1+(\omega/\omega_2')^2} - \frac{1/\omega_1}{1+(\omega/\omega_1)^2} - \frac{1/\omega_2}{1+(\omega/\omega_2)^2}\right],$$

therefore, for $\omega = \omega_u = (\omega_1'\omega_2')^{1/2} = (\omega_1\omega_2)^{1/2}$:

$$\left[\frac{d}{d\log\omega}\arg C(j\omega)\right]_{\omega=\omega_u} = 4.6\omega_u\, m\left[\frac{1}{\omega_1' + \omega_2'} - \frac{1}{\omega_1 + \omega_2}\right].$$

Given that (Figure P13.1)

$$\omega_1' = \alpha\omega_1,$$

$$\omega_2' = \eta\omega_1' = \alpha\eta\omega_1$$

and

$$\omega_2 = \alpha\omega_2' = \alpha^2\eta\omega_1,$$

we obtain:

$$\omega_u = (\omega_1'\omega_2')^{1/2} = \alpha\eta^{1/2}\omega_1,$$

$$\omega_1' + \omega_2' = \alpha(1+\eta)\omega_1$$

and

$$\omega_1 + \omega_2 = (1+\alpha^2\eta)\omega_1.$$

We then have:

$$\left[\frac{d}{d\log\omega}\arg C(j\omega)\right]_{\omega=\omega_u} = 4.6m\,\alpha\eta^{1/2}\left[\frac{1}{\alpha(1+\eta)} - \frac{1}{1+\alpha^2\eta}\right].$$

5) The phase slope of the plant at the frequency, ω_u, admits the expression:

$$\left[\frac{d}{d\log\omega}\arg G(j\omega)\right]_{\omega=\omega_u} = -2.3\frac{\omega_u/\omega_0}{1+(\omega_u/\omega_0)^2} = -1.15 = p_u.$$

6) The equation ensuring a phase locking in open loop around the frequency, ω_u, expresses the opposition of the phase slopes of the regulator and the plant at the frequency, ω_u, that is:

$$\left[\frac{d}{d\log\omega}\arg C(j\omega)\right]_{\omega=\omega_u} = -\left[\frac{d}{d\log\omega}\arg G(j\omega)\right]_{\omega=\omega_u} = -p_u,$$

or:

$$4.6m\,\alpha\eta^{1/2}\left[\frac{1}{\alpha(1+\eta)} - \frac{1}{1+\alpha^2\eta}\right] = -p_u.$$

7.1) Figure P13.2 illustrates the influence of m and α on the phase slope of the regulator around the frequency, ω_u.

7.2) Quick analysis of the equation obtained in response to Question 6 shows that, in calculation terms, the first strategy is certainly the simplest.

7.3) On the contrary, where feasibility is concerned, the second strategy is the simplest.

8) From the equation ensuring the phase locking, we obtain the literal expression of m, that is:

$$m = -\frac{p_u}{4.6\alpha\eta^{1/2}\left[\frac{1}{\alpha(1+\eta)} - \frac{1}{1+\alpha^2\eta}\right]},$$

the numerical value corresponding to $\alpha = \eta = 10$ being:

$$m = 0.977\,.$$

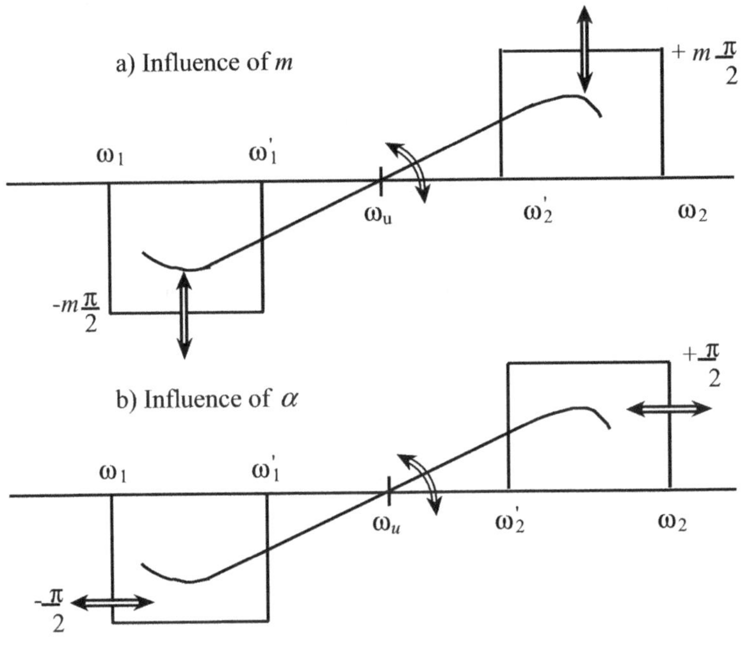

Figure P13.2. *(a) Influence of m and (b) influence of α on the phase slope of $C(j\omega)$ around ω_u*

9) The equation ensuring the phase locking can be written as:

$$\frac{1}{\alpha(1+\eta)} - \frac{1}{1+\alpha^2\eta} = -\frac{p_u}{4.6m\ \alpha\eta^{1/2}},$$

that is:

$$\frac{1}{1+\eta} - \frac{\alpha}{1+\alpha^2\eta} = -\frac{p_u}{4.6m\ \eta^{1/2}},$$

or:

$$\frac{1}{1+\eta} + \frac{p_u}{4.6m\ \eta^{1/2}} = \frac{\alpha}{1+\alpha^2\eta},$$

or indeed, by introducing the function, $g(p_u,\eta,m)$, as defined in the presentation:

$$g(p_u,\eta,m) = \frac{\alpha}{1+\alpha^2\eta},$$

this relation making it possible to establish the second-degree equation in α:

$$\eta\, g(p_u,\eta,m)\alpha^2 - \alpha + g(p_u,\eta,m) = 0,$$

or, for $m = 1$ and $\eta = 10$:

$$0.119\alpha^2 - \alpha + 0.0119 = 0,$$

from which we obtain:

$$\alpha = 8.4.$$

10) The n order of the control satisfies the second-degree equation,

$$an^2 + bn + c - D_1' = 0,$$

in which

$$a = 79.195;\quad b = -138.507;\quad c = 59.528.$$

n is therefore given by:

$$n = \frac{-b \pm \sqrt{b^2 - 4a\left(c - D_1'\right)}}{2a},$$

that is, numerically:

$$n = \frac{138.507 \pm 59.116}{158.39},$$

from which we obtain:

$$n = 1.25 \quad \text{and} \quad n = 0.5,$$

of which we retain only:

$$n = 1.25$$

since n must be comprised between 1 and 2.

11) In the case of a narrow frequency domain, the principle of a constant-phase controller arises from the smoothing of the asymptotic phase diagram of a double phase-lead controller. Figure P13.3 gives a distribution of the transitional frequencies corresponding to the different zeros and poles of this double controller.

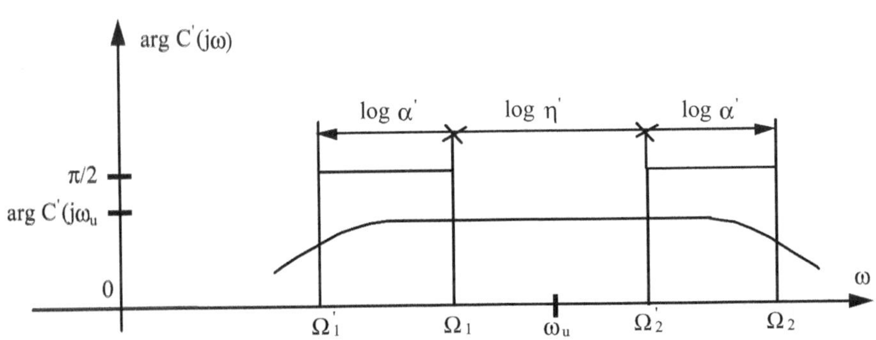

Figure P13.3. *Smoothing of the asymptotic phase diagram of a double phase-lead controller*

With the η' ratio being set, the phase around the frequency, ω_u, is adjusted to the desired value, $\arg C'(j\omega_u)$, by varying the α' ratio.

A variation of the transitional frequencies, Ω_1' and Ω_2 (Figure P13.4), indeed enables us to act on the phase at the frequency, ω_u, this being an increasing function of α'. Besides the simplicity of the structure, a notable property of this regulator is the quasi-constancy of the phase around ω_u (on around one decade for $\eta' = 5$).

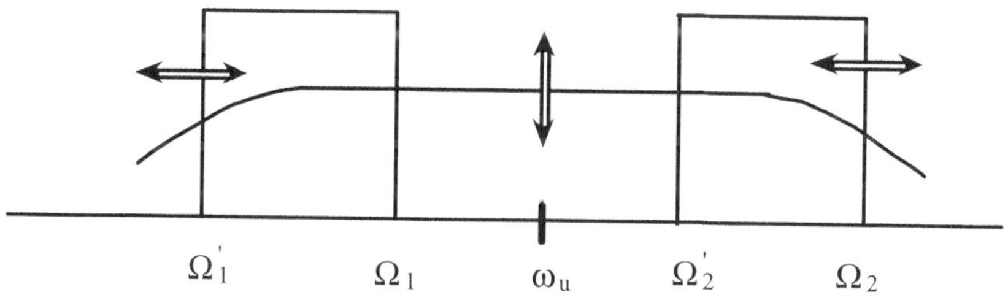

Figure P13.4. *Modification of the phase around ω_u through a variation of the zero, Ω_1', and the pole, Ω_2*

To determine the phase at the frequency, ω_u, it is sufficient to make use of the phase diagram's symmetry with respect to ω_u, expressing that the contribution of Ω_1' and Ω_1 on the phase at the frequency, ω_u, is the same as that of the frequencies, Ω_2' and Ω_2, therefore:

$$\arctan\frac{\omega_u}{\Omega_1'} - \arctan\frac{\omega_u}{\Omega_1} = \frac{1}{2}\arg C'(j\omega_u),$$

that is, given that

$$\Omega_1' = \omega_u / \alpha'\eta'^{1/2} \text{ and } \Omega_1 = \omega_u / \eta'^{1/2}:$$

$$\arctan \alpha'\eta'^{1/2} - \arctan \eta'^{1/2} = \frac{1}{2}\arg C'(j\omega_u),$$

from which we obtain:

$$\alpha' = \eta'^{-1/2} \tan\left[\frac{1}{2}\arg C'(j\omega_u) + \arctan \eta'^{1/2}\right],$$

this equation enables the value of α' to be determined, setting that of η', considering that $\arg C'(j\omega_u)$ constitutes a datum, which is deduced from the following equations:

$$\arg \beta(j\omega_u) = \arg C(j\omega_u) + \arg C'(j\omega_u) + \arg G(j\omega_u)$$

and

$$\arg \beta(j\omega_u) = -n\frac{\pi}{2},$$

from which we indeed obtain:

$$\arg C'(j\omega_u) = -n\frac{\pi}{2} - \arg C(j\omega_u) - \arg G(j\omega_u),$$

that is, numerically:

$$\arg C'(j\omega_u) = -(1.25)90° - 0° + 135° = 22.5°.$$

Considering that $\eta' = 5$, the ratio, α', is calculated in compliance with:

$$\alpha' = \frac{1}{\sqrt{5}} \tan\left[\frac{1}{2}22.5° + \arctan\sqrt{5}\right],$$

therefore:

$$\alpha' = 1.96.$$

The transitional frequencies are then:

$$\Omega_1 = \frac{\omega_u}{\sqrt{\eta}} = 4.47 \text{ rd/s and } \Omega_2' = \sqrt{\eta}\ \omega_u = 22.36 \text{ rd/s};$$

$$\Omega_1' = \frac{\Omega_1}{\alpha'} = 2.28 \text{ rd/s and } \Omega_2 = \alpha'\Omega_2' = 43.83 \text{ rd/s}.$$

With regard to the static gain, C_0', it verifies the equation:

$$|\beta(j\omega_u)| = |C(j\omega_u)||C'(j\omega_u)||G(j\omega_u)| = 1,$$

or, given that $|C(j\omega_u)| = |G(j\omega_u)| = 1$:

$$|C'(j\omega_u)| = 1\,,$$

therefore:

$$C'_0 \frac{\sqrt{1+\left(\frac{\omega_u}{\Omega'_1}\right)^2}\sqrt{1+\left(\frac{\omega_u}{\Omega'_2}\right)^2}}{\sqrt{1+\left(\frac{\omega_u}{\Omega_1}\right)^2}\sqrt{1+\left(\frac{\omega_u}{\Omega_2}\right)^2}} = 1\,,$$

from which we obtain C'_0:

$$C'_0 = \frac{\sqrt{1+\left(\frac{\omega_u}{\Omega_1}\right)^2}\sqrt{1+\left(\frac{\omega_u}{\Omega_2}\right)^2}}{\sqrt{1+\left(\frac{\omega_u}{\Omega'_1}\right)^2}\sqrt{1+\left(\frac{\omega_u}{\Omega'_2}\right)^2}}\,,$$

that is, numerically:

$$C'_0 = 0.51\,.$$

12) In the synthesis of a narrow-band vertical template, the function of the regulator, $C(j\omega)$, is to compensate locally (around ω_u) for the plant phase variation. Its function is therefore a *compensation function*.

In the synthesis of a narrow-band generalized template, the function of the regulator, $C(j\omega)$ (around ω_u), is no longer compensation, but *undercompensation* or *overcompensation*, depending on whether the open-loop phase *decreases* or *increases* with the frequency, the imaginary integration order, b, then being either *negative* or *positive*.

Concerning the strategy that makes it possible to satisfy this new function, it is appropriate to start with the configuration ensuring a compensation, namely:

$$m = 0.977 \quad \text{for} \quad \alpha = \eta = 10$$

and

$$\alpha = 8.4 \text{ for } m = 1 \text{ and } \eta = 10,$$

these configurations being obtained in response to Questions 8 and 9, respectively.

Considering that the phase slope of $C(j\omega)$ at the frequency, ω_u, increases with m or α:

– a value of m below 0.977 or a value of α below 8.4 ensure an undercompensation;

– a value of m above 0.977 or a value of α above 8.4 ensure an overcompensation.

APPENDICES

Appendix 1

From Regulation Function to Active Noise Control

A1.1. An active principle based on an anti-noise

Let there be a *noise* (acoustic, ambient or background) resulting from a spatially limited sound environment, such as that associated with a car interior or with an acoustic chamber in an experimental context.

Let there also be another noise, called *anti-noise*, coming from a loudspeaker, which, by its sign, partially or completely opposes the noise, depending on its intensity: a complete (or total) opposition corresponds to the canceling of the output of an omnidirectional microphone simultaneously receiving noise and anti-noise (Figure A1.1), thus canceling the sum of the sound waves at the input of the microphone (this input materializing the pickup area).

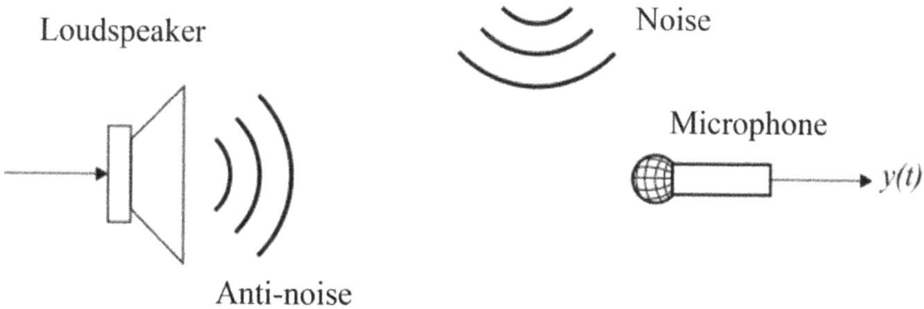

Figure A1.1. *Electro-acoustic configuration for active noise control*

The output, $y(t)$, of the microphone is thus the sum of two very distinct electrical components (Figure A1.2):

– one, denoted as $y_n(t)$, representative of noise, in situ natural noise or noise coming from a loudspeaker in the case of an experimental study [LOI 15];

– the other, denoted as $y_a(t)$, representative of the anti-noise to be generated in accordance with the following section.

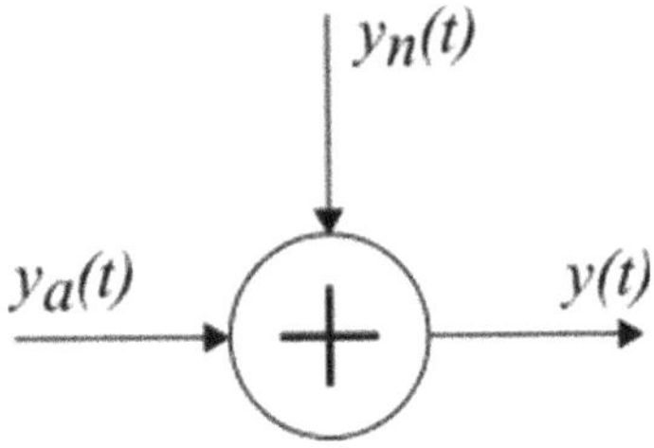

Figure A1.2. *Breaking down of the output, $y(t)$, of the microphone into two components, $y_n(t)$ and $y_a(t)$, representative of noise and anti-noise*

A1.2. On the generation of anti-noise

The idea of the generation of anti-noise is based on the principle of disturbance rejection, which defines the regulation function of the control loop (the reference, $r(t)$, then being nil, Figure A1.3).

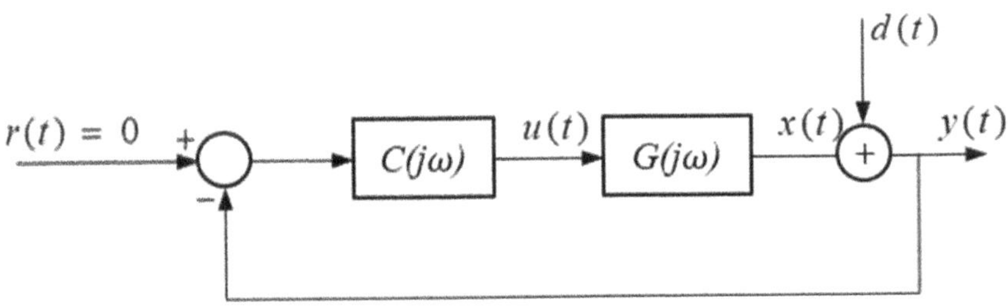

Figure A1.3. *Elementary control loop used in regulation ($r(t) = 0$), and in which $x(t)$ denotes the signal delivered by the transfer, $G(j\omega)$*

The signal, $x(t)$, stemming from the transfer, $G(j\omega)$, plays a major role here through *its relation to the disturbance*, $d(t)$, which helps establish the manner in which anti-noise is generated.

This relation is established by determining $X(j\omega)$ using the equation:

$$X(j\omega) = C(j\omega)G(j\omega)\left[-\left(X(j\omega) + D(j\omega)\right)\right], \qquad \text{[A1.1]}$$

That is, given that

$$C(j\omega)G(j\omega) = \beta(j\omega): \qquad \text{[A1.2]}$$

$$X(j\omega) = -\beta(j\omega)X(j\omega) - \beta(j\omega)D(j\omega), \qquad \text{[A1.3]}$$

or, by factoring $X(j\omega)$:

$$X(j\omega)\left[1+\beta(j\omega)\right] = -\beta(j\omega)D(j\omega), \qquad \text{[A1.4]}$$

from which we obtain:

$$X(j\omega) = -\frac{\beta(j\omega)}{1+\beta(j\omega)}D(j\omega), \qquad \text{[A1.5]}$$

or, for a spectrum of $d(t)$, significantly below the frequency, ω_u, therefore at low frequencies ($\omega << \omega_u$), for which $|\beta(j\omega)| >> 1$:

$$X(j\omega) \simeq -D(j\omega), \qquad \text{[A1.6]}$$

or, in the time domain:

$$x(t) \simeq -d(t), \qquad \text{[A1.7]}$$

a remarkably simple result expressing that $x(t)$ is practically the opposite of $d(t)$ or, in terms that anticipate the solution to our problem, $x(t)$ is practically the *total-opposition anti-disturbance*, $-d(t)$.

Thus, in the functional diagram of Figure A1.3, by replacing the disturbance, $d(t)$, with the noise, $y_n(t)$ (or component $y_n(t)$, which represents the noise), $x(t)$ practically becomes the *total-opposition anti-noise*, $-y_n(t)$, a general representation of which is given by the component, $y_a(t)$, in Figure A1.2. Given that the product, $C(j\omega)G(j\omega)$, is none other than $\beta(j\omega)$, the elementary control loop in

Figure A1.3 is reduced to that of Figure A1.4, then, in more technological terms, to that of Figure A1.5.

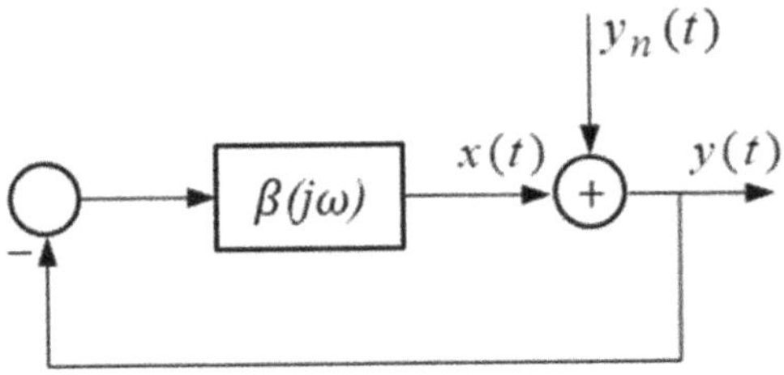

Figure A1.4. *Elementary control loop, in which the signal, $x(t)$, offers an achievement of the component, $y_a(t)$*

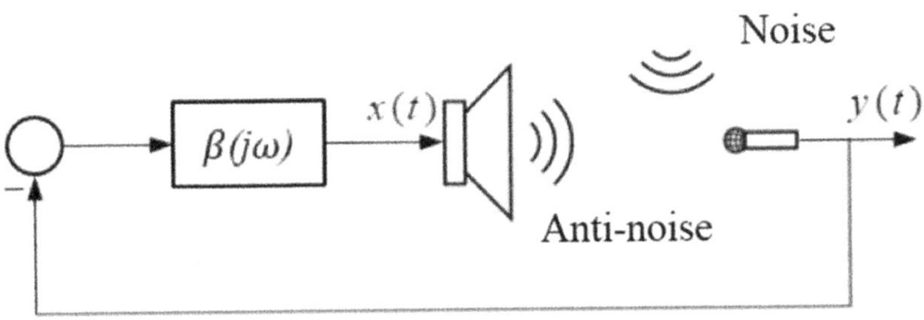

Figure A1.5. *Simplified technological representation of anti-noise generation*

NOTE.– If electro-acoustic conversion between the loudspeaker input and the microphone input does not enable the reduction of the transfer between $x(t)$ and $y_a(t)$ to a transfer factor, this transfer factor needs to be generalized by a transfer function denoted by $\beta''(s)$: in Figure A1.5, $\beta(j\omega)$ must then be replaced with $\beta'(j\omega)$ such that

$$\beta(j\omega) = \beta'(j\omega)\beta''(j\omega) . \qquad [A1.8]$$

A1.3. A remark dictated by technology

As at zero frequency and at neighboring frequencies the loudspeaker and the microphone do not provide any transmission, the control loop of Figure A1.5 is found to be open at these frequencies (therefore it is continuous and at very low frequencies).

If we return to the frequency response in open loop used in Chapter 1, defined by

$$\beta(j\omega) = \left(\frac{\omega_u}{j\omega}\right)^n \text{, with } 1 < n < 2, \qquad \text{[A1.9]}$$

it is then convenient (to avoid any offset drift) to truncate $\beta(j\omega)$ at low frequency, in compliance with the new response,

$$\beta(j\omega) = \frac{\beta_0}{\left(1 + j\frac{\omega}{\omega_b}\right)^n}, \qquad \text{[A1.10]}$$

which, in order to simplify the calculations, can be replaced with

$$\beta(j\omega) = \frac{\beta_0}{1 + \left(j\frac{\omega}{\omega_b}\right)^n}, \qquad \text{[A1.11]}$$

a response that is only different around the transitional frequency, ω_b (significantly lower than the transition frequency, ω_u, for $\beta_0 >> 1$).

The frequency response in regulation is then written as:

$$F_d(j\omega) = \frac{1}{1 + \beta(j\omega)} = \frac{1}{1 + \frac{\beta_0}{1 + \left(j\frac{\omega}{\omega_b}\right)^n}}, \qquad \text{[A1.12]}$$

or:

$$F_d(j\omega) = \frac{1 + \left(j\frac{\omega}{\omega_b}\right)^n}{1 + \beta_0 + \left(j\frac{\omega}{\omega_b}\right)^n} = \frac{1 + \left(j\frac{\omega}{\omega_b}\right)^n}{(1 + \beta_0)\left[1 + \left(j\frac{\omega}{\omega_b}\right)^n \frac{1}{1 + \beta_0}\right]}, \qquad \text{[A1.13]}$$

or:

$$F_d(j\omega) = \frac{1+\left(j\frac{\omega}{\omega_b}\right)^n}{(1+\beta_0)\left[1+\left(j\frac{\omega}{\omega_b'}\right)^n\right]}, \text{ with } \omega_b' = (1+\beta_0)^{1/n}\,\omega_b, \qquad [A1.14]$$

or indeed, for $\beta_0 >> 1$:

$$F_d(j\omega) \simeq \frac{1}{\beta_0}\frac{1+\left(j\frac{\omega}{\omega_b}\right)^n}{1+\left(j\frac{\omega}{\omega_u}\right)^n}, \text{ with } \omega_u \simeq \beta_0^{1/n}\omega_b, \qquad [A1.15]$$

a frequency response whose asymptotic gain diagram is given in Figure A1.6.

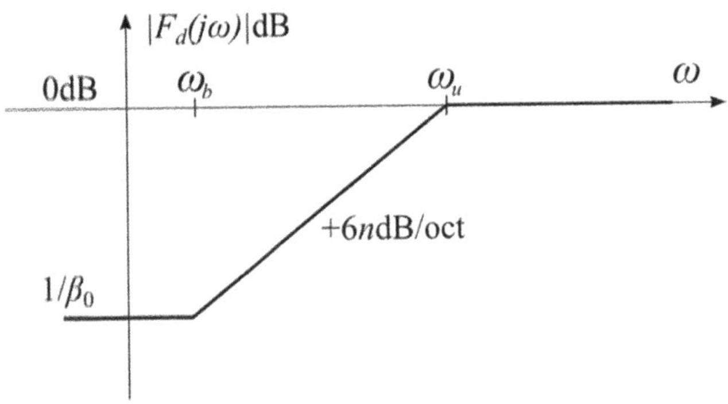

Figure A1.6. *Asymptotic gain diagram showing that, for a given value of n, the attenuation of the noise spectrum is all the better if the frequency, ω_u, is high*

A1.4. Quantifying the attenuation of a spectral noise component

Among the set of (sinusoidal) components of a noise spectrum, which can be a broadband spectrum, thus exceeding the frequency, ω_u, let us consider a component of frequency, ω_0, such that ω_0 is the geometric mean of the transitional frequencies, ω_b and ω_u, that is, $\omega_0 = (\omega_b\omega_u)^{1/2}$.

For $n = 3/2$ and for a ω_u/ω_b ratio of 10^4, which leads to $\omega_0 = \omega_u/100$, the aim of this section is to determine, literally and numerically, the attenuation of the component of frequency, ω_0.

In sinusoidal steady state, the response of the output, $y(t)$, of the control loop to the component of frequency, ω_0 ω_0,

$$c(t) = c_m \cos\left(\omega_0 t + \varphi_c\right), \tag{A1.16}$$

is also a sinusoidal vibration of the form:

$$y(t) = y_m \cos\left(\omega_0 t + \varphi_y\right), \tag{A1.17}$$

the ratio of the amplitudes, y_m and c_m, being given by the gain in regulation, that is:

$$\frac{y_m}{c_m} = \left|\frac{1}{1+\beta(j\omega_0)}\right|, \tag{A1.18}$$

from which we obtain:

$$y_m = \left|\frac{1}{1+\beta(j\omega_0)}\right| c_m. \tag{A1.19}$$

For

$$\beta(j\omega_0) = \frac{\beta_0}{1+\left(j\dfrac{\omega_0}{\omega_b}\right)^n} \simeq \frac{\beta_0}{\left(j\dfrac{\omega_0}{\omega_b}\right)^n} \text{ since } \omega_0/\omega_b = 100, \tag{A1.20}$$

the expression of y_m is particularized in accordance with:

$$y_m \simeq \left|\frac{1}{1+\dfrac{\beta_0}{\left(j\dfrac{\omega_0}{\omega_b}\right)^n}}\right| c_m = \left|\frac{1}{1+\dfrac{\beta_0 \omega_b{}^n}{(j\omega_0)^n}}\right| c_m, \tag{A1.21}$$

or:

$$y_m \simeq \left| \frac{1}{1+\frac{\left(\beta_0{}^{1/n}\omega_b\right)^n}{(j\omega_0)^n}} \right| c_m = \left| \frac{1}{1+\left(\frac{\omega_u}{j\omega_0}\right)^n} \right| c_m, \qquad [A1.22]$$

or:

$$y_m \simeq \left| \left(j\frac{\omega_0}{\omega_u} \right)^n \right| c_m \text{ since } \omega_u/\omega_0 = 100, \qquad [A1.23]$$

or indeed:

$$y_m \simeq \left(\frac{\omega_0}{\omega_u} \right)^n c_m, \qquad [A1.24]$$

that is, numerically, given that $n = 3/2$ and $\omega_0/\omega_u = 1/100$:

$$y_m \simeq \left(\frac{1}{100} \right)^{3/2} c_m = \left[\left(\frac{1}{100} \right)^{\frac{1}{2}} \right]^3 c_m = \frac{c_m}{1{,}000}, \qquad [A1.25]$$

hence an *attenuation of a factor 1,000* of the component of frequency, ω_0.

In view of this result, it is important to note that the attenuation would only be a factor 100 for $n = 1$, which clearly shows, here, the interest of a 3/2 order.

Appendix 2

Closed-Loop Behavior and Dynamic Performance of Second-Generation CRONE Control

A2.1. Description of closed-loop dynamic behavior

Closed-loop dynamic behavior is exclusively linked to the open-loop behavior close to the unit gain frequency. So, the study of such behavior can be based on transmittances in tracking, $\tilde{F}_r(s)$, and regulation, $\tilde{F}_d(s)$, determined from the transmittance in open loop, $\beta(s)$, reduced to the template description transmittance, $\tilde{\beta}(s)$, that is:

$$\beta(s) = \tilde{\beta}(s) \text{ for } \omega \in [\omega_A, \omega_B], \tag{A2.1}$$

with

$$\tilde{\beta}(s) = \left(\frac{\omega_u}{s}\right)^n, \qquad n \in [1, 2]. \tag{A2.2}$$

$\tilde{F}_r(s)$ and $\tilde{F}_d(s)$ admit the following expressions:

$$\tilde{F}_r(s) = K\frac{\tilde{\beta}(s)}{1+\tilde{\beta}(s)} = \frac{K}{1+\left(\dfrac{s}{\omega_u}\right)^n} \tag{A2.3}$$

and

$$\tilde{F}_d(s) = \frac{1}{1+\tilde{\beta}(s)} = \frac{1}{1+\left(\frac{\omega_u}{s}\right)^n}. \qquad [A2.4]$$

A2.2. Closed-loop dynamic performances

A2.2.1. *Frequency dynamic performances*

The results presented in this section are demonstrated in section A2.3.

The gains, $\left|\tilde{F}_r(j\omega)\right|$ and $\left|\tilde{F}_d(j\omega)\right|$, reach a maximum at the *resonance frequencies* (in tracking and regulation) given by:

$$\omega_r = \left(-\cos n\frac{\pi}{2}\right)^{\frac{1}{n}} \omega_u \qquad [A2.5]$$

in tracking, and

$$\omega_r = \left(-\cos n\frac{\pi}{2}\right)^{-\frac{1}{n}} \omega_u \qquad [A2.6]$$

in regulation (with the resonance frequencies notated ω_{r_r} and ω_{r_d} when it is necessary to differentiate between them); these results express the existence of a resonance when $\cos(n\pi/2) < 0$, that is, for $1 < n < 3$, therefore for the CRONE control, since $1 < n < 2$.

The *resonance ratios* (in tracking and regulation) are expressed by:

$$Q = Q(n) = \frac{\left|\tilde{F}_r(j\omega_r)\right|}{\left|\tilde{F}_r(j0)\right|} = \frac{1}{\sin n\frac{\pi}{2}} \qquad [A2.7]$$

in tracking, and

$$Q = Q(n) = \frac{\left|\tilde{F}_d(j\omega_r)\right|}{\left|\tilde{F}_d(j\infty)\right|} = \frac{1}{\sin n\dfrac{\pi}{2}} \qquad \text{[A2.8]}$$

in regulation (with the resonance ratios notated Q_r and Q_d when it is necessary to differentiate between them); these results convey *the same resonance ratio, Q, which is exclusively linked to the n order of the control*, thus enabling the notion of *robust resonance* to be introduced.

The set of four formulas presented enable us to analytically determine the link between the frequency dynamic performances in tracking and regulation, that is:

$$\left(\omega_{r_r}\,\omega_{r_d}\right)^{1/2} = \omega_u \qquad \text{[A2.9]}$$

and

$$Q_r = Q_d. \qquad \text{[A2.10]}$$

The relation between the resonance frequencies expresses that *the resonance frequencies in tracking and regulation are distributed geometrically with respect to the unit gain frequency in open loop.*

The relation between the resonance ratios, which expresses *the identity of the resonance ratios in tracking and regulation*, is moreover confirmed graphically by the tangency of the vertical template with two amplitude contours presenting the same parameter, one of the Nichols chart and the other of the inverted Nichols chart.

A2.2.2. *Time dynamic performances*

Given the expressions of $\tilde{F}_r(s)$ and $\tilde{F}_d(s)$, written with the same denominator, that is,

$$\tilde{F}_r(s) = \frac{K}{1+\left(\dfrac{s}{\omega_u}\right)^n} \qquad \text{[A2.11]}$$

and

$$\tilde{F}_d(s) = \frac{\left(\dfrac{s}{\omega_u}\right)^n}{1+\left(\dfrac{s}{\omega_u}\right)^n}, \qquad \text{[A2.12]}$$

the poles in closed loop (zeros of the denominator or roots of the denominator equaled to zero) satisfy the characteristic equation:

$$1+\left(\frac{s}{\omega_u}\right)^n = 0, \qquad \text{[A2.13]}$$

or:

$$\left(\frac{s}{\omega_u}\right)^n = -1 = e^{\pm j\pi}, \qquad \text{[A2.14]}$$

with the argument of -1 thus chosen (i.e. $\pm\pi$), complying with the determination, $\left]-\pi,+\pi\right[$, of the argument of s, which is imposed by the *non-defined nature* of s^n for $s \in \mathbb{R}^-$; s^n indeed results from an infinite number of alternating zeros and poles, recursively distributed on $\mathbb{R}^-$ and infinitely close.

There exist, exclusively, two poles, that is:

$$s = \omega_u e^{\pm j\pi/n}; \qquad \text{[A2.15]}$$

for $1 < n < 2$, their argument is indeed compliant with the determination, $\left]-\pi,+\pi\right[$. They are conjugate complex, forming an angle at the center, 2Θ, with $\Theta = (\pi - \pi/n)$ (see Figure A2.1). They are at the origin of an oscillatory mode. As ω_u depends on the plant parameters, the poles move with a constant angle (set by the n order of the control) when the parameters vary. They are therefore two half-lines forming the same angle, Θ, with respect to the axis of real values, which are known as *isodamping half-straight lines*, illustrating robustness in the s plane.

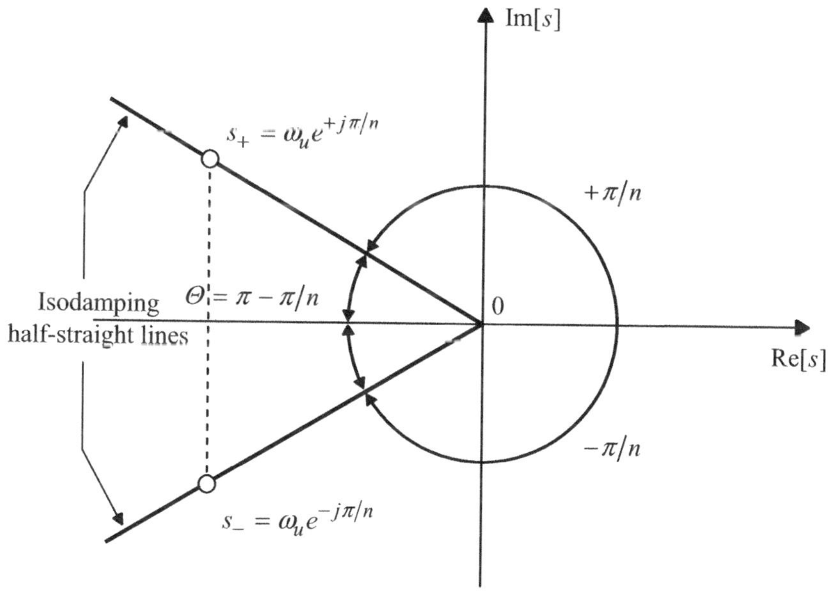

Figure A2.1. *Locus of the poles in closed loop illustrating robustness in the s plane*

The *natural frequency* (projection of s_+ on the imaginary axis) and the *damping ratio* (cosine of the half-angle at the center, Θ) are directly deduced from the poles through their modulus, ω_u, and the half-angle at the center, Θ, that they form:

$$\omega_p = \omega_u \sin\Theta = \omega_u \sin\left(\pi - \frac{\pi}{n}\right) = \omega_u \sin\frac{\pi}{n} \qquad \text{[A2.16]}$$

and

$$\zeta = \zeta(n) = \cos\Theta = \cos\left(\pi - \frac{\pi}{n}\right) = -\cos\frac{\pi}{n} \qquad \text{[A2.17]}$$

This result expresses that the damping ratio, ζ, is exclusively linked to the n order of the control, thus enabling the introduction of the concept of a robust oscillatory mode, whereas the natural frequency, ω_p, is not robustified (which is an advantage in terms of natural frequency).

It is worth specifying that a *non-dominant, aperiodic mode* is added to this dominant oscillatory mode, due to an infinity of infinitely close poles on $\mathbb{R}^-$ and is thus called an *aperiodic multimode*.

A2.3. Demonstrations

Given the expressions of $\tilde{F}_r(s)$ and $\tilde{F}_d(s)$, that is,

$$\tilde{F}_r(s) = \frac{K}{1+\left(\dfrac{s}{\omega_u}\right)^n} \qquad \text{[A2.18]}$$

and

$$\tilde{F}_d(s) = \frac{1}{1+\left(\dfrac{\omega_u}{s}\right)^n}, \qquad \text{[A2.19]}$$

the corresponding frequency responses are written as:

$$\tilde{F}_r(j\omega) = \frac{K}{1+\left(j\dfrac{\omega}{\omega_u}\right)^n} \qquad \text{[A2.20]}$$

and

$$\tilde{F}_d(j\omega) = \frac{1}{1+\left(\dfrac{\omega_u}{j\omega}\right)^n}. \qquad \text{[A2.21]}$$

By replacing j with $e^{j\frac{\pi}{2}}$ in $\tilde{F}_r(j\omega)$ and $1/j = -j$ with $e^{-j\frac{\pi}{2}}$ in $\tilde{F}_d(j\omega)$, we can write:

$$\tilde{F}_r(j\omega) = \frac{K}{1+\left(\dfrac{\omega}{\omega_u}\right)^n e^{jn\frac{\pi}{2}}} \qquad \text{[A2.22]}$$

and

$$\tilde{F}_d(j\omega) = \frac{1}{1+\left(\frac{\omega_u}{\omega}\right)^n e^{-jn\frac{\pi}{2}}}, \qquad \text{[A2.23]}$$

or:

$$\tilde{F}_r(j\omega) = \frac{K}{1+\left(\frac{\omega}{\omega_u}\right)^n \left(\cos n\frac{\pi}{2} + j\sin n\frac{\pi}{2}\right)} \qquad \text{[A2.24]}$$

and

$$\tilde{F}_d(j\omega) = \frac{1}{1+\left(\frac{\omega_u}{\omega}\right)^n \left(\cos n\frac{\pi}{2} - j\sin n\frac{\pi}{2}\right)}, \qquad \text{[A2.25]}$$

or:

$$\tilde{F}_r(j\omega) = \frac{K}{1+\left(\frac{\omega}{\omega_u}\right)^n \cos n\frac{\pi}{2} + j\left(\frac{\omega}{\omega_u}\right)^n \sin n\frac{\pi}{2}} \qquad \text{[A2.26]}$$

and

$$\tilde{F}_d(j\omega) = \frac{1}{1+\left(\frac{\omega_u}{\omega}\right)^n \cos n\frac{\pi}{2} - j\left(\frac{\omega_u}{\omega}\right)^n \sin n\frac{\pi}{2}}, \qquad \text{[A2.27]}$$

or, by writing $a = \omega/\omega_u$:

$$\tilde{F}_r(j\omega) = \frac{K}{1+a^n \cos n\frac{\pi}{2} + ja^n \sin n\frac{\pi}{2}} \qquad \text{[A2.28]}$$

and

$$\tilde{F}_d(j\omega) = \frac{1}{1 + a^{-n}\cos n\frac{\pi}{2} - ja^{-n}\sin n\frac{\pi}{2}}. \quad \text{[A2.29]}$$

The modulus of $\tilde{F}_r(j\omega)$ and $\tilde{F}_d(j\omega)$ take the form:

$$\left|\tilde{F}_r(j\omega)\right| = \frac{K}{\left[\left(1 + a^n\cos n\frac{\pi}{2}\right)^2 + a^{2n}\sin^2 n\frac{\pi}{2}\right]^{1/2}} \quad \text{[A2.30]}$$

and

$$\left|\tilde{F}_d(j\omega)\right| = \frac{1}{\left[\left(1 + a^{-n}\cos n\frac{\pi}{2}\right)^2 + a^{-2n}\sin^2 n\frac{\pi}{2}\right]^{1/2}}, \quad \text{[A2.31]}$$

or:

$$\left|\tilde{F}_r(j\omega)\right| = \frac{K}{\left[1 + 2a^n\cos n\frac{\pi}{2} + a^{2n}\cos^2 n\frac{\pi}{2} + a^{2n}\sin^2 n\frac{\pi}{2}\right]^{1/2}} \quad \text{[A2.32]}$$

and

$$\left|\tilde{F}_d(j\omega)\right| = \frac{1}{\left[1 + 2a^{-n}\cos n\frac{\pi}{2} + a^{-2n}\cos^2 n\frac{\pi}{2} + a^{-2n}\sin^2 n\frac{\pi}{2}\right]^{1/2}}, \quad \text{[A2.33]}$$

or indeed:

$$\left|\tilde{F}_r(j\omega)\right| = \frac{K}{\left[1 + 2a^n\cos n\frac{\pi}{2} + a^{2n}\right]^{1/2}} \quad \text{[A2.34]}$$

and

$$\left|\tilde{F}_d(j\omega)\right| = \frac{1}{\left[1+2a^{-n}\cos n\dfrac{\pi}{2}+a^{-2n}\right]^{1/2}}. \quad \text{[A2.35]}$$

$\left|\tilde{F}_r(j\omega)\right|$ and $\left|\tilde{F}_d(j\omega)\right|$ pass through a maximum when the quantities under the radicals are minimal, that is, when

$$\frac{d}{da}\left[1+2a^{n}\cos n\frac{\pi}{2}+a^{2n}\right]=0 \quad \text{[A2.36]}$$

and

$$\frac{d}{da}\left[1+2a^{-n}\cos n\frac{\pi}{2}+a^{-2n}\right]=0, \quad \text{[A2.37]}$$

that is, for:

$$a^{n}+\cos n\frac{\pi}{2}=0 \quad \text{[A2.38]}$$

and

$$a^{-n}+\cos n\frac{\pi}{2}=0. \quad \text{[A2.39]}$$

The resonance frequencies thus satisfy the relations:

$$\left(\frac{\omega_r}{\omega_u}\right)^{n}+\cos n\frac{\pi}{2}=0 \quad \text{[A2.40]}$$

and

$$\left(\frac{\omega_r}{\omega_u}\right)^{-n}+\cos n\frac{\pi}{2}=0, \quad \text{[A2.41]}$$

or:

$$\left(\frac{\omega_r}{\omega_u}\right)^n = -\cos n\frac{\pi}{2} \qquad \text{[A2.42]}$$

and

$$\left(\frac{\omega_r}{\omega_u}\right)^{-n} = -\cos n\frac{\pi}{2}, \qquad \text{[A2.43]}$$

or:

$$\frac{\omega_r}{\omega_u} = \left(-\cos n\frac{\pi}{2}\right)^{\frac{1}{n}} \qquad \text{[A2.44]}$$

and

$$\frac{\omega_r}{\omega_u} = \left(-\cos n\frac{\pi}{2}\right)^{-\frac{1}{n}}, \qquad \text{[A2.45]}$$

from which we obtain:

$$\omega_r = \omega_r(n, \omega_u) = \left(-\cos n\frac{\pi}{2}\right)^{\frac{1}{n}} \omega_u \qquad \text{[A2.46]}$$

in tracking, and

$$\omega_r = \omega_r(n, \omega_u) = \left(-\cos n\frac{\pi}{2}\right)^{-\frac{1}{n}} \omega_u \qquad \text{[A2.47]}$$

in regulation.

Defined by

$$Q = \frac{\left|\tilde{F}_r(j\omega_r)\right|}{\left|\tilde{F}_r(j0)\right|} \qquad \text{[A2.48]}$$

in tracking, and

$$Q = \frac{\left|\tilde{F}_d(j\omega_r)\right|}{\left|\tilde{F}_d(j\infty)\right|} \qquad \text{[A2.49]}$$

in regulation, the resonance ratios are expressed by:

$$Q = \frac{1}{\left[1+2\left(\frac{\omega_r}{\omega_u}\right)^n \cos n\frac{\pi}{2} + \left(\frac{\omega_r}{\omega_u}\right)^{2n}\right]^{1/2}} \qquad \text{[A2.50]}$$

in tracking, and

$$Q = \frac{1}{\left[1+2\left(\frac{\omega_r}{\omega_u}\right)^{-n} \cos n\frac{\pi}{2} + \left(\frac{\omega_r}{\omega_u}\right)^{-2n}\right]^{1/2}} \qquad \text{[A2.51]}$$

in regulation.

Given that

$$\left(\frac{\omega_r}{\omega_u}\right)^n = -\cos n\frac{\pi}{2} \text{ in tracking} \qquad \text{[A2.52]}$$

and

$$\left(\frac{\omega_r}{\omega_u}\right)^{-n} = -\cos n\frac{\pi}{2} \text{ in regulation,} \qquad \text{[A2.53]}$$

the resonance ratios become:

$$Q = \frac{1}{\left[1 + 2\left(-\cos n\frac{\pi}{2}\right)\left(\cos n\frac{\pi}{2}\right) + \left(-\cos n\frac{\pi}{2}\right)^2\right]^{1/2}} \qquad \text{[A2.54]}$$

in tracking, and

$$Q = \frac{1}{\left[1 + 2\left(-\cos n\frac{\pi}{2}\right)\left(\cos n\frac{\pi}{2}\right) + \left(-\cos n\frac{\pi}{2}\right)^2\right]^{1/2}} \qquad \text{[A2.55]}$$

in regulation, with the identical results expressing *the same resonance ratio in tracking and regulation*, that is,

$$Q = \frac{1}{\left[1 - 2\cos^2 n\frac{\pi}{2} + \cos^2 n\frac{\pi}{2}\right]^{1/2}}, \qquad \text{[A2.56]}$$

or:

$$Q = \frac{1}{\left[1 - \cos^2 n\frac{\pi}{2}\right]^{1/2}} = \frac{1}{\left[\sin^2 n\frac{\pi}{2}\right]^{1/2}}, \qquad \text{[A2.57]}$$

therefore, ultimately:

$$Q = Q(n) = \frac{1}{\sin n\frac{\pi}{2}}. \qquad \text{[A2.58]}$$

After these demonstrations, two limit scenarios are worth analyzing, where $n = 1$ and $n = 2$.

For $n = 1$, the frequencies and the resonance ratio are reduced to:

$$\omega_{r_r} = 0, \quad \omega_{r_d} = \infty \quad \text{and} \quad Q = 1. \qquad \text{[A2.59]}$$

Appendix 3

Iso-overshoot Contours and Isodamping Contours

A3.1. On performance contours

In accordance with Chapter 2 (sections 2.4 and 2.7), the dynamic performances quantifying the stability degree admit the following decomposition:

– the time dynamic performances, namely

- the reduced first overshoot, D_1',

- the damping ratio, ζ;

– the frequency dynamic performances, namely

- the resonance ratio in tracking, Q_r, which parameterizes (in decibels) a Nichols amplitude contour,

- the resonance ratio in regulation, Q_d, which parameterizes (in decibels) an inverted Nichols amplitude contour.

Within the framework of the time-frequency relations necessary for the synthesis of a control in the frequency domain, it is convenient to identify the frequency dynamic performances that are significant of the time dynamic performances (in this case, D_1' and ζ), given that the performance specifications imposed by the designer come under the time domain.

Whereas the resonance ratio in tracking, Q_r, is significant for the reduced first overshoot, D_1', a Nichols amplitude contour thus creating an *iso-overshoot contour* (as shown in section A3.4), the resonance ratio in regulation, Q_d, on the other hand, is not significant for any time dynamic performance. Notably, while it parameterizes a circle centered on the critical point, $(-1,0)$, and thus measuring a veritable distance to this point, it is nonetheless not significant for the damping ratio, ζ.

Moreover, in order for the time dynamic performances, D_1' and ζ, to each have their equivalent in the frequency domain, it is convenient to seek a new frequency dynamic performance or a new contour coming under the frequency domain significant for the damping ratio, ζ. In fact, such a contour is none other than an *isodamping contour*, as introduced in section 2.4.2 and whose construction (intuitive then deductive) is discussed in section A3.5.

Legitimately interpretable as *performance contours*, the iso-overshoot and isodamping contours notably consolidate the arsenal of frequency tools for linear automation. Indeed, they enable an open-loop Nichols locus satisfying the overshoot and damping specifications to be systematically synthesized.

With the performance contours thus being defined, notably as for their complementarity, the purpose of this appendix is, first, to *validate* the Nichols amplitude contours as iso-overshoot contours, and, second, to *build* the isodamping contours.

A3.2. Covering of generalized templates as a performance contour

The most immediate geometric approach for validating or building a performance contour consists of using the covering technique. The contour is then defined as the cover formed by a family of straight-line segments tangent to it (Figure A3.1).

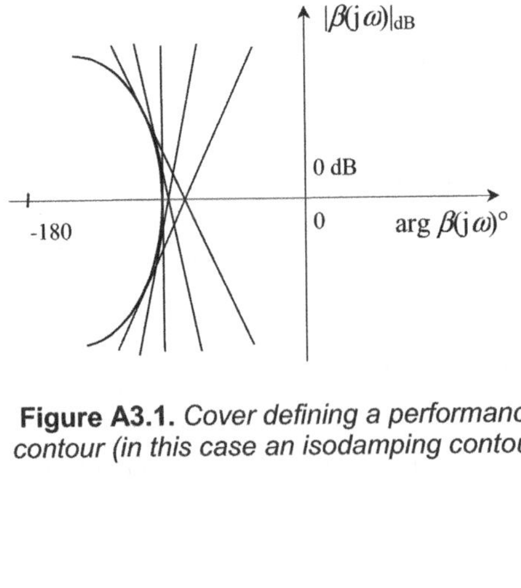

Figure A3.1. *Cover defining a performance contour (in this case an isodamping contour)*

Within the framework of a representation in the Nichols plane, each straight-line segment of the family thus defined can be interpreted as the rectilinear section of an open-loop Nichols locus that ensures, in closed loop, the overshoot or damping corresponding to the contour. Moreover, such a rectilinear section around the unit gain frequency in open loop, ω_u, is none other than the generalized template characterizing the third-generation CRONE control.

Considering that the closed-loop dynamic behavior is exclusively linked to the open-loop behavior close to the unit gain frequency, ω_u, it is sufficient to determine a transmittance likely to describe the generalized template in order to obtain the overshoot through the transmittance in tracking (or in regulation) and the damping through the characteristic equation. As the following developments show, the description transmittance of the generalized template is based on complex, non-integer integration.

A3.3. Generalized template and complex, non-integer integration

The vertical template is characterized by a real, non-integer integration order, n, which determines its phase placement at $-n90°$ (Figure A3.2).

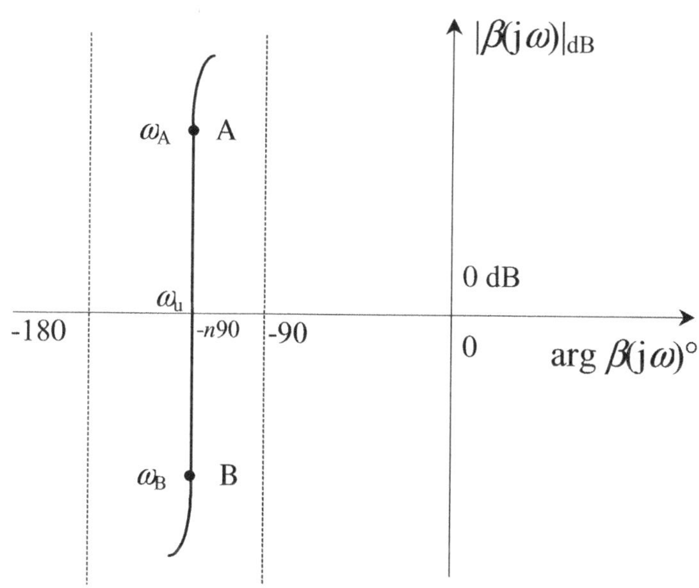

Figure A3.2. *Representation of the vertical template by means of a vertical straight-line segment in the Nichols plane*

The vertical template ($\omega \in [\omega_A, \omega_B]$) is indeed described by the transmittance of a real, non-integer integrator, that is:

$$\tilde{\beta}(s) = \left(\frac{\omega_u}{s}\right)^n \text{ with } n \in \mathbb{R}^+. \quad \text{[A3.1]}$$

Within the context of a hypothesis, it is fair to assume, by extension of the description of the vertical template, that the generalized template can be characterized by a complex, non-integer integration order, n, the real part of which determines its phase placement at the frequency, ω_u, that is $-\text{Re}(n)90°$, and whose imaginary part then determines its incline to the vertical (Figure A3.3).

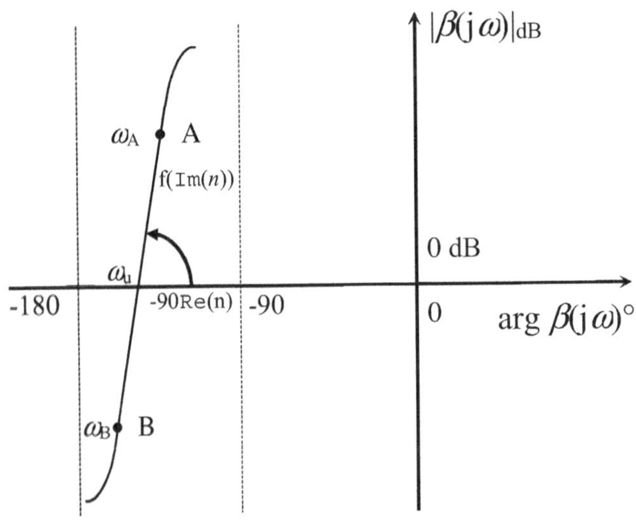

Figure A3.3. *Representation of the generalized template by means of a straight-line segment of any direction in the Nichols plane*

The generalized template ($\omega \in [\omega_A, \omega_B]$) is then, a priori, described by the restriction, in the operational plane, $\mathbb{C}_j$, of the transmittance of a complex, non-integer integrator, that is:

$$\tilde{\beta}(s) = \left[\left(\frac{\omega_u}{s}\right)^n\right]_{\mathbb{C}_j} \text{ with } n \in \mathbb{C}_i. \quad \text{[A3.2]}$$

The pure imaginary unit, i, featuring in the integration order, n $(n = a + ib,\ a > 0)$, is independent of the pure imaginary unit, j, featuring in the operational variable, s $(s = \sigma + j\omega)$.

As the transmittance of a complex, non-integer integrator is broken down in compliance with relation

$$\left(\frac{\omega_u}{s}\right)^n = \left(\frac{\omega_u}{s}\right)^a \left(\cos\left(b\ln\left(\frac{\omega_u}{s}\right)\right) + i\sin\left(b\ln\left(\frac{\omega_u}{s}\right)\right)\right),\ a > 0, \qquad \text{[A3.3]}$$

its restriction in $\mathbb{C}_j$ is expressed by:

$$\left[\left(\frac{\omega_u}{s}\right)^n\right]_{\mathbb{C}_j} = \left(\frac{\omega_u}{s}\right)^a \cos\left(b\ln\left(\frac{\omega_u}{s}\right)\right). \qquad \text{[A3.4]}$$

The frequency behavior of $\tilde{\beta}(s)$ close to the frequency, ω_u, is characterized by the set of relations:

$$\left|\tilde{\beta}(j\omega_u)\right| = \cosh\left(b\frac{\pi}{2}\right), \qquad \text{[A3.5]}$$

$$\arg\tilde{\beta}(j\omega_u)^\circ = -a90, \qquad \text{[A3.6]}$$

$$\frac{d}{d\log\omega}\left|\tilde{\beta}(j\omega)\right|\mathrm{dB}\Big|_{\omega=\omega_u} = -20a \qquad \text{[A3.7]}$$

and

$$\frac{d}{d\log\omega}\arg\tilde{\beta}(j\omega)^\circ\Big|_{\omega=\omega_u} = -\frac{180}{\pi}\ln(10)b\tanh\left(b\frac{\pi}{2}\right). \qquad \text{[A3.8]}$$

These relations express that at the frequency, ω_u:

– the gain must be readjusted in compliance with relation [A3.5] in order to ensure $\left|\tilde{\beta}(j\omega_u)\right| = 1$;

– the phase placement of the Nichols locus of $\tilde{\beta}(j\omega)$ (relation [A3.6]) is exclusively linked to the real integration order, a;

– the gain slope (relation [A3.7]) is not affected by the imaginary integration order, b;

– the phase slope (relation [A3.8]) is an even function of b and is therefore always negative.

Thus, a posteriori and in order for the imaginary integration order, b, to act not only on the absolute value of the Nichols locus slope of $\tilde{\beta}(j\omega)$ at the frequency, ω_u, but also on its sign, it is convenient to describe the generalized template by the transmittance:

$$\tilde{\beta}(s) = \left(\cosh\left(b\frac{\pi}{2}\right)\right)^{sign(b)} \left(\frac{\omega_u}{s}\right)^a \left(\cos\left(b\ln\left(\frac{\omega_u}{s}\right)\right)\right)^{-sign(b)}, \qquad [A3.9]$$

which we can write in the form:

$$\tilde{\beta}(s) = \left(\cosh\left(b\frac{\pi}{2}\right)\right)^{sign(b)} \left(\frac{\omega_u}{s}\right)^a \left(\left[\left(\frac{\omega_u}{s}\right)^{ib}\right]_{\mathbb{C}_j}\right)^{-sign(b)}. \qquad [A3.10]$$

The parameter n ($n = a + ib$) only represents (strictly speaking) the complex integration order of the transmittance, $\tilde{\beta}(s)$ ([A3.10]) when the parameter, b, is negative. Nevertheless, in order to lessen subsequent developments, we will agree to conserve the "integration order" designation when the b parameter is positive.

Aside from the values of b of less than 1, for which very low gain and phase undulations occur, Figure A3.4 reveals the excellent linearity of the generalized templates, the transmittance [A3.10] thus proving an excellent description transmittance of the generalized template. For a given real integration order, that is, $a = 1.5$, this figure represents the generalized templates corresponding to different values of the imaginary integration order, b, in this case: –2; –1.5; –1; –0.5; 0; 0.5; 1; 1.5; 2.

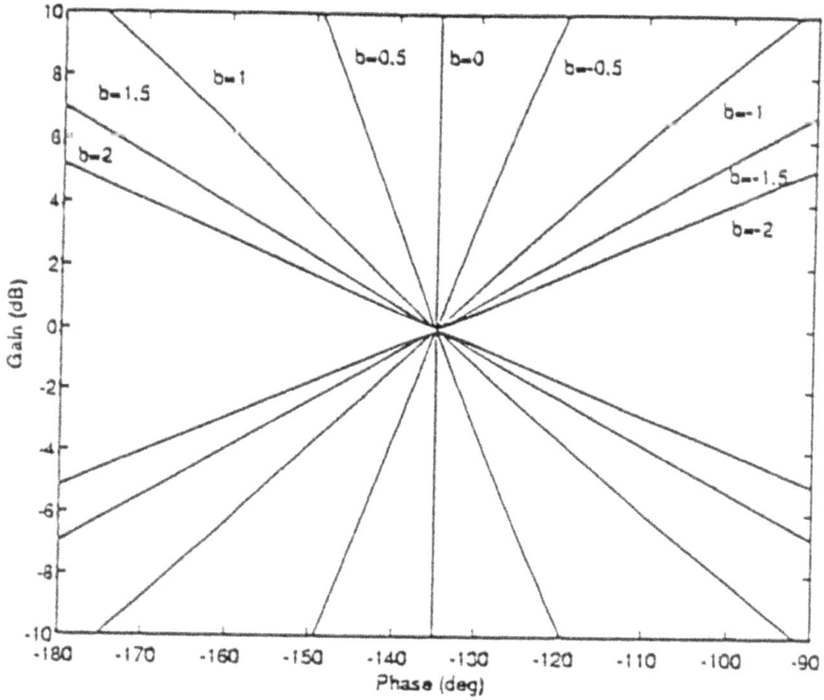

Figure A3.4. *Generalized templates for a given real integration order, in this case a = 1.5, and different imaginary integration orders*

A3.4. Validation of amplitude contours as iso-overshoot contours

The *validation of the Nichols amplitude contours as iso-overshoot contours* occurs based on curves representing the variations of the reduced first overshoot as a function of the imaginary integration order obtained for different values of the real integration order (Figure A3.5). These curves are deduced from the step responses in tracking, determined using standard simulation tools, after synthesizing, through a recursive distribution of zeros and complex poles [OUS 95a, OUS 00], the complex non-integer integration of the description transmittance of the generalized template, $\tilde{\beta}(s)$. The fractured nature of the curves in Figure A3.5 is simply due to an insufficient number of step tests, each of them being, moreover, at the origin of a unit step response of the form:

$$y_r(t) = \mathcal{L}^{-1}\left(\frac{\tilde{\beta}(s)}{1+\tilde{\beta}(s)}\frac{1}{s}\right). \qquad [A3.11]$$

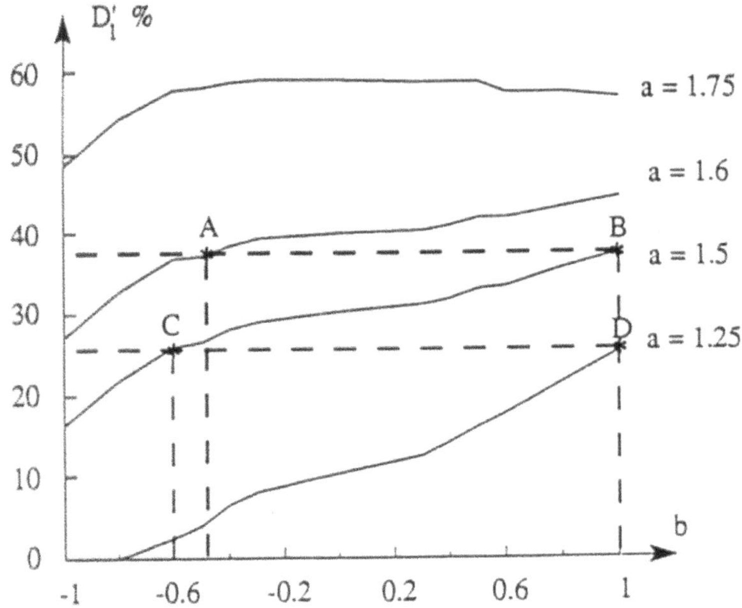

Figure A3.5. *Variations of the reduced first overshoot D_1' as a function of b for different values of a*

The intersection points between a horizontal and the overshoot curves (parameterized in a and represented as a function of b) determine the couples (a and b) corresponding to the same reduced first overshoot. Two examples are presented by way of illustration:

For $D_1' = 37.5\%$, the intersection points, A and B, define the following couples:

$$\text{A } (a = 1.6; b = -0.46) \text{ and B } (a = 1.5; b = 1).$$

For $D_1' = 25\%$, the intersection points, C and D, define the following couples:

$$\text{C } (a = 1.5; b = -0.6) \text{ and D } (a = 1.25; b = 1).$$

For $D_1' = 37.5\%$, we can observe that the generalized templates defined by the couples, $(a = 1.6; b = -0.46)$ and $(a = 1.5; b = 1)$, are tangent to the same amplitude contour, 5 dB, of the Nichols chart (Figure A3.6).

For $D_1' = 25\%$, we can observe that the generalized templates defined by the couples, $(a = 1.5; b = -0.6)$ and $(a = 1.25; b = 1)$, are tangent to the same amplitude contour, 3.1 dB, of the Nichols chart (Figure A3.7).

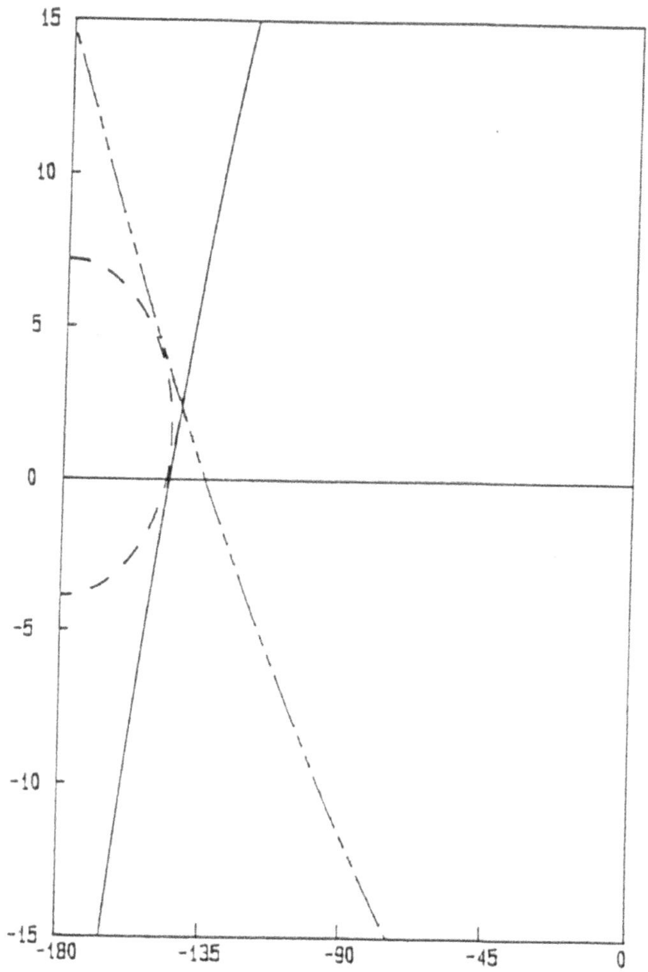

Figure A3.6. *The generalized templates defined by the couples (a and b) corresponding to $D_1' = 37.5\%$ are tangent to the same Nichols amplitude contour (in this case the amplitude contour, 5 dB)*

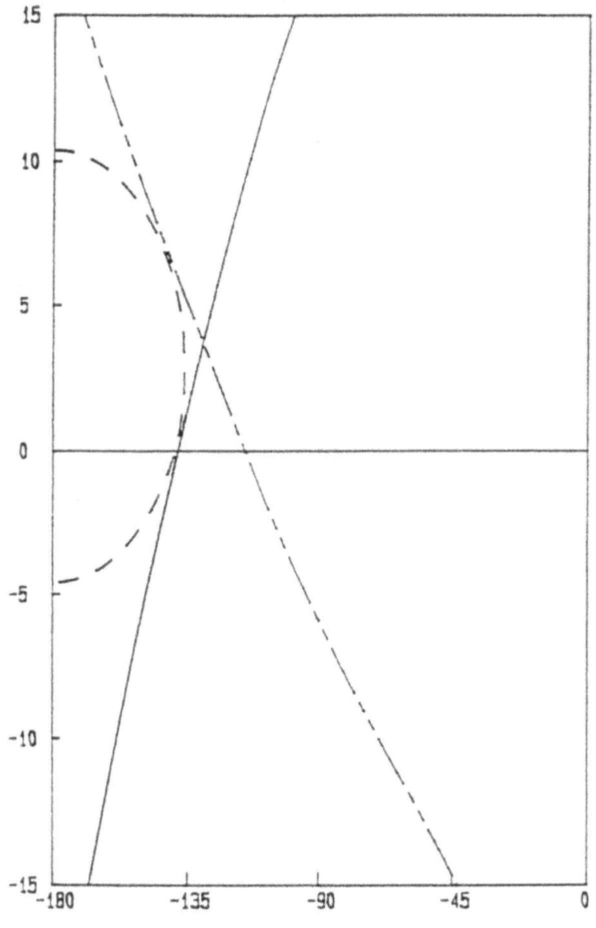

Figure A3.7. *The generalized templates defined by the couples (a and b) corresponding to* $D_1' = 25\%$ *, are tangent to the same Nichols amplitude contour (in this case the amplitude contour, 3.1 dB)*

Thus, by tracing different classes of generalized templates corresponding to different overshoots, it is possible to build (using the covering technique) iso-overshoot contours which, through the two examples presented, prove to be the Nichols amplitude contours (Figure A3.8).

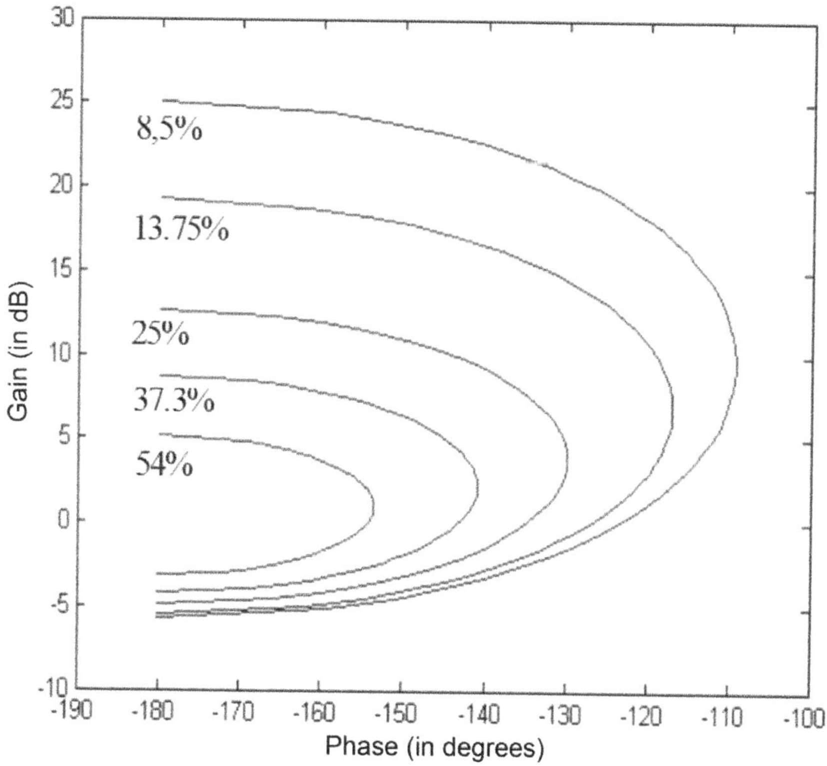

Figure A3.8. *Iso-overshoot contours*

The parameterization of a Nichols amplitude contour by the corresponding reduced first overshoot can be conducted in a similar manner to that of an isodamping contour, as illustrated by Figure A3.10(b). The "overshoot as a percentage" parameter, between 0 and 1, which thus references the contour, can be given by the reduced first overshoot, assured, in closed loop, by the Nichols locus of a generalized integrator tangent to this contour (Figure A3.9), that is [OUS 91]:

$$D_1' = an^2 + bn + c\,, \qquad \text{[A3.12]}$$

the coefficients having the following numerical values:

$$a = 79.195; \quad b = -138.507 \quad \text{and} \quad c = 59.528\,, \qquad \text{[A3.13]}$$

the order of integration, n (between 1 and 2), being such that $-n\pi/2$ is the abscissa of the tangency point.

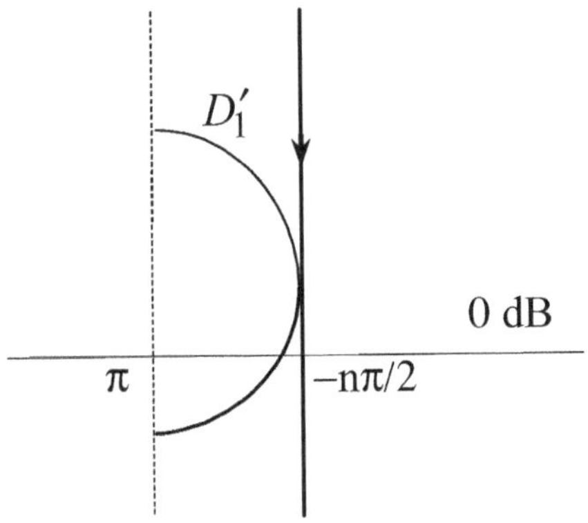

Figure A3.9. *Parameterization of an iso-overshoot contour*

Moreover, knowing that an n-order generalized integrator ensures, in closed loop, a resonance ratio in tracking of the form [OUS 91]

$$Q_r = \frac{1}{\sin \frac{n\pi}{2}}, \qquad \text{[A3.14]}$$

it is possible to express n as a function of Q_r, namely

$$n = \frac{2}{\pi} \arcsin \frac{1}{Q_r}, \qquad \text{[A3.15]}$$

with

$$\frac{\pi}{2} < \arcsin \frac{1}{Q_r} = n\frac{\pi}{2} < \pi \text{ since } 1 < n < 2, \qquad \text{[A3.16]}$$

then to apply this expression of n in that of the reduced first overshoot, that is:

$$D_1' = a\left(\frac{2}{\pi} \arcsin \frac{1}{Q_r}\right)^2 + b\left(\frac{2}{\pi} \arcsin \frac{1}{Q_r}\right) + c. \qquad \text{[A3.17]}$$

We thus obtain an expression of D_1' as a function of Q_r, enabling parameterization the in overshoot of a Nichols amplitude contour, not based on the parameter, n, but directly based on the parameter, Q_r, which (expressed in decibels) ordinarily references it.

A3.5. Building isodamping contours

As mentioned in section 2.4.2, an isodamping contour is referenced by a parameter between 0 and 1, which directly expresses the damping ratio in tracking and regulation, ζ. It is built according to two separate graphical methods, the first based on amplitude contours of the same parameter, and the second based on the covering technique.

1) *Intuitive method*

The first method (corresponding to the initial idea) is an intuitive, somewhat empirical, method, which nevertheless provides positive results. It uses the symmetry with the axis of arguments of the Nichols amplitude contours and the inverted contours of the same parameter [OUS 91]. Using two symmetrical amplitude contours, one a Nichols contour and the other its dual, the method consists of plotting a symmetrical contour whose ordinate of each point is none other than the algebraic mean of the ordinates of the points of the same abscissa of the two amplitude contours (Figure A3.10(a)). The idea behind this building method results from an analysis highlighting the following property: two open-loop Nichols loci which lap, respectively (one above and the other below), the contour thus obtained, lead to the same closed-loop damping ratio, moreover, identical (give or take several percent) to that ensured by the vertical template of the CRONE control tangent to the same contour. The parameter that references this contour can be given by the damping ratio ensured, in closed loop, by the Nichols locus of a generalized integrator tangent to this contour (Figure A3.10(b)), that is, (given relation [A2.17]) $\zeta = -\cos(\pi/n)$.

2) *Deductive method*

The second method is a deductive method that uses the covering technique [OUS 95b]. An isodamping contour is then defined as the cover formed by a family of straight-line segments. These straight-line segments of any direction around the unit gain frequency in open loop, known as generalized templates, constitute the rectilinear sections of open-loop Nichols loci that ensure, in closed loop, the same damping ratio.

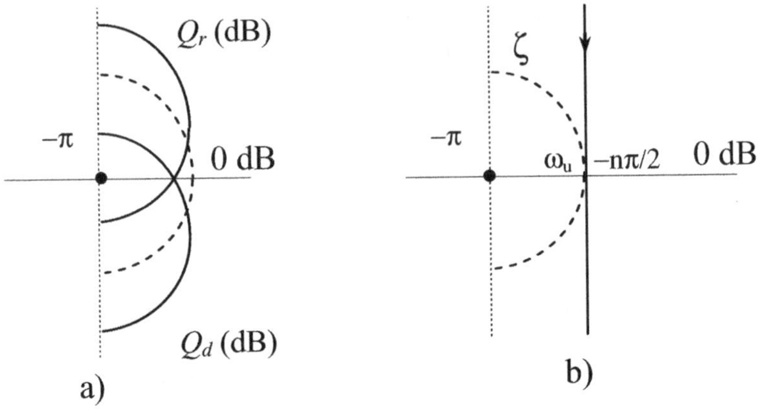

Figure A3.10. *Building of an isodamping contour (a) and principle of its parameterization (b)*

The purpose of this section is therefore to determine a family of description transmittances of the type [A3.10], which notably present the same unit gain frequency, ω_u, and which ensure the same damping ratio in closed loop, then use the linearity of the corresponding Nichols loci to build an isodamping contour using the covering technique.

The pair of complex-conjugate poles, from which the oscillatory mode in closed loop results, satisfies the characteristic equation:

$$1+\tilde{\beta}(s)=0\,, \qquad \text{[A3.18]}$$

or, in the case of a normalization of the poles with respect to the frequency, ω_u, that is, $s' = s/\omega_u$:

$$1+\tilde{\beta}(s')=0\,, \qquad \text{[A3.19]}$$

or, given the expression of $\tilde{\beta}(s)$ (relation [A3.9]):

$$1+\left(\cosh\left(b\frac{\pi}{2}\right)\right)^{sign(b)} s'^{-a}\left(\cos\left(b\ln s'\right)\right)^{-sign(b)}=0\,. \qquad \text{[A3.20]}$$

For the operational variable, s', expressed in polar coordinates, that is, $s' = \rho e^{i\theta}$, the normalized characteristic equation becomes:

$$F(a,b,\rho,\theta) = 0, \qquad \text{[A3.21]}$$

with

$$F(a,b,\rho,\theta) = \rho^{-sign(b)a} e^{-jsign(b)a\theta} \cosh\left(b\frac{\pi}{2}\right) + \cos\left(b\ln\rho + jb\theta\right). \qquad \text{[A3.22]}$$

As the frequency, ω_u, is set, the pair of poles, the solution to this equation, depends solely on the integration order, n, of the transmittance, $\tilde{\beta}(s)$ (relation [A3.10]).

When the pair of poles describes the two isodamping half-straight lines represented by Figure A3.11 in $\mathbb{C}_j$, the integration order, n, associated with the generalized template, describes in $\mathbb{C}_i$ the locus, Γ_ζ, of the integration orders ensuring the same damping ratio ζ (defined by the cosine of the half-angle at the center formed by the pair of poles), that is:

$$\Gamma_\zeta = \left\{n \in \mathbb{C}_i, n = a + ib; \rho \in \mathbb{R}^+, \theta = \theta_0 = \pi - \arccos\zeta, F(a,b,\rho,\theta) = 0\right\}. \qquad \text{[A3.23]}$$

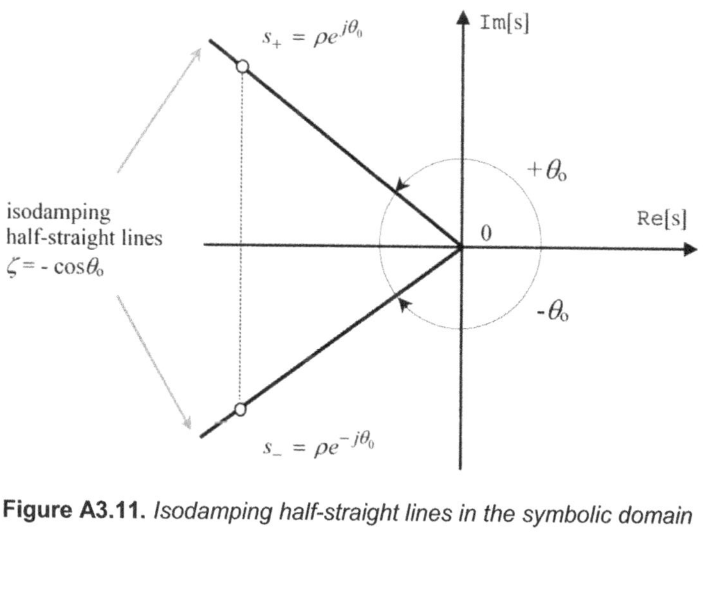

Figure A3.11. *Isodamping half-straight lines in the symbolic domain*

Given the property defined by the equality

$$\mathrm{F}(a,b,\rho,\theta) = \mathrm{F}(a,-b,\rho^{-1},\theta), \tag{A3.24}$$

the locus, Γ_ζ, presents symmetry with the real axis of the complex plane, $\mathbb{C}_i$. This property makes it possible to limit the study to the negative imaginary integration orders, b.

Such a reduction of the study domain enables the determination of the elements of Γ_ζ to be reduced to resolving a system of two nonlinear equations with separable variables, resulting from the breaking down into real and imaginary parts of $\mathrm{F}(a,b,\rho,\theta)$, that is:

$$\begin{cases} f_1(a) = g_1(b) \\ f_2(a) = g_2(b) \end{cases}, \tag{A3.25}$$

where the functions, f_1, f_2, g_1 and g_2 are defined, respectively, by:

$$f_1(a) = \rho^2 \cos(a\theta_0), \tag{A3.26}$$

$$f_2(a) = \rho^2 \sin(a\theta_0), \tag{A3.27}$$

$$g_1(b) = \frac{-\cosh(b\theta_0)\cos(b\ln\rho)}{\cosh\left(b\frac{\pi}{2}\right)} \tag{A3.28}$$

and

$$g_2(b) = \frac{\sinh(b\theta_0)\sin(b\ln\rho)}{\cosh\left(b\frac{\pi}{2}\right)}. \tag{A3.29}$$

The solution of such a system is obtained by the minimization (using the conjugate gradient method) of a quadratic criterion, that is:

$$J = \left(f_1(a) - g_1(b)\right)^2 + \left(f_2(a) - g_2(b)\right)^2. \tag{A3.30}$$

It should be noted that the locus, Γ_ζ, admits as an element the real integration order,

$$n = -\frac{\pi}{\theta_0}\left(\mathrm{Im}[n] = 0\right), \qquad [A3.31]$$

which corresponds to the verticality of the generalized template.

Figure A3.12 represents the locus, Γ_ζ, corresponding to the value, $\zeta = 0.1$.

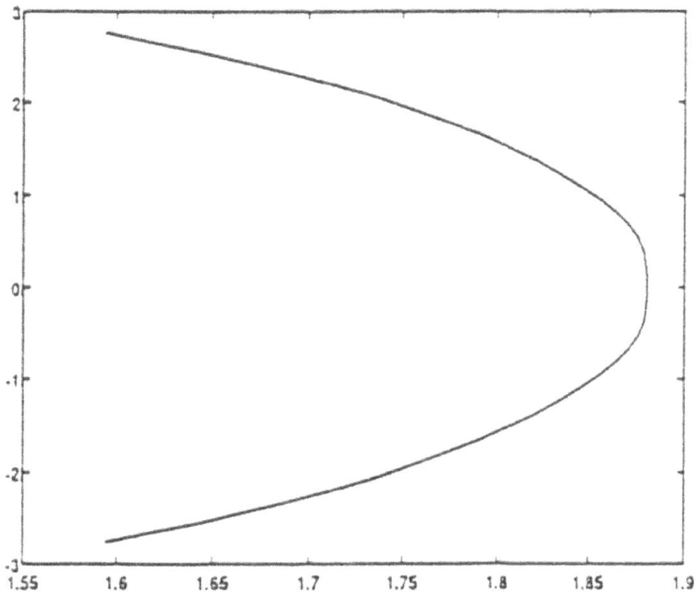

Figure A3.12. *Locus Γ_ζ of integration orders ensuring the same damping ratio, in this case $\zeta = 0.1$*

Given the excellent linearity of the Nichols locus of $\tilde{\beta}(s)$ ([A3.10]) around the frequency, ω_u, a legitimate approximation consists of replacing it, around this frequency, with its tangent at the frequency point, ω_u, Δ_n, of equation

$$\Delta_n : Y = A_n X + B_n , \qquad [A3.32]$$

with

$$A_n = -\frac{sign(b)a\pi}{9\ln(10)b\tanh\left(b\frac{\pi}{2}\right)} \qquad [A3.33]$$

and

$$B_n = -\frac{sign(b)a^2 10\pi}{\ln(10)b\tanh\left(b\frac{\pi}{2}\right)}. \qquad [A3.34]$$

Using $I_{n_1n_2}$ to denote the intersection point of two generalized templates, restrictions of two straight lines, Δ_{n_1} and Δ_{n_2}, the isodamping contour, C_ζ, is defined as the locus of the points, $I_{n_1n_2}$, such that the orders, n_1 and n_2, are two elements of the locus, Γ_ζ, that are different but infinitely close, that is:

$$C_\zeta = \left\{ M \in P / n_1, n_2 \in \Gamma_\zeta, M = \lim_{\substack{n_1 - n_2 \to 0 \\ n_1 \neq n_2}} I_{n_1,n_2} = \left\{\Delta_{n_1} \cap \Delta_{n_2}\right\} \right\}. \qquad [A3.35]$$

In other words, the isodamping contour is defined as the cover of the family of straight lines, Δ_n, corresponding to orders, n, belonging to the same locus, Γ_ζ.

Figure A3.13 illustrates the building of the isodamping contour corresponding to a damping ratio, ζ, of 0.1.

The contour thus plotted shows that an isodamping contour is only subject to a partial representation. This is as a result of the limitations of the building method. Indeed, the validity of such a method requires that within the framework of its application, certain major principles of automation are observed.

1) Generally speaking, in order for the closed-loop dynamics (determined by the open-loop behavior close to the unit gain frequency, ω_u) to be representative of the transient of the step response in tracking or in regulation, it is notably necessary for the open-loop gain to increase sufficiently quickly when the frequency decreases below ω_u.

2) In the case of a step response in tracking or in regulation comprising an oscillatory mode, in order for the natural frequency and the damping ratio of such a mode to quantify the rapidity and damping of the dynamics, this mode needs to be sufficiently dominant. The verification of this requirement then ensures that of the first condition.

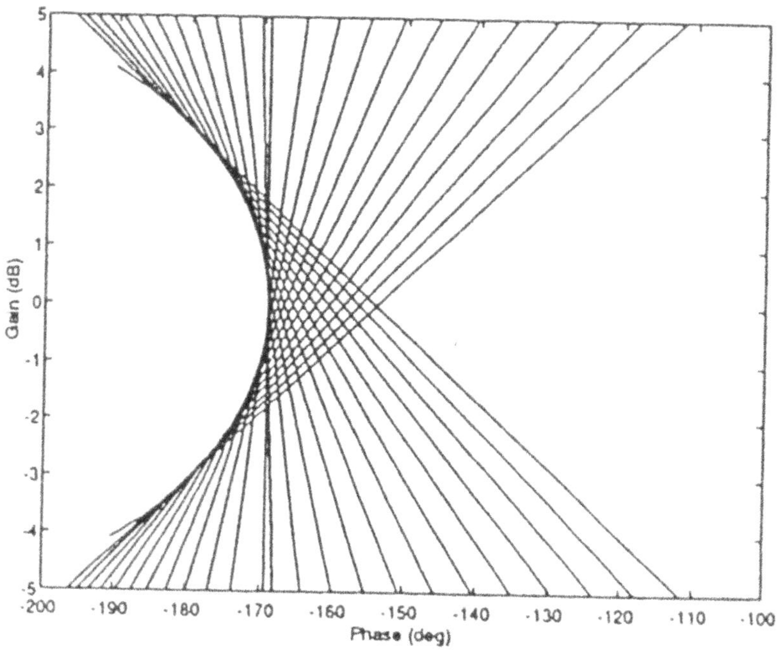

Figure A3.13. *Building of the isodamping contour corresponding to the damping ratio,* $\zeta = 0.1$

In the case of a generalized integration in open loop, where the gain slope is $-20a\text{dB/dec}$ around the frequency, ω_u, with a denoting the real integration order, it is therefore necessary for a to remain sufficiently large, in this case greater than one, to ensure a sufficiently rapid increase in the gain, with a decrease in the frequency below ω_u; such a condition then imposes a maximal limit to the phase placement of the templates, in this case at $-\pi/2$, for the building of isodamping contours using the covering technique.

Figure A3.14 proposes a network of isodamping contours built according to this technique and corresponding to values of the damping ratio, ζ, comprised between 0.1 and 1, in steps of 0.1.

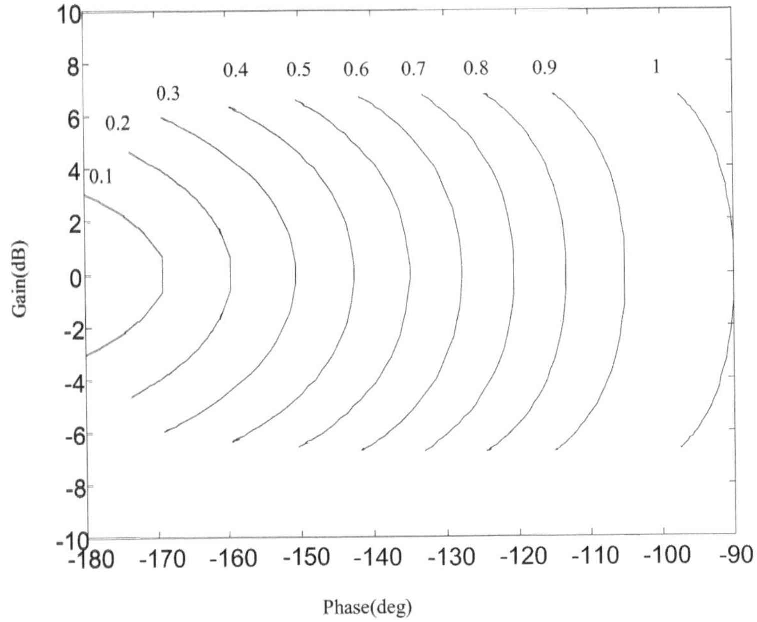

Figure A3.14. *Network of isodamping contours*

A3.6. Conclusion

Starting from a rectilinear open-loop behavior in the Nichols plane, the operational description of which is based on complex, non-integer integration (section A3.3). First, we have *validated* the Nichols amplitude contours as *iso-overshoot contours* (section A3.4), and these contours signify the reduced first overshoot of the step responses in tracking and regulation; second, we have *built* new contours, namely *isodamping contours* (section A3.5), which signify the damping ratio of the oscillatory mode of the step responses in tracking and regulation [OUS 95a, OUS 95b, OUS 99b].

The interest of these contours lies in the fact that their field of application goes beyond the case study for validation or building. Indeed, we have shown that they remain valid, first, in the case of limited rectilinear behavior around the unit gain frequency in open loop, and, second, in the case of curvilinear behavior that is also limited and presents a permissible degree of concavity.

References

[AOU 93] AOUCHICHE, K., D'ANDRÉA-NOVEL, B., "Nonlinear dynamic output feedback for an equilibrist robot", *Proceedings of the IEEE SMC'93 Conference*, Le Touquet, France, vol. 4, pp. 39–44, 1993.

[BAL 56] BALCHEN, J.G., "A graphical method for determination of closed loop frequency response curves for feedback control systems containing nonlinear elements", *Tekniske Skrifter*, Oslo, Norway, 1956.

[BAL 92] BALLOUK A., La dérivation non entière réelle et complexe : synthèse et applications dans les sciences de l'ingénieur, Thesis, Université Bordeaux 1, France, 1992.

[BEN 79] BENNETT, S., *A History of Control Engineering 1800–1930*, IEE Control Engineering Series, vol. 8, Peter Peregrinus Ltd, Hitchin, UK, 1979.

[BEN 93] BENNETT, S., *A History of Control Engineering 1930–1955*, IEE Control Engineering Series, vol. 47, Peter Peregrinus Ltd, Stevenage, UK, 1993.

[CHA 91] CHARLET, B., LÉVINE, J., MARINO, R., "Sufficient conditions for dynamic state feedback linearization", *SIAM Journal of Control and Optimization*, vol. 29, pp. 38–57, 1991.

[DAN 88] D'ANDRÉA-NOVEL, B., *Commande non linéaire des robots*, Hermès, Paris, 1988.

[DAN 91] D'ANDRÉA-NOVEL, B., MARTIN, P., SÉPULCHRE, R., "Full linearization of a class of mechanical systems via dynamic state feedback", *Proceedings of MTNS'91*, Kobe, Japan, 1991.

[DAN 92] D'ANDRÉA-NOVEL, B., CAMPION, G., BASTIN, G., "Dynamic feedback linearization of nonholonomic wheeled mobile robots", *Proceedings of the IEEE Conference on Robotics and Automation*, Nice, France, 1992.

[DEL 09] DE LARMINAT, P., *Automatique Appliquée*, Hermès Lavoisier, Paris, France, 2009.

[DEM 91] DEMIRCIOGLU, H., GAWTHROP, P.J., "Continuous-time generalized predictive control (CGPC)", *Automatica*, vol. 27, no. 1, pp. 55–74, 1991.

[DES 85] DESCUSSE, J., MOOG, C.H., "Decoupling with dynamic compensation for strong invertible affine nonlinear systems", *International Journal of Control*, vol. 42, pp. 1387–1398, 1985.

[FLI 95] FLIESS, M., LÉVINE, J., MARTIN, P. *et al.*, "Flatness and defect of nonlinear systems: Introductory theory and applications", *International Journal of Control*, vol. 61, pp. 1327–1361, 1995.

[GIB 63] GIBSON, J.E., *Nonlinear Automatic Control,* McGraw-Hill Book Co., New York, USA, 1963.

[HIL 43] HALL, A.C., *The Analysis and Synthesis of Linear Servomechanisms*, The Technology Press, Cambridge, MA, USA, 1943.

[HIL 62] HILL, J.C., "Closed loop response of nonlinear systems by a functional transformation approach", *IRE Transactions on Automatic Control*, vol. AC-7, no. 4, pp. 39–45, 1962.

[ISI 95] ISIDORI, A., *Nonlinear Control Systems*, 3rd edition, Springer-Verlag, London, UK, 1995.

[JAM 47] JAMES, H.M., NICHOLS, N.B., PHILLIPS, R.S., *Theory of Servomechanisms*, McGraw-Hill Book Co., New York, USA, 1947.

[LAN 94] LANUSSE P., De la commande CRONE de première génération à la commande CRONE de troisième génération, Thesis, Université Bordeaux 1, France, 1994.

[LOI 15] LOISEAU, P., CHEVREL, P., YAGOUBI, M. *et al.*, "Broadband active noise control design through nonsmooth H_∞ synthesis", *8th IFAC Symposium on Robust Control Design*, Bratislava, Slovak Republic, July, 2015.

[MAR 86] MARINO, R., "On the largest feedback linearizable subsystem", *Systems and Control Letters*, vol. 6, pp. 345–351, 1986.

[MAT 94] MATHIEU, B., OUSTALOUP, A., LEVRON, F., "Une solution globale à l'identification paramétrique dans le domaine fréquentiel", Réunion du pôle SARTA-G.R. Automatique CNRS, Paris, France, November 7–8, 1994.

[OUS 75] OUSTALOUP, A., Etude et réalisation d'un système d'asservissement d'ordre 3/2 de la fréquence d'un laser à colorant continu, Thesis, Université Bordeaux 1, France, 1975.

[OUS 78a] OUSTALOUP, A., "Etude de la fonction régulation d'un système asservi au Moyen d'un nouvel abaque", *Onde Electrique*, vol. 58, nos 8–9, pp. 543–546, 1978.

[OUS 78b] OUSTALOUP, A., Oscillateur sinusoïdal d'ordre demi-entier, Patent 7835728, INPI, 1978.

[OUS 79] OUSTALOUP, A., "Systèmes de modulation de fréquence BF à grande linéarité", *L'onde electrique*, vol. 59, nos 8–9, pp. 102–107, 1979.

[OUS 81a] OUSTALOUP, A., "Fractional order sinusoidal oscillators: Optimization and their use in highly linear FM modulation", *IEEE Transactions on Circuits and Systems*, vol. 28, no. 10, pp. 1007–1009, 1981.

[OUS 81b] OUSTALOUP A., "Linear feedback control systems of fractional order between 1 and 2", *IEEE International Symposium on Circuits and Systems*, Chicago, IL, USA, April 27–29, 1981.

[OUS 81c] OUSTALOUP, A., Systèmes asservis linéaires d'ordre fractionnaire, Thesis, Université Bordeaux 1, France, 1981.

[OUS 83a] OUSTALOUP, A., *Fractional Order Control*, North-Holland Publishing Company, Amsterdam, The Netherlands, 1983.

[OUS 83b] OUSTALOUP, A., *Systèmes asservis linéaires d'ordre fractionnaire*, Masson, Paris, France, 1983.

[OUS 89] OUSTALOUP, A., "From fractality to non integer derivation through recursivity, a property common to these two concepts: A fundamental idea for a new process control strategy", *Computing and Computers for Control Systems*, Editorial Office, The Netherlands, pp. 421–426, 1989.

[OUS 91] OUSTALOUP, A., *La commande CRONE*, Hermès, Paris, France, 1991.

[OUS 92a] OUSTALOUP, A., *The CRONE Control: Template and Real Non-integer Derivation*, Elsevier Sciences Publishers, Amsterdam, The Netherlands, 1992.

[OUS 92b] OUSTALOUP A., LANUSSE, P., *New Trends in CRONE Control: Generalized Template and Complex Non-integer Derivation*, Elsevier Sciences Publishers, Amsterdam, The Netherlands, 1992.

[OUS 95a] OUSTALOUP, A., *La dérivation non entière : théorie, synthèse et applications*, Hermès, Paris, France, 1995.

[OUS 95b] OUSTALOUP, A., MATHIEU, B., LANUSSE, P., "Intégration non entière complexe et contours d'isoamortissement", *RAIRO-APII*, vol. 29, no. 2, pp. 177–203, 1995.

[OUS 95c] OUSTALOUP, A., MATHIEU, B., LANUSSE, P., "The CRONE control of resonant plants: Application to a flexible transmission", *European Journal of Control*, vol. 1, no. 2, pp. 113–121, 1995.

[OUS 99a] OUSTALOUP, A., SABATIER, J., LANUSSE, P., "From fractal robustness to the CRONE control", *Fractional Calculus and Applied Analysis (FCAA): An International Journal for Theory and Applications*, vol. 2, no. 1, pp. 1–30, 1999.

[OUS 99b] OUSTALOUP, A., MATHIEU, B., *La commande CRONE : du scalaire au multivariable*, Hermès, Paris, France, 1999.

[OUS 00] OUSTALOUP, A., LEVRON, F., MATHIEU, B., *et al.*, "Frequency-band complex noninteger differentiator: Characterization and synthesis", *IEEE Transactions on Circuits and Systems I: Fundamental Theory and Applications*, vol. 47, pp. 25–39, 2000.

[OUS 02] OUSTALOUP, A., LANUSSE, P., LEVRON, F., "Frequency domain synthesis of a filter using Viète root functions", *IEEE Transactions on Automatic Control*, vol. 47, no. 5, pp. 837–841, 2002.

[OUS 14] OUSTALOUP, A., *Diversity and Non-integer Differentiation for System Dynamics*, ISTE Ltd, London, UK, and John Wiley and Sons, Hoboken, NJ, USA, 2014.

[ROU 94] ROUCHON, P., "Necessary condition and genericity of dynamic feedback linearization", *Journal of Mathematical Systems, Estimation and Control*, vol. 5, no. 3, pp. 345–358, 1994.

Index

A, C

D, E, F

I

L

M, N

O, P

R

S

T

U, W

Other titles from

ISTE

in

Systems and Industrial Engineering – Robotics

2024

ANDRÉ Jean-Claude
Knowledge Production Modes between Science and Applications 1: Concepts
Knowledge Production Modes between Science and Applications 2: Applications

BRIFFAUT Jean-Pierre
Complexities 2: Various Approaches in the Field of Social and Human Sciences
(Systems of Systems Complexity Set – Volume 6)

YALAOUI Farouk, NGUYEN Nhan-Quy, OUAZENE Yassine, ZEMZAMI Maria, MERGHEM-BOULAHIA Leïla
Connected Innovation and Technology X.0 1: Advancing Digital Transformation: Innovational Strategies for Smart Industry and Sustainable Innovation

2023

BERRAH Lamia, CLIVILLÉ Vincent
Decision and Decision-maker in an Industrial Environment

BRIFFAUT Jean-Pierre
Complexities 1: Various Approaches in the Field of Techno-scientific Knowledge
(Systems of Systems Complexity Set – Volume 5)

CHAINEAUX Jacques
The Mechanisms of Explosions: 27 Case Studies for their Understanding

DIALLO Saikou, BOGUI Jean-Jacques Maomra, ATCHOUA Julien
Technologies and African Societies in Pandemic Times: Using Technology to Survive and Thrive During the Covid-19 Era
(Africa Digital Age Set – Volume 2)

JAEGLER Anicia, ROQUES Thierry
Sustainable Supply Chain Management: Sustainable Decision Support

PRINTZ Jacques
Organization and Pedagogy of Complexity: Systemic Case Studies and Prospects
(Systems of Systems Complexity Set – Volume 4)

2022

AMARA Yacine, BEN AHMED Hamid, GABSI Mohamed
Hybrid Excited Synchronous Machines: Topologies, Design and Analysis

BOURRIÈRES Jean-Paul, PINÈDE Nathalie, TRAORÉ Mamadou Kaba, ZACHAREWICZ Grégory
From Logistic Networks to Social Networks: Similarities, Specificities, Modeling, Evaluation

DEMOLY Frédéric, ANDRÉ Jean-Claude
4D Printing 1: Between Disruptive Research and Industrial Applications
4D Printing 2: Between Science and Technology

HAJJI Rafika, JARAR OULIDI Hassane
Building Information Modeling for a Smart and Sustainable Urban Space

KROB Danicl
Model-based Systems Architecting: Using CESAM to Architect Complex Systems
(Systems of Systems Complexity Set – Volume 3)

LOUIS Gilles
Dynamics of Aircraft Flight

2020

BRON Jean-Yves
System Requirements Engineering

KRYSINSKI TOMASZ, MALBURET FRANÇOIS
Energy and Motorization in the Automotive and Aeronautics Industries

PRINTZ Jacques
System Architecture and Complexity: Contribution of Systems of Systems to Systems Thinking

2019

ANDRÉ Jean-Claude
Industry 4.0: Paradoxes and Conflicts

BENSALAH Mounir, ELOUADI Abdelmajid, MHARZI Hassan
Railway Information Modeling RIM: The Track to Rail Modernization

BLUA Philippe, YALAOU Farouk, AMODEO Lionel, DE BLOCK Michaël, LAPLANCHE David
Hospital Logistics and e-Management: Digital Transition and Revolution

BRIFFAUT Jean-Pierre
From Complexity in the Natural Sciences to Complexity in Operations Management Systems
(Systems of Systems Complexity Set – Volume 1)

BUDINGER Marc, HAZYUK Ion, COÏC Clément
Multi-Physics Modeling of Technological Systems

FLAUS Jean-Marie
Cybersecurity of Industrial Systems

JAULIN Luc
Mobile Robotics – Second Edition Revised and Updated

KUMAR Kaushik, DAVIM Paulo J.
Optimization for Engineering Problems

TRIGEASSOU Jean-Claude, MAAMRI Nezha
Analysis, Modeling and Stability of Fractional Order Differential Systems 1: The Infinite State Approach
Analysis, Modeling and Stability of Fractional Order Differential Systems 2: The Infinite State Approach

VANDERHAEGEN Frédéric, MAAOUI Choubeila, SALLAK Mohamed, BERDJAG Denis
Automation Challenges of Socio-technical Systems

2018

BERRAH Lamia, CLIVILLÉ Vincent, FOULLOY Laurent
Industrial Objectives and Industrial Performance: Concepts and Fuzzy Handling

GONZALEZ-FELIU Jesus
Sustainable Urban Logistics: Planning and Evaluation

GROUS Ammar
Applied Mechanical Design

LEROY Alain
Production Availability and Reliability: Use in the Oil and Gas Industry

MARÉ Jean-Charles
Aerospace Actuators 3: European Commercial Aircraft and Tiltrotor Aircraft

MAXA Jean-Aimé, BEN MAHMOUD Mohamed Slim, LARRIEU Nicolas
Model-driven Development for Embedded Software: Application to Communications for Drone Swarm

MBIHI Jean
Analog Automation and Digital Feedback Control Techniques
Advanced Techniques and Technology of Computer-Aided Feedback Control

MORANA Joëlle
Logistics

SIMON Christophe, WEBER Philippe, SALLAK Mohamed
Data Uncertainty and Important Measures
(Systems Dependability Assessment Set – Volume 3)

TANIGUCHI Eiichi, THOMPSON Russell G.
City Logistics 1: New Opportunities and Challenges
City Logistics 2: Modeling and Planning Initiatives
City Logistics 3: Towards Sustainable and Liveable Cities

ZELM Martin, JAEKEL Frank-Walter, DOUMEINGTS Guy, WOLLSCHLAEGER Martin
Enterprise Interoperability: Smart Services and Business Impact of Enterprise Interoperability

2017

ANDRÉ Jean-Claude
From Additive Manufacturing to 3D/4D Printing 1: From Concepts to Achievements
From Additive Manufacturing to 3D/4D Printing 2: Current Techniques, Improvements and their Limitations
From Additive Manufacturing to 3D/4D Printing 3: Breakthrough Innovations: Programmable Material, 4D Printing and Bio-printing

ARCHIMÈDE Bernard, VALLESPIR Bruno
Enterprise Interoperability: INTEROP-PGSO Vision

CAMMAN Christelle, FIORE Claude, LIVOLSI Laurent, QUERRO Pascal
Supply Chain Management and Business Performance: The VASC Model

FEYEL Philippe
Robust Control, Optimization with Metaheuristics

MARÉ Jean-Charles
Aerospace Actuators 2: Signal-by-Wire and Power-by-Wire

POPESCU Dumitru, AMIRA Gharbi, STEFANOIU Dan, BORNE Pierre
Process Control Design for Industrial Applications

RÉVEILLAC Jean-Michel
Modeling and Simulation of Logistics Flows 1: Theory and Fundamentals
Modeling and Simulation of Logistics Flows 2: Dashboards, Traffic Planning and Management
Modeling and Simulation of Logistics Flows 3: Discrete and Continuous Flows in 2D/3D

2016

ANDRÉ Michel, SAMARAS Zissis
Energy and Environment
(Research for Innovative Transports Set – Volume 1)

AUBRY Jean-François, BRINZEI Nicolae, MAZOUNI Mohammed-Habib
Systems Dependability Assessment: Benefits of Petri Net Models
(Systems Dependability Assessment Set – Volume 1)

BLANQUART Corinne, CLAUSEN Uwe, JACOB Bernard
Towards Innovative Freight and Logistics
(Research for Innovative Transports Set – Volume 2)

COHEN Simon, YANNIS George
Traffic Management
(Research for Innovative Transports Set – Volume 3)

MARÉ Jean-Charles
Aerospace Actuators 1: Needs, Reliability and Hydraulic Power Solutions

REZG Nidhal, HAJEJ Zied, BOSCHIAN-CAMPANER Valerio
Production and Maintenance Optimization Problems: Logistic Constraints and Leasing Warranty Services

TORRENTI Jean-Michel, LA TORRE Francesca
Materials and Infrastructures 1
(Research for Innovative Transports Set – Volume 5A)
Materials and Infrastructures 2
(Research for Innovative Transports Set – Volume 5B)

WEBER Philippe, SIMON Christophe
Benefits of Bayesian Network Models
(Systems Dependability Assessment Set – Volume 2)

YANNIS George, COHEN Simon
Traffic Safety
(Research for Innovative Transports Set – Volume 4)

2015

AUBRY Jean-François, BRINZEI Nicolae
Systems Dependability Assessment: Modeling with Graphs and Finite State Automata

BOULANGER Jean-Louis
CENELEC 50128 and IEC 62279 Standards

BRIFFAUT Jean-Pierre
E-Enabled Operations Management

MISSIKOFF Michele, CANDUCCI Massimo, MAIDEN Neil
Enterprise Innovation

2014

CHETTO Maryline
Real-time Systems Scheduling
Volume 1 – Fundamentals
Volume 2 – Focuses

DAVIM J. Paulo
Machinability of Advanced Materials

ESTAMPE Dominique
Supply Chain Performance and Evaluation Models

FAVRE Bernard
Introduction to Sustainable Transports

GAUTHIER Michaël, ANDREFF Nicolas, DOMBRE Etienne
Intracorporeal Robotics: From Milliscale to Nanoscale

MICOUIN Patrice
Model Based Systems Engineering: Fundamentals and Methods

MILLOT Patrick
Designing Human–Machine Cooperation Systems

NI Zhenjiang, PACORET Céline, BENOSMAN Ryad, RÉGNIER Stéphane
Haptic Feedback Teleoperation of Optical Tweezers

OUSTALOUP Alain
Diversity and Non-integer Differentiation for System Dynamics

REZG Nidhal, DELLAGI Sofien, KHATAD Abdelhakim
Joint Optimization of Maintenance and Production Policies

STEFANOIU Dan, BORNE Pierre, POPESCU Dumitru, FILIP Florin Gh.,
EL KAMEL Abdelkader
Optimization in Engineering Sciences: Metaheuristics, Stochastic Methods and Decision Support

2013

ALAZARD Daniel
Reverse Engineering in Control Design

ARIOUI Hichem, NEHAOUA Lamri
Driving Simulation

CHADLI Mohammed, COPPIER Hervé
Command-control for Real-time Systems

DAAFOUZ Jamal, TARBOURIECH Sophie, SIGALOTTI Mario
Hybrid Systems with Constraints

FEYEL Philippe
Loop-shaping Robust Control

FLAUS Jean-Marie
Risk Analysis: Socio-technical and Industrial Systems

FRIBOURG Laurent, SOULAT Romain
Control of Switching Systems by Invariance Analysis: Application to Power Electronics

GROSSARD Mathieu, RÉGNIER Stéphane, CHAILLET Nicolas
Flexible Robotics: Applications to Multiscale Manipulations

GRUNN Emmanuel, PHAM Anh Tuan
Modeling of Complex Systems: Application to Aeronautical Dynamics

HABIB Maki K., DAVIM J. Paulo
Interdisciplinary Mechatronics: Engineering Science and Research Development

HAMMADI Slim, KSOURI Mekki
Multimodal Transport Systems

JARBOUI Bassem, SIARRY Patrick, TEGHEM Jacques
Metaheuristics for Production Scheduling

KIRILLOV Oleg N., PELINOVSKY Dmitry E.
Nonlinear Physical Systems

LE Vu Tuan Hieu, STOICA Cristina, ALAMO Teodoro,
CAMACHO Eduardo F., DUMUR Didier
Zonotopes: From Guaranteed State-estimation to Control

MACHADO Carolina, DAVIM J. Paulo
Management and Engineering Innovation

MORANA Joëlle
Sustainable Supply Chain Management

SANDOU Guillaume
Metaheuristic Optimization for the Design of Automatic Control Laws

STOICAN Florin, OLARU Sorin
Set-theoretic Fault Detection in Multisensor Systems

2012

AÏT-KADI Daoud, CHOUINARD Marc, MARCOTTE Suzanne, RIOPEL Diane
Sustainable Reverse Logistics Network: Engineering and Management

BORNE Pierre, POPESCU Dumitru, FILIP Florin G., STEFANOIU Dan
Optimization in Engineering Sciences: Exact Methods

CHADLI Mohammed, BORNE Pierre
Multiple Models Approach in Automation: Takagi-Sugeno Fuzzy Systems

DAVIM J. Paulo
Lasers in Manufacturing

DECLERCK Philippe
Discrete Event Systems in Dioid Algebra and Conventional Algebra

DOUMIATI Moustapha, CHARARA Ali, VICTORINO Alessandro, LECHNER Daniel
Vehicle Dynamics Estimation using Kalman Filtering: Experimental Validation

GUERRERO José A, LOZANO Rogelio
Flight Formation Control

HAMMADI Slim, KSOURI Mekki
Advanced Mobility and Transport Engineering

MAILLARD Pierre
Competitive Quality Strategies

MATTA Nada, VANDENBOOMGAERDE Yves, ARLAT Jean
Supervision and Safety of Complex Systems

POLER Raúl, CARNEIRO Luis Maia, JASINSKI Thomas, ZOLGHADRI Marc, PEDRAZZOLI Paolo
Intelligent Non-hierarchical Manufacturing Networks

TROCCAZ Jocelyne
Medical Robotics

YALAOUI Alice, CHEHADE Hicham, YALAOUI Farouk, AMODEO Lionel
Optimization of Logistics

ZELM Martin, SANCHIS Raquel, POLER Raúl, DOUMEINGTS Guy
Enterprise Interoperability: I-ESA'12 Proceedings

2011

CANTOT Pascal, LUZEAUX Dominique
Simulation and Modeling of Systems of Systems

DAVIM J. Paulo
Mechatronics

DAVIM J. Paulo
Wood Machining

GROUS Ammar
Applied Metrology for Manufacturing Engineering

KOLSKI Christophe
Human–Computer Interactions in Transport

LUZEAUX Dominique, RUAULT Jean-René, WIPPLER Jean-Luc
Complex Systems and Systems of Systems Engineering

ZELM Martin, VAN SINDEREN Marten, DOUMEINGTS Guy, JOHNSON Pontus
Enterprise Interoperability: IWEI2011 Proceedings

2010

BOTTA-GENOULAZ Valérie, CAMPAGNE Jean-Pierre, LLERENA Daniel, PELLEGRIN Claude
Supply Chain Performance: Collaboration, Alignment and Coordination

BOURLÈS Henri, GODFREY K.C. Kwan
Linear Systems

BOURRIÈRES Jean-Paul
Proceedings of CEISIE'09

CHAILLET Nicolas, REGNIER Stéphane
Microrobotics for Micromanipulation

DAVIM J. Paulo
Sustainable Manufacturing

GIORDANO Max, MATHIEU Luc, VILLENEUVE François
Product Life-Cycle Management: Geometric Variations

LOZANO Rogelio
Unmanned Aerial Vehicles: Embedded Control

LUZEAUX Dominique, RUAULT Jean-René
Systems of Systems

VILLENEUVE François, MATHIEU Luc
Geometric Tolerancing of Products

2009

DIAZ Michel
Petri Nets: Fundamental Models, Verification and Applications

OZEL Tugrul, DAVIM J. Paulo
Intelligent Machining

PITRAT Jacques
Artificial Beings

2008

ARTIGUES Christian, DEMASSEY Sophie, NÉRON Emmanuel
Resources–Constrained Project Scheduling

BILLAUT Jean-Charles, MOUKRIM Aziz, SANLAVILLE Eric
Flexibility and Robustness in Scheduling

DOCHAIN Denis
Bioprocess Control

LOPEZ Pierre, ROUBELLAT François
Production Scheduling

THIERRY Caroline, THOMAS André, BEL Gérard
Supply Chain Simulation and Management

2007

DE LARMINAT Philippe
Analysis and Control of Linear Systems

DOMBRE Etienne, KHALIL Wisama
Robot Manipulators

LAMNABHI-LAGARRIGUE Françoise, LAGHROUCHE Salah, LORIA Antonio, PANTELEY Elena
Taming Heterogeneity and Complexity of Embedded Control

LIMNIOS Nikolaos
Fault Trees

2006

FRENCH COLLEGE OF METROLOGY
Metrology in Industry

NAJIM Kaddour
Control of Continuous Linear Systems

Printed and bound by CPI Group (UK) Ltd, Croydon, CR0 4YY

16/09/2026

14963079-0004